D0882413

MONOCLONAL ANTIBODIES: PRINCIPLES AND APPLICATIONS

MONOCLONAL ANTIBODIES: PRINCIPLES AND APPLICATIONS

Editors

J. R. BIRCH
Celltech Biologics
Slough, United Kingdom

E. S. LENNOX
Henley-on-Thames, United Kingdom

A JOHN WILEY & SONS, INC., PUBLICATION
New York • Chichester • Brisbane • Toronto • Singapore

Samford University Library

Address All Inquiries to the Publisher
Wiley-Liss, Inc., 605 Third Avenue, New York, NY 10158-0012

Copyright © 1995 Wiley-Liss, Inc.

Printed in the United States of America.

Under the conditions stated below the owner of copyright for this book hereby grants permission to users to make photocopy reproductions of any part or all of its contents for personal or internal organizational use, or for personal or internal use of specific clients. This consent is given on the condition that the copier pay the stated per-copy fee through the Copyright Clearance Center, Incorporated, 27 Congress Street, Salem, MA 01970, as listed in the most current issue of "Permissions to Photocopy" (Publisher's Fee List, distributed by CCC, Inc.), for copying beyond that permitted by sections 107 or 108 of the US Copyright Law. This consent does not extend to other kinds of copying, such as copying for general distribution, for advertising or promotional purposes, for creating new collective works, or for resale.

While the authors, editors, and publisher believe that drug selection and dosage and the specifications and usage of equipment and devices, as set forth in this book, are in accord with current recommendations and practice at the time of publication, they accept no legal responsibility for any errors or omissions, and make no warranty, express or implied, with respect to material contained herein. In view of ongoing research, equipment modifications, changes in governmental regulations and the constant flow of information relating to drug therapy, drug reactions and the use of equipment and devices, the reader is urged to review and evaluate the information provided in the package insert or instructions for each drug, piece of equipment or device for, among other things, any changes in the instructions or indications of dosage or usage and for added warnings and precautions.

Library of Congress Cataloging-in-Publication Data

Monoclonal antibodies : principles and applications / editors, J.R. Birch, E.S. Lennox.
p. cm.
Includes bibliographical references and index.
ISBN 0-471-05147-0
1. Monoclonal antibodies—Biotechnology. I. Birch, J.R. (John R.) II. Lennox, E.S. (Edwin Samuel), 1920– .
[DNLM: 1. Antibodies, Monoclonal—diagnostic use. 2. Antibodies, Monoclonal—therapeutic uses. QW 575 M75135 1995]
TP248.65.M65M66 1995
660'.63—dc20
DNLM/DLC
for Library of Congress 94-39791
10 9 8 7 6 5 4 3 2 CIP

The text of this book is printed on acid-free paper.

TP
248.65
.M65
M66
1995

CONTENTS

CONTRIBUTORS

C.R. Bebbington, Celltech Ltd., Slough, Berkshire SL1 4EN, UK **[137]**

J.R. Birch, Celltech Biologics, Slough, Berkshire SL1 4EN, UK **[231]**

J. Bonnerjea, Celltech Biologics, Slough, Berkshire SL1 4EN, UK **[231]**

Paul B. Chapman, Leukemia and Clinical Immunology Services, Department of Medicine, Memorial Sloan-Kettering Cancer Center, New York, NY 10021 **[45]**

Mike Clark, Department of Pathology, Immunology Division, Cambridge University, Cambridge CB2 1QP, UK **[1]**

Patrick Crawley, Patents and Trademarks Division, Sandoz Technology Ltd., CH-4002 Basel, Switzerland **[299]**

Allan Darling, Q-One Biotech Ltd., West of Scotland Science Park, Glasgow G20 OXA, UK **[267]**

S. Flatman, Celltech Biologics, Slough, Berkshire SL1 4EN, UK **[231]**

C.R. Hill, Celltech Biologics, Slough, Berkshire SL1 4EN, UK **[121]**

Lois Hinman, Medical Research Division, Lederle Laboratories, American Cyanamid Company, Pearl River, NY 10965-1299 **[187]**

Gillian Lees, Q-One Biotech Ltd., West of Scotland Science Park, Glasgow G20 OXA, UK **[267]**

Arnold Oronsky, InterWest Partners, Menlo Park, CA 94025-7112 **[187]**

M.J. Perry, Celltech Ltd., Slough, Berkshire SL1 4EN, UK **[107]**

David A. Scheinberg, Leukemia and Clinical Immunology Services, Department of Medicine, Memorial Sloan-Kettering Cancer Center, New York, NY 10021 **[45]**

Janis Upeslacis, Medical Research Division, Lederle Laboratories, American Cyanamid Company, Pearl River, NY 10965-1299 **[187]**

The numbers in brackets are the opening page numbers of the contributors' articles.

S. Vranch, Celltech Biologics, Slough, Berkshire SL1 4EN, UK; present address: Jacobs Engineering, Croydon, Surrey CR0 6SR, UK **[231]**

E. Sally Ward, Department of Microbiology, Cancer Immunobiology Center, University of Texas Southwestern Medical Center, Dallas, TX 75235-8576 **[137]**

PREFACE

This book arises out of the experience of the two editors in the uses and manufacture of monoclonal antibodies in both academic and commercial settings. We have together been involved in the processes from the selection of cells producing antibodies with desired specificity through the steps for optimizing their productivity, including the redesign of their properties for specific diagnostic and therapeutic purposes. That process is long, has many steps, and needs varied skills. Careful planning is required because decisions made early on in this long path can have serious consequences later on.

Although much has been written on particular aspects of the uses and production of monoclonal antibodies, we felt that there was a need for a book dealing with the many steps between the original selection of an antibody for a particular purpose and its path through those steps that allow it to be tested as a candidate therapeutic in the clinic or as a diagnostic product. This is our attempt to create such a book, concerned with antibodies in their role as research tools, diagnostic reagents, and therapeutics, with an emphasis on the latter two roles.

We focused mainly on monoclonal antibodies as products of biotechnology, and that determined the choice of subjects for each of the chapters and of the experts we asked to write them. We attempted to have the text flow smoothly from the introduction through the subjects of applications, molecular redesign, methods of production, regulatory considerations, and patent issues. In particular we wanted to emphasize that through the combination of conventional antibody induction and selection methods with rDNA techniques, and the ability to introduce chemical modifications, it is possible to improve the properties of antibodies to address a wide range of potential uses. We also wanted to emphasize that it takes a great deal of research to know what properties have to be designed into these molecules. In many instances that research has not yet been done.

In addition, there may be manufacturing or regulatory consequences of altering antibodies or of expressing them in different organisms. For those in commercial organizations or in academic ones with an eye towards later commercial applications (which is normal these days), these considerations, as well as patent issues, are important. In the design of antibodies as products their intended use must be kept in mind at all stages. For example, an early poor choice of cell line may subsequently lead to manufacturing difficulties or regulatory concerns. It is essential, then, that these potential technical problems and the patent status of the product and proposed process are considered at the earliest possible stage in a research program. We hope that this book will contribute usefully to the understanding of these issues.

J. R. Birch
E. S. Lennox

CHAPTER 1

GENERAL INTRODUCTION

MIKE CLARK
Department of Pathology, Immunology Division, Cambridge University, Cambridge CB2 1QP, UK

1.1. INTRODUCTION

In this chapter various aspects of the technology and the background to the technology of monoclonal antibodies will be introduced and discussed at a basic level. This will give a baseline for the following chapters, which will then concentrate on particular issues in greater detail. Throughout this chapter, various critical aspects of the issues discussed will be raised, including likely problems, advantages, or areas of current uncertainty with the technologies. Where appropriate, the reader will be referred to the other relevant chapters in this volume for a more complete and in-depth coverage of the topic in question. This is not meant to be an authoritative "cookbook"; very many excellent compilations of detailed descriptions of techniques of relevance to the technology already exist, only a few of which are referred to within this chapter.

1.2. THE IMMUNE RESPONSE

When challenged with a pathogen or other antigen, the immune system of most higher vertebrates responds by making an immune response that is aimed at the specific elimination of the pathogen or antigen from the body [1–3]. A major antigen-specific component of this response is the production of antigen-specific immunoglobulins, otherwise termed antibodies, which have affinity for and bind to the antigen. The immune response in an animal is polyclonal in nature, and a mixture of many antibodies with different specificities and affinities for the antigen is made. This polyclonal response to antigen is a result of the activation of a large number of different clones of B cells, each clone encoding a single immunoglobulin type with a single specificity for antigen, interacting and cooperating together with

Monoclonal Antibodies: Principles and Applications, pages 1–43
© 1995 Wiley-Liss, Inc.

other cells of the immune system, including the thymus-derived T lymphocytes and antigen-presenting cells (often abbreviated to APCs). In 1975 Kohler and Milstein [4] described a technique for the specific immortalization of the individual B cells from an animal that are responsible for the production of antibodies; during culture and propagation, these immortalized B cells can be grown as single clones of cells, each of which secretes a monoclonal antibody.

In order to set the background for some of the later arguments that are raised in this chapter and discussed at length in other chapters—particularly concerning the recent advances in the derivation of recombinant antibodies—it is necessary that the reader have a basic knowledge of our current understanding of how the immune system decides what antibody response to make toward a given antigen. Figure 1.1 shows schematically a grossly oversimplified view of the B-cell immune response, which illustrates some key features. During their development and in a random and antigen-independent manner, B cells rearrange their immunoglobulin genes such that each clone expresses a surface immunoglobulin molecule of a unique specificity (see sects. 1.3.1 and 1.3.2 below for more detail). These clones of "virgin" B cells have a rapid turnover in the immune system and hence form the basis for the total B-cell repertoire, but they do not secrete their immunoglobulins in quantity. Upon encountering antigen, these clones of B cells that possess a surface antibody with affinity for antigen are able to bind and internalize the antigen. In the case of complex protein or protein-associated antigens, the internalized complex is de-

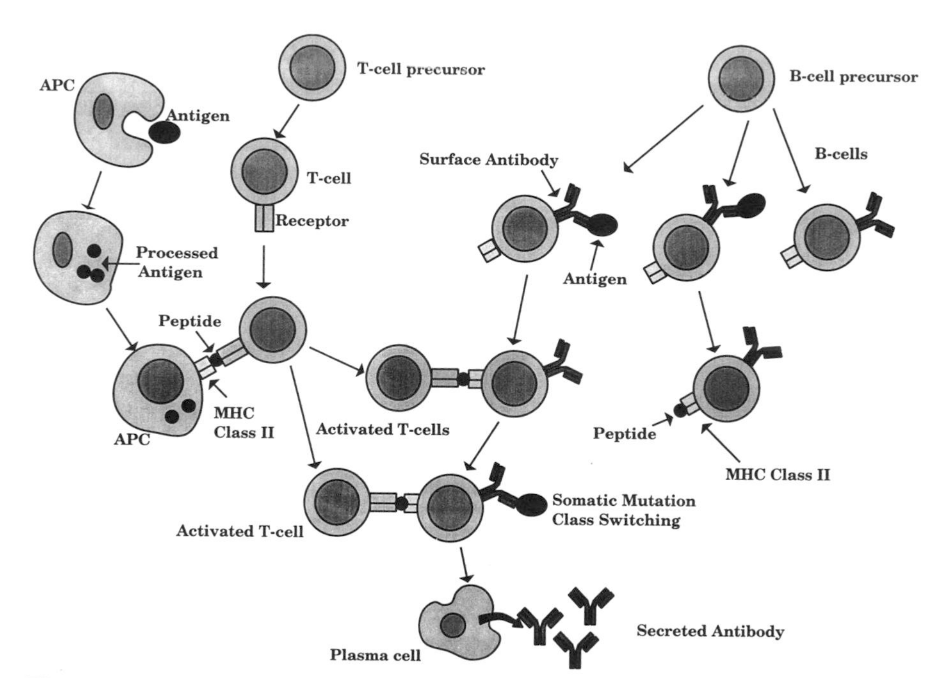

Fig. 1.1. Shown schematically is a simplified representation of the differentiation of B cells in response to antigen. The involvement of antigen-presenting cells and cooperation between antigen-specific T cells and B cells are represented.

graded and certain unique peptides may become associated with the major histocompatibility complex (MHC) class II molecules. The MHC class II molecules then display the peptides on the surface of the B cell, where they can be recognized by the antigen-specific receptors of T cells. If an appropriate T-cell recognition takes place, then the T cell may decide to "help" the B cell by delivering appropriate signals, which trigger the proliferation and differentiation of the B cell. Thus the antigen, and T-cell selected, B cells may go on to differentiate into antibody-secreting plasma cells, or to become memory B cells which can participate in a response to subsequent antigen encounter. Also during the differentiation process, the B cells may produce isotype class switches of their immunoglobulin genes (see sect. 1.3.2) or somatically mutate their immunoglobulin V region genes to give altered binding affinity.

Although this description of the B-cell response is grossly oversimplified, it should be noted that a key feature is that the production of antibody is controlled by "help" from T cells, in addition to the binding of antigen by the B cell. During their development in the thymus, T cells that respond to peptides generated from self molecules are eliminated, and so the mature peripheral T cells selectively only "help" B cells that are not producing antibody cross-reacting on self molecules.

As well as the B cells and T cells (lymphocytes), there are many other types of blood cells—including macrophages, neutrophils, eosinophils, platelets, and mast cells—which cooperate and play a role in the recognition and elimination of antigens and pathogens. These other cell types possess invariant receptors which enable recognition of the antigen. These invariant receptors include receptors for certain carbohydrate structures and for immune complexes containing antibody and complement components (Fc receptors and complement receptors). Each of the cell types has a different combination of receptors and possesses different mechanisms for assisting in the elimination of the antigen or pathogen. The complexity of the interactions of these cells with each other and with immune complexes and antibody is not fully understood but nevertheless is of importance, particularly with regard to the potential use of monoclonal antibodies as *in vivo* therapeutic agents.

1.3. ANTIBODY STRUCTURE AND PROPERTIES

1.3.1. Protein Structure

The immunoglobulin structure is well adapted to the functional role that an antibody molecule (i.e., an antigen-specific immunoglobulin) has to perform. Antibodies can be regarded as universal adapter molecules [1–3,5–9]. As mentioned above, their purpose is to direct appropriate mechanisms of the cellular and humoral immune response toward any given antigen. To accomplish this, it is necessary that first they interact in a precise and predictable way with the cellular and humoral effector mechanisms (i.e., Fc receptors and complement) and, second, that they be adaptable to the binding of any given antigen. To achieve this end, immunoglobulin molecules are built up from discrete units of genes encoding for variable segments and constant segments. The variable segments differ markedly from one antibody to another and are responsible for the differences in antigen binding, while the constant segments determine the basic antibody structure and its interaction with effector mechanisms. The basic structure of an antibody molecule as it is usually schematically represented, as a linearized N-terminal to C-terminal sequence, is given in Figure 1.2. This structure relates more specifically to the IgG, class although it can more generally be used to represent other antibody classes. It may be useful to compare the schematic representation in Figure 1.2 with an alternative schematic representation of the structure of an antibody in Figure 1.3, which attempts to convey the way the chains fold into globular domains and relate to each other in three dimensions.

Figure 1.2 and 1.3 show the basic protein structure of an antibody, which consists of two immunoglobulin heavy chains and two immunoglobulin light chains. The chains have overall structures divided into domains that share a similar repeating structure. The light (L) chains have two such domains: a variable region domain (V_L) and and a constant region domain (C_L). There are two classes of light chains found in human, mouse, and rat, called κ and λ. Thus we have V_κ, C_κ and V_λ, C_λ domains. The heavy chains have more domains, with the exact number depending upon the class of immunoglobulin; the details of these classes are summarized in Tables 1.1–1.3. All three of the species human, mouse, and rat have the classes

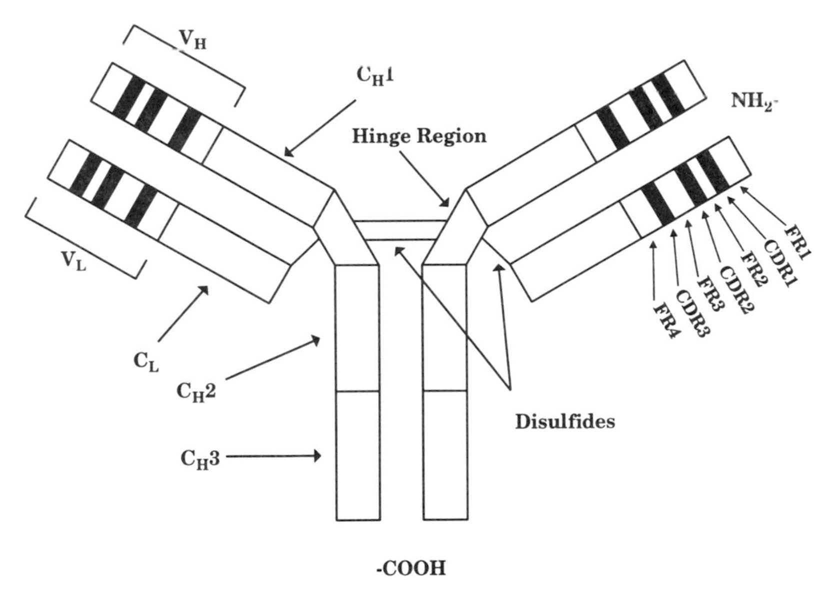

Fig. 1.2. Schematic generalized and linear representation of an antibody molecule (in particular, IgG) that has two identical heavy chains and two identical light chains covalently associated by disulphide bonds. There is also a single conserved intradomain disulphide in each of the globular V-region and C-region domains. In this linear representation, the relative positions of the recognized framework regions (FR) and complementaritiy determining regions (CDR) are marked for the V-region domains.

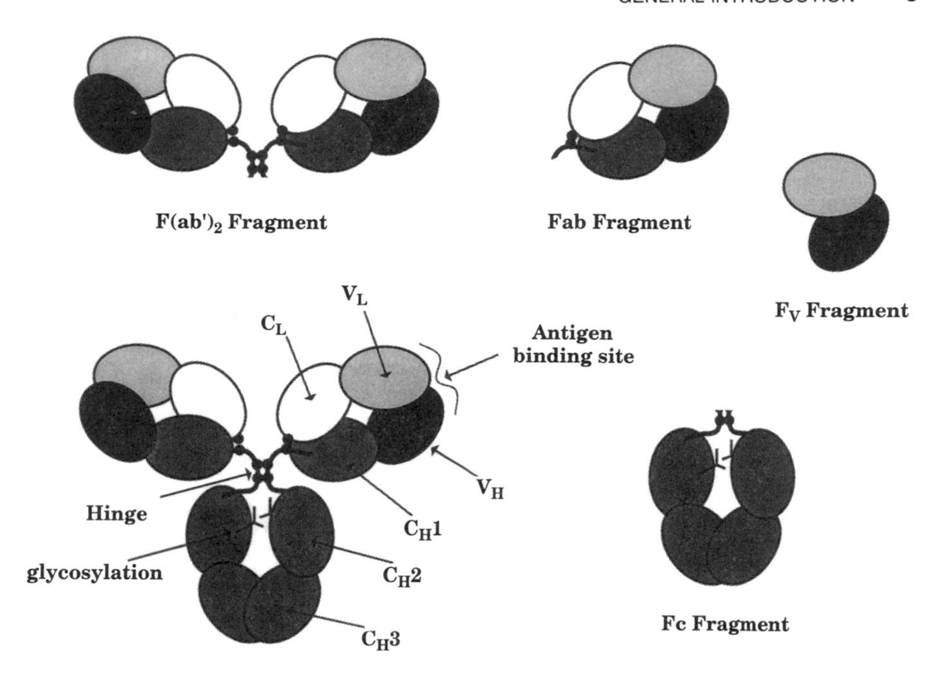

Fig. 1.3. Schematic representations of an antibody molecule (IgG) and a number of the more commonly used fragments derived from the basic structure. In this diagram, an attempt has been made to convey the three-dimensional organization of the globular domains, and this illustrates that there is a pseudo diad axis of symmetry about the vertical plane rather than the implied mirror symmetry in the more commonly used representation such as shown in Figure 1.2.

IgM, IgD, IgG, IgA, and IgE, which have heavy chains called μ δ, γ, α, and ϵ. In human but not rat or mouse there are two IgA subclasses called IgA1 and IgA2, and in all three species there are four subclasses of IgG. These are called IgG1, IgG2, IgG3, and IgG4 in human, IgG1, IgG2a, IgG2b, and IgG3 in mouse, and IgG1, IgG2a, IgG2b, and IgG2c in rat. Note that while there is an overall structural and sequence homology for these classes between species, this does not imply identity at the subclass level [10]. It is interesting to note that a recent paper describes a subclass of IgG found in camels that has no light chain but still binds to antigen [11]. A study of the camel heavy chains may be valuable in understanding how single Fv type antibodies might be derived [11].

In the case of IgG, there is a variable domain (V_H) followed by the first constant domain (C_H1), then a less homologous region of poorly defined structure called the

TABLE 1.1. Antibody Isotypes

Mouse	IgM	IgD	IgG1	IgG2a	IgG2b	IgG3	IgA		IgE
Rat	IgM	IgD	IgG1	IgG2a	IgG2b	IgG2c	IgA		IgE
Human	IgM	IgD	IgG1	IgG2	IgG3	IgG4	IgA1	IgA2	IgE

TABLE 1.2. Human Antibody Properties and Functions by Isotype[1]

	IgM	IgD	IgG1	IgG2	IgG3	IgG4	IgA1	IgA2	IgE
Properties									
Heavy chain	μ	δ	$\gamma1$	$\gamma2$	$\gamma3$	$\gamma4$	$\alpha1$	$\alpha2$	ϵ
H-chain domains	5	4	4	4	4	4	4	4	5
$M_r/10^3$	970	~200	146	146	165	146	160	160	~196
Serum concentration (mg/ml)	1.5	0.03	9	3	1	0.5	3	0.5	10^{-4}
Half-life (days)	5.1	2.8	21	21	7.1	21	5.8	5.8	2.3
Protein A binding	−	−	+++	+++	−	+++	−	−	−
Protein G binding	−	−	+++	+++	+++	+++	−	−	−
Functions									
Complement activation	+++	−	+++	+	++	−	+/−	+/−	−
Placental transfer	−	−	+	−	+	+	−	−	−
Muscosal transfer	+	−	−	−	−	−	+	+	−
$Fc_\gamma RI$ (CD64)	−	−	+++	−	+++	+	−	−	−
$Fc_\gamma RII$ (CD32)	−	−	+	(+)	+	−	−	−	−
$Fc_\gamma RIII$ (CD16)	−	−	++	−	++	−	−	−	−
$Fc_\epsilon RI$	−	−	−	−	−	−	−	−	+++
$Fc_\epsilon RII$	−	−	−	−	−	−	−	−	+
$Fc_\alpha R$	−	−	−	−	−	−	+	+	−
$Fc_\mu R$	+	−	−	−	−	−	−	−	−

[1]Data have been compiled from multiple sources, some of which are contradictory in the information given [1–3, 5–9, 24–39]. Some differences may be due to undocumented variation between allotypes of immunoglobulins and receptors (see recent review by van de Winkel and Capel [37]). An indication of the relative strength of binding is given by the number of plus signs.

TABLE 1.3. Mouse and Rat Antibody Functions by Isotype[1]

	IgM	IgG1	IgG2a	IgGb (mouse)/ IgG2b (rat)	IgG3 (mouse)/ IgG2c (rat)
Mouse					
Heavy chain	μ	γ1	γ2a	γ2b	γ3
Protein A binding	−	+	+++	+++	++
Protein G binding	−	+++	+++	++	++
Human FcγRI	−	−	++	−	++
Human FcγRII	−	+?	+	+	−
Human FcγRIII	−	−	+	+	++
Rat					
Heavy chain	μ	γ1	γ2a	γ2b	γ2c
Protein A binding	+/−	+/−	+/−	+/−	++
Protein G Binding	−	+	+++	++	++
Complement activation (human)	+++	−	+	+++	+
Human FcγRI	−			+++	
Human FcγRIII	−	+	−	++	−

[1]This table shows the functions of mouse and rat monoclonal antibodies that may be of importance in human therapeutic use. As for Table 1.2, the data above are collated from multiple and sometimes contradictory sources [32, 35–37, 40–44]. Comparative information on all the functions for all isotypes is not available, and some entries are absent or incomplete.

hinge, and finally two more constant domains (C_H2 and C_H3). The inter–heavy chain disulfide bridges are found within the hinge region, and the heavy and light chain disulfide bridges are found within the hinge or C_H1 region, depending on class. IgM does not have a hinge region but has instead four constant region domains, C_H1 to C_H4. In addition, secreted IgM consists primarily of a polymerized covalently associated (pentameric) complex of five of the basic 2H + 2L subunits. IgA and IgE also differ in detail from this basic structure of domains, and IgA can be found as covalent polymers of two or more subunits. The immunoglobulin molecule contains conserved N-linked glycosylation sites within the heavy chain constant region domains (e.g., position 297 in the C_H2 domain of all IgG molecules), and this post-translational modification is critical for many antibody effector functions.

X-ray crystallographic techniques have been used to determine the structures of fragments of some of the classes and subclasses [12–15]. Thus the Fab fragment structure (V_LC_L + V_HC_H1) and Fc fragment structure (2 × C_H2C_H3) for human IgG1 is known, but the whole structure—including the hinge with two Fab arms attached to the FC—is not known, although a complete crystal structure of mouse IgG2a has been described [15].

Sequence analysis at the protein level of the variable regions defines three regions of high variability between antibodies, called the "hypervariable regions," that are within four regions of more limited variation, which are the "framework regions" (FR1 to FR4) (see Fig. 1.2)[10)]. From the crystallographic data it appears that these hypervariable regions are principally (but not exclusively) responsible for contact with the antigen. These regions are also called complementarity-determining regions (or CDR1 to CDR3).

1.3.2. Antibody Gene Organization

The basic protein structure of a variable region attached to a constant region is brought about by the rearrangement of one of a large number of variable region encoding gene segments to one of a limited number of constant region encoding segments (see Fig. 1.4 [1–3,16–18]. The variable regions are formed by the re-arrangement of V_L segments with junctional segments J_L to form a unit with the adjacent C_L in the case of light chains and V_H with diversity segments D_H and junctional segments J_H to form a unit with the adjacent C_H encoding segments in the case of heavy chains (cf. Figs. 1.4 and 1.2). The rearrangement of these gene segments is controlled by signal sequences. The number of different V_H and V_L gene segments in the genome determines the variability in the FR1, CDR1, FR2, CDR2, and FR3 sequences. The CDR3 region is more variable and is encoded by the junction of V_L and J_L (light chains) and by junction of V_H with D_H and J_H (heavy chains). The FR4 is determined by the J region. Additional variability is introduced because of imprecision of joining of the segments as well as introduction of extra nucleotides at the junctions. These regions of extra nucleotides not encoded in the genome arc often referred to as N-region diversity, and in Figure 1.4 they would

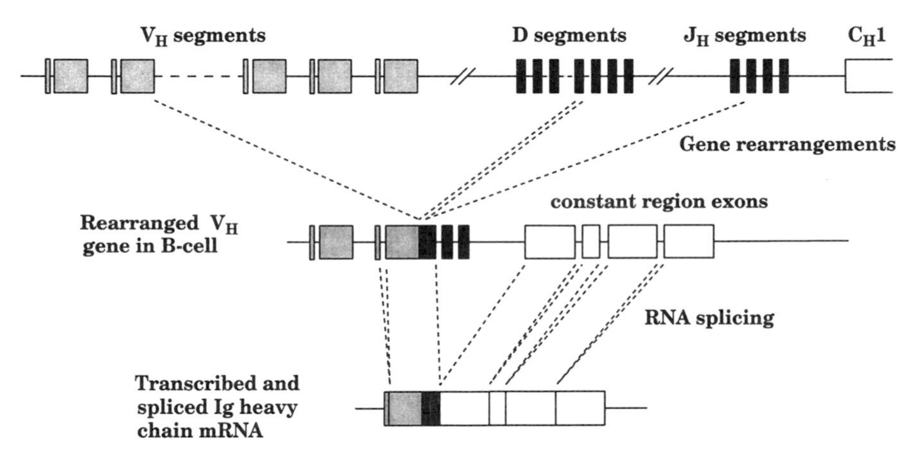

Fig. 1.4. Schematic representation of the organization of the immunoglobulin heavy chain locus. During B-cell development, a rearrangement of the separate DNA segments takes place to give rise to a conventional gene structure involving a number of exons separated by introns. After RNA transcription, these introns are spliced out to give the contiguous mRNA sequence. The light chain genes are similar in organization, but they lack the D segments, and the V segments therefore rearrange directly to the J segments.

occur at the junction of V_H with D and D with J_H. During B-cell development, there is also somatic mutation of the whole V region, followed by antigen-driven selection. This differentiation of B cells is in part regulated by antigen-specific T cells and is therefore said to be "T-cell-dependent" [1–3].

The constant region segments adjacent to the rearranged variable region segments determine the class of chain being encoded. Initially, in the case of heavy chains, this is μ giving rise to IgM production. However, during B-cell differentiation, there are further possible rearrangements whereby constant region genes are deleted such that a new constant region segment is adjacent [1–3,16].

In human, rat, and mouse, the heavy chains are encoded by a gene locus found on one chromosome and the two light chain classes κ and λ are each encoded on different chromosomes. This ensures that the segments can only rearrange appropriately within a class. In addition, there are precise feedback control mechanisms to ensure that any given B cell only encodes one heavy chain allele and one allele from one of the two light chain classes. The mechanism for expression of only a single immunoglobulin allele is commonly referred to as "allelic exclusion" and for a single heavy chain and light chain class as "isotypic exclusion." The result is that every clone of B cells encodes a single unique antibody molecule.

1.3.3. Antibody Protein Fragments

Various protein fragments of antibodies can be derived that may be of practical use in different circumstances (see Fig. 1.3; Chaps. 2, 4) [6,19–23]. These fragments can conveniently be derived by enzyme proteolysis. In general, the Fab region is quite resistant to proteolysis, whereas the Fc and particularly the hinge region are comparatively susceptible. Depending upon which protease is used and the particular antibody isotype (and the animal species from which the antibody is derived), the proteolytic cleavage may occur at the hinge region. If the cleavage occurs on the C-terminal side of the interchain disulfide bridges, then $F(ab')_2$ fragments are generated if on the N-terminal side Fab fragments are generated. Alternatively, mildly reducing conditions can be used to separate the $F(ab')_2$ fragment into two F(ab') fragments.

Similar fragments can also be expressed using recombinant DNA techniques (see below and Chap. 3). Other recombinant products such as Fv fragments contain only the V-region domains, which are not covalently associated and could be considered as the smallest unit of antibody that should still be capable of antigen binding with the original single site affinity. As discussed above, the existence in camels of an IgG-like subclass with heavy chains and no light chains could allow the derivation of smaller antibody fragments with binding specificity.

All of these small fragments of antibodies are considerably useful because they are still capable of binding to antigen but have lost the ability to bind to Fc receptors and to activate the complement cascade. Their smaller size can in certain situations improve their diffusion and penetration properties, particularly, for example, when they are used for staining of tissue sections *in vitro* or for targeting of cellular antigens *in vivo*.

TABLE 1.4. Sequence Homologies of Mouse, Rat, and Human IgG[1]

	Mouse γ1	Mouse γ2a	Mouse γ2b	Mouse γ3	Human γ1
Human γ1	65	65	63	66	—
Human γ2	67	64	63	65	94
Human γ3	65	64	62	66	95
Human γ4	65	64	62	63	93
Rat γ1	80	70	66	66	65
Rat γ2a	82	69	67	68	66
Rat γ2b	71	78	76	70	67
Rat γ2c	69	72	71	81	65
Mouse γ1	—	72	69	69	65
Mouse γ2a		—	80	72	65
Mouse γ2b			—	70	63
Mouse γ3				—	66

[1]The sequence homologies were determined for the predicted amino acid sequences of the constant regions excluding the hinge regions, which are not well conserved [10].

1.3.4. Antibody Properties

The following points of discussion relate to some of the properties of an antibody that are likely to be critical to the intended use, together with a few examples of the situations where these may be important. An attempt has been made to summarize a large amount of data on antibody properties in Tables 1.1–1.5. These data have been compiled from a number of independent sources [24–44], and in some cases the information is incomplete or even contradictory. The data should therefore not be regarded as definitive, particularly as there are often many factors that have an

TABLE 1.5. Fc Receptor Expression on Human Cells[1]

	$Fc_{\gamma}RI$	$Fc_{\gamma}RII$	$Fc_{\gamma}RIII$	$Fc_{\epsilon}RI$	$Fc_{\epsilon}RII$	$Fc_{\alpha}R$	$Fc_{\alpha}R$
Monocytes	(+,i)	+	(+,i)	–	+	+	–
Macrophages	+	+	+	–	+	+	+
Neutrophils	(+,i)	+	+	–	+	+	–
Eosinophils	(+,i)	+	(+,i)	–	+	+	–
Mast cells	–	–	–	+	–	–	–
Basophils	–	+	–	+	–	–	–
Platelets	–	+	–	–	+	–	–
B Lymphocytes	–	+	–	–	+	+	+
T Lymphocytes	–	(+,i)	(+,i)	–	+	+	+
LGLs (K/NK)	–	–	+	–	–	–	–

[1]The expression of Fc receptors on different cells is shown. The expression of some of these Fc receptors on certain cell types is contraversial (e.g., on T lymphocytes) and may be due to differences in activation and or induction [32–38]. Also, some of these receptors may only be expressed on subpopulations of the cell types concerned. (+,i) = Receptors that are not normally expressed but that have been reported to be inducible.

influence on the quantitative as well as the qualitative aspects of the measured properties, particularly effector functions, of individual monoclonal antibodies. Despite these caveats and limitations, the summarized data do help to illustrate the arguments of how the different properties of monoclonal antibodies can influence their usefulness for different intended uses.

Affinity or Avidity. For many uses, both *in vivo* and *in vitro,* the affinity of the antibody for antigen is critical [44–46]. Strictly, the affinity of an antibody for its antigen (association constant or K_a expression in units of M^{-1}) is a measure of the ratio of the concentrations of bound antibody-antigen complex to free antibody and antigen at a thermodynamic equilibrium. It assumes that the interaction with antigen is of single valency, which is more likely to be the case only for very simple antigens or for antibody Fab or Fv fragments.

$$[Ab] + [Ag] \underset{k_{back}}{\overset{k_{forward}}{\rightleftharpoons}} [AbAg]$$

$$K_a = \frac{k_{forward}}{k_{back}} = \frac{[AbAg]}{[Ab][Ag]}$$

Often the interaction of a bivalent (e.g., whole IgG) or multivalent (e.g., IgM) antibody with multivalent antigen (e.g., a cell surface or antigen immobilized on a solid surface) is the critical parameter, and this functional affinity of an antibody with multivalent antigen (e.g., a cell surface or antigen immobilized on a solid surface) is the critical parameter, and this functional affinity of an antibody is referred to as the "avidity." The affinity or avidity of an antibody for antigen is also related to the ratio of the rates of the forward reaction for formation of the complex to back reaction for decay of the complex. Two antibodies may have a similar affinity for antigen measured at equilibrium, but one may have a much slower on-rate ($k_{forward}$) and of course a proportionally slower off-rate (k_{back}). For many uses, the antibody will not be used under conditions of thermodynamic equilibrium; for example, when using an antibody to affinity-purify an antigen or when using antibodies in immunometric assays, the antibody is usually in excess and a faster rate of the forward reaction may then be desirable. In the use of radiolabeled antibodies for the radioimaging of tumors in *in vivo,* the antibody needs first to circulate through the body and then to diffuse and penetrate through the tissues before it even has a chance to interact with the antigen. The stability of the antibody on the tumor once it is bound (affinity and off-rate) as well as the diffusion rates of the antibody in tissues (a product of the antibody or fragment size) are both factors that determine the suitability of one antibody versus another.

Specificity. Particularly in an operational sense, the specificity of an antibody for its antigen is only in part related to its affinity or avidity. It is possible—and also highly likely—that an antibody will have a spectrum of affinities for a range of different antigens [47]. Sometimes these antigens may be completely unrelated, while more often they may share related structural features; for example, many different peptide hormones and growth factors often have a subunit or peptide chain in common with other "family" members. Another example is that many different complex carbohydrate structures share features in common. Different monoclonal antibodies to the same antigen may therefore show different functional cross-reactions on other antigens. Clearly, if the intended use of the antibody is to discriminate between the different antigens in a complex mixture, then the cross-reactions of the antibody are as critical as the avidity for the correct antigen (for example, in the immunometric determination of hormone levels in a serum sample). However, if the antibody is to be used in a situation where the "alternative" antigens are not likely to be encountered; For example, in the affinity purification of an antigen product from a batch culture process, any cross-reactions are irrelevant. In the use of antibodies for *in vivo* therapy or diagnostics, there are so many different tissue antigens that unexpected cross-reactions of the antibody on an irrelevant tissue may frequently be a complicating factor in the development of a monoclonal antibody-based product. The observation of the cross-reaction of an antibody on a second antigen is, of course, related to the avidity of the antibody for that antigen and the sensitivity of the assay being used to measure the interaction. This can lead to the situation where an apparent improvement in the sensitivity of an assay leads to a deterioration in the specificity of the assay. Similarly, the selection of monoclonal antibodies or the engineering of antibodies to produce higher affinities for an antigen may also increase the affinities for the alternative cross-reacting antigens and again lead to a reduction in "specificity" [48]. The solution is to search for a single monoclonal antibody that does not show undesirable cross-reactions in the system used or, alternatively, to blend a number of monoclonal antibodies together such that each of the individual cross-reactions is diluted out (a situation that exists in natural polyclonal antisera).

For the therapeutic use of monoclonal antibodies, it is the operational specificity that is of absolute importance. For example, it may be acceptable to use an antibody that does cross-react on some normal tissue antigens if the ratio of damage to a tumor tissue results in benefits to the patient that outweigh the small amount of damage caused to the normal tissue.

Species and Isotype. Monoclonal antibodies can be prepared from a range of different species using techniques described later (see sec. 1.4). The vast majority of monoclonal antibodies are of mouse or rat origin and particularly of the IgM or IgG classes. Increasingly, through the adoption of a number of different technologies, there is also a growing trend toward the production of human monoclonal antibodies. Within each of these species the antibodies are grouped within a number of classes and subclasses, as summarized in Table 1.1. While there is structural and sequence homology between species for each of the classes, there is no absolute correlation for the subclasses (see Table 1.4). Thus the four subclasses of human IgG are all very closely related by sequence (except for the hinge region), having greater than 90% amino acid homology, while the IgG subclasses of mouse are less homologous to each other, having 69%–80% homology (Table 1.4) [10]. Comparing human with mouse or human with rat gives homologies in the region of 65%

[10]. Thus some of the subclasses of mouse are almost as homologous to the four human subclasses as they are to other mouse subclasses. Obviously, in each of the species the immune system has evolved with its own set of immunoglobulins required to interact with its own effector systems (as summarized for human in Tables 1.2 and 1.5). While there is some conservation and homology for these effector systems, again the correlation is not absolute [49]. In the use of monoclonal antibodies *in vitro,* for example, in affinity purification of antigens or in immunoassays, the species and/or isotype may not be of great importance. In such situations, the species/isotype may sometimes have some advantages or disadvantages; if, for example, antibody fragments are to be used, then different enzymes or conditions may have to be used for their preparation. Problems might be encountered if rat or mouse monoclonal antibodies were used in assays of antigens present in human samples such as serum, because the samples might also contain naturally occurring antimouse or antirat Ig antibodies [50].

When used *in vivo,* additional problems arise. First, the mouse or rat Ig becomes a target for an antiglobulin response, which usually takes about 7 to 10 days to be detected and rises to a peak at about 20 to 30 days, thus severely limiting the useful window within which an antibody can be administered. Second, depending upon the species/isotype, the antibody will interact differently with human effector systems such as complement and Fc receptors. Table 1.3 attempts to summarize some of the data available on interactions of mouse and rat monoclonal antibodies with human complement and human Fc receptors. Some items of data within Table 1.3 are omitted because the properties have not been studied or reported on in any detail. These areas of uncertainty could be crucial to an understanding of why certain isotypes and antibodies have shown therapeutic potential while others have not (see sec. 1.8.3; Chap. 2). The fact that rat IgG2b is very good at activating human complement and at activating antibody-dependent cell-mediated cytotoxicity through $Fc_{\gamma}RIII$, and that rat IgG2a is not [43], may be a significant factor in the success of the therapeutic rat IgG2b antibody Campath-1G (CDw52) [51,52]. Alternatively, for an antibody that is to be used in radioimaging or in targeting, the binding of the antibody to Fc receptors or activation of complement is likely to be undesirable (see Chap. 2).

When making recombinant antibodies, it is possible to engineer into the final construct any species/isotype, and so a choice can be made of the appropriate properties and functions required [25,28–30,43]. For therapeutic applications, it now seems likely that most mouse and rat monoclonal antibodies will need to be humanized by using recombinant DNA technology [28,53–55]. At present a reasonable approach in the early stages of such projects is to produce a matched set of antibodies of different isotypes and then to compare them to determine the most useful isotype [26,29–31,39,43]. Using such an approach, the rat IgG2b antibody Campath-1G (CDw52) was humanized to give the human IgG1 antibody Campath-1H [28,39,56].

Allotypes. Within each species studied, some of the isotypes also exist in different allelic forms. Some of these alleles can be distinguished serologically, giving rise to recognized "allotypes" [57–60]. At present there are very few data on the properties and possible functional differences that might be determined by these allelic differences. An example of the kind of differences that can occur is provided by the human IgG3 subclass. The most common allotypes of IgG3 found within the white population fail to bind protein A (see Table 1.2), but an alternative allele found at higher frequency in the East Asian population binds strongly. Also, in a study of a matched set of hapten-specific chimeric antibodies, which included two IgG3 allotypes (b and g), quantitative differences were seen in complement activation [25,26,39]. Similarly, functional differences between recombinant IgG1 allotypes in complement-mediated lysis can be detected under certain circumstances [39]. The allotype of a therapeutic antibody may also be important, not only because of possible functional differences, but also because of the likelihood of an antiallotypic immune response in those individual patients who differ in allotype. One possibility is to generate the same antibody in multiple allotypes and to match the patient's allotype; however, this may not be a reasonable commercial aspiration (see Chap. 6). An alternative approach might be to attempt to generate antibodies in which the allotypes had been mutated out to give "null allotypes" [39,60].

Glycosylation. All classes and subclasses of immunoglobulin contain conserved sites for N-linked glycosylation. In addition, it is sometimes found that the variable regions are also glycosylated as a result of the presence of a suitable sequence (Asn-X-Ser or Asn-X-Thr where X is not Pro). The carbohydrate attached is composed of mixtures of complex branched chain structures, the ratios of each of the individual structures to each other varying with the isotype and species [5–9,61]. All IgG molecules contain a single conserved asparagine (297) that is glycosylated in the C_H2 domain (see sec. 1.3.1; Figs. 1.2, 1.4). The removal of the carbohydrate using enzymes or prevention of glycosylation by mutating the sequence or by using metabolic inhibitors such as tunicamycin results in IgG molecules with grossly impaired functions with regard to activation of complement and binding to Fc receptors [39,62–64]. In addition, there are carbohydrate receptors found *in vivo* that can bind to carbohydrate structures which do not contain a terminal sialic acid residue, and this can result in a shorter *in vivo* half-life. This is highly relevant for monoclonal antibody production, because the state of glycosylation of the antibody is very dependent upon the method of production [65–67]. Bacteria are unable to attach *N*-linked carbohydrate and also have difficulty in forming complex disulphide bridges, and so the generation of fully functional antibodies in such systems is not a viable prospect and limits these cells to the production of antibody fragments (see Chaps. 3, 5) [68]. Expression of antibodies in mammalian cells including hybridomas, plasmacytomas, and nonlymphoid cells such as Chinese hamster ovary (CHO) cells results in glycosylated product. However, the exact structures attached to the antibody seem to vary with both the particular cell line and the growth conditions [65–67] (see Chap. 5). The therapeutic antibody Campath-1H (CDw52) has been prepared from the rat hybridoma cell line Y0 and from CHO cells, and both forms of the antibody have been used successfully in patient treatment [28,56,69–72]. Some naturally occurring forms of "aberrant" glycosylation have been associated with disease in a human [73–76]. More work is still needed to clarify the importance of individual types of glycosylation in antibody function.

Disulfide Bonds. Antibodies contain conserved disulfide bonds as intrachain bonds within each domain, as well as interchain bonds between the heavy and light chains and between the heavy chains [5–9]. The exact positions and numbers of the interchain bonds do vary with the species and isotype. In addition, the multiple subunit structures of IgM and IgA involve disulfide bonds to each other and to J chain. Antibodies are remarkably stable proteins, and no doubt the disulfide bonds assist in this. However, treatment of antibodies with mildly reducing conditions (e.g., 20 mM mercaptoethanol) does not produce a major breakdown of the structures. Under such conditions, the subunits of IgM will dissociate from each other but still remain intact and an associated drop in the avidity of binding antigen is usually observed. The inter–heavy chain and the heavy and light chain bonds within the hinge region of IgG can be disrupted, resulting in a possible increase in flexibility and sometimes a consequent increase in avidity [77], but again the overall association of chains remains intact. The functional activity of IgG in binding Fc receptors is impaired by reduction of disulfide bonds in the hinge region. These observations are again critical to the production, purification, and end use of the antibody (see Chaps. 2, 4, 5). For example, bacteria have trouble in correctly disulfide-bonding large proteins and so are more successfully used in the expression of smaller fragments of antibodies (see Chap. 5) [68].

1.4. DERIVATION OF MONOCLONAL ANTIBODIES

1.4.1. Somatic Cell Fusion

Conventionally, monoclonal antibody-secreting cell lines have been derived by immortalizing immune B cells by somatic cell fusion with a suitable tissue culture cell line [4,78]. For reasons that are not fully understood, this technique has proven to be most successful for a limited range of species, particularly for mouse- and rat-derived monoclonal antibodies of the IgM and IgG classes [79], although other species such as sheep, hamsters, and humans have been used. A range of different myeloma (or plasmacytoma) cell lines have been established and used for the derivation of hybridomas [4,79–84], but, for example, many of the mouse cell lines have been derived from a single original plasmacytoma cell line, P3-X63Ag8 (MOPC-21). Different cell lines appear to have differing efficiencies in immortalization of cells [79].

A large number of variations in the methods for cell fusion and subsequent selection of hybridomas exist, and these are well documented in textbooks and reviews devoted to the methodology [44,46,78,85–87]. Originally Kohler and Milstein used Sendai virus to induce cell fusion [4]. This virus was then replaced with polyethylene glycol (PEG) [78], which is still used widely, although some protocols have now adopted electrofusion techniques [88]. For mouse and rat hybridomas, the efficiencies of these procedures are all high, and typically several hundred to thousands of individual hybridoma clones can be obtained from one animal spleen [79,89].

1.4.2. Cell Transformation

Although limited success in the production of human monoclonal antibodies has been reported using cell fusion with mouse myelomas as well as a number of human myelomas, the efficiency of the process is extremely poor compared to mouse or rat monoclonal antibodies [90–92]. Human monoclonal antibodies have in addition been derived by transformation of human B lymphocytes with a suitable virus such as the Epstein-Barr virus (EBV) [90,92]. Common problems with this technique are that the transformed cells are not stable for production of the antibody at reasonable levels and there is a constant requirement to reclone and rescreen to maintain a suitable culture. Attempts to solve this problem include the somatic cell fusion of EBV-transformed clones with plasmacytoma cell lines [90,92] or, alternatively, the rescue of the antigen-specific antibody by cloning the immunoglobulin chains and then expressing them as recombinant products in another cell line [93].

Using somatic cell fusion or cell transformation either alone or in combination, it has proved very difficult to make human monoclonal antibodies. The time and effort expended is far, far greater than that needed to produce the equivalent mouse or rat monoclonal antibodies. It is this difficulty in the production of human monoclonal antibodies that has driven forward the strategies for rescue of human antibodies using recombinant DNA technology (see sec. 1.6; Chap. 3).

1.4.3. Immunization

Immunization of the animal is a critical factor in the derivation of any monoclonal antibody. As discussed in section 1.2 and show schematically in Figure 1.1, the antibody response to most antigens is a T-cell-dependent B-cell response [1–3]. The antigen is taken up by specialized presenting cells (which include macrophages and B cells) and processed to give rise to T-cell epitopes that are short linear peptides expressed in association with MHC class II molecules. Thus, for a good response to be generated, the antigen complex should possess both T- and B-cell epitopes. The use of adjuvants and carrier antigens to initiate antigen processing and in providing this T-cell help for the B cells is often crucial [1–3,44,46]. In general, although antibodies are generated by B cells, which randomly rearrange and then somatically mutate their immunoglobulin genes, the immune response in an animal is directed toward those parts of an antigen that are dissimilar to self antigens. Thus there is very strong selection during an immune response in favor of some B-cell clones and against others, and this is in part determined by the complex genetics and ontogony of the T-cell repertoire and antigen presentation on polymorphic MHC molecules. The important point is that it may prove very difficult to immunize for some specificities in a given strain or species. If difficulty is encountered with raising an antibody response of a required specificity, then the method of immunization, adjuvant, form of antigen, animal strain, and species can all be varied [44,78,87].

Even given a good antibody response there are still some complicating factors with regard to their selection and perhaps of relevance to their proposed practical and commercial use. In addition to the specificities, the isotypes of an antibody made to any given antigen are also subject to regulation by processes that are currently only poorly understood. Thus the form of the antigen, route of immunization, carrier, adjuvant, timing, strain, and species may all play a major role in biasing the immune response to produce a restricted pattern of immunoglobulin isotypes. The isotype of an antibody will influence its effector and other properties, thus determining the applicability of the individual antibodies for the intended purposes (see Tables 1.1–1.3, 1.5).

In addition to the above complications, in order to derive the required monoclonal antibodies it is necessary to immortalize the responding clones of B cells. Unfortunately, B-cell responses to antigens can occur in different lymphoid organs, and this again is influenced by the nature of the antigen and route of administration.

Efficient methods for *in vitro* immunization of cells is one avenue of research that is currently being pursued to tackle some of these problems [94,95]. This is particularly relevant to humans, where immunization is both practically and ethically difficult to achieve for a majority of antigens. In addition, even where it might be practical to immunize or to find a naturally immune human donor, it is difficult to obtain a source of the responding antigen-specific B cells, as these are likely to be localized in a lymphoid organ and not circulating in the blood.

1.4.4. Screening and Assays

Screening of the cultures is a critical part of the process in successfully generating the required monoclonal antibodies [44,46,78,85,87]. A desirable approach is to derive an assay that tests the individual antibodies under the circumstances in which they are to be finally used. This is an ideal that often cannot be achieved for very practical reasons. An important lesson that has been learned time and time again by those involved in monoclonal antibody production is that what you get at the end often meets the requirements of your assays but is not suitable for the intended end use. It is therefore very critical to understand what the intended use of the antibody is, what properties this demands (see sec. 1.3.4), and then to design the simplest assays that meet the requirements for testing these properties. It is necessary to apply the screening assays to large numbers of cultures and to obtain the results quickly enough to satisfy the tissue culture demands of maintaining and recloning stable cell lines from the heterogeneous and polyclonal starting cultures.

Many assay procedures—e.g., enzyme-linked immunoadsorbent assays (ELISA), radioimmunoassays, and immunofluorescence assays—test the binding properties of the antibody [44,46,87]. Using indirect procedures involving isotype-specific second antibodies or reagents such as protein A and protein G (see Tables 1.2, and 1.3) allows for the selection of additional properties of the antibody at the same time. In many of these procedures, it is the multivalent avidity of the monoclonal antibody that is being tested rather than the single Fab affinity, and this may be critical to the desired end use (see sec. 1.3.4).

Functional properties of an antibody such as virus neutralization, complement, and cell-mediated cytotoxicity are all properties of an antibody that are isotype-dependent (see Tables 1.2, 1.3) and that can be assayed for *in vitro* [25,28–30,86].

Using such *in vitro* cytotoxicity assays (both complement and cell-mediated) on a range of different cell lines, tissues, and cell types often gives a pattern of cell killing that does not correlate exactly with the binding spectrum as measured in ELISA or immunofluorescence assays for the same cells. This is because the cytotoxicity of an antibody for a given cell type is influenced by a large number of factors, many of which are only poorly understood [96–98]. Thus, although the antigen density may be one factor, the cell type is also important, as many cells possess an array of cell-associated molecules and functions that protect them from lysis through activation of their own effector systems [99]. Also, some antigens may not be as readily accessible to an antibody in one tissue as in another, although it is very difficult to predict how an antibody might react *in vivo*. The valency of interaction of the antibody may also be important; for example, monovalent antibodies are sometimes many times more potent than bivalent antibodies (see below) [100–102]. Monoclonal antibodies are often less potent and do not exhibit the same properties as polyclonal antisera. Blending mixtures of monoclonal antibodies may be a solution to this problem, as often two monoclonal antibodies may behave synergistically if they bind to different epitopes of the same antigen [103,104]. This synergy may be exhibited as an improved binding avidity or ability to precipitate the antigen [105] with a corresponding increase in the apparent specificity of reaction, improved virus neutralization, or a greatly increased lytic activity against cell targets. A synergistic pair of rat IgG2b antibodies to the CD45 antigen, which is also known as the leucocyte common antigen, have been used therapeutically to treat kidney allograft graft rejection by removal of passenger leucocytes (which express the antigen at high levels) from the graft using an antibody containing perfusion prior to grafting [106]. The commercial exploitation of such products may be more difficult in terms of the regulatory issues as discussed in Chapter 6.

Results from animal models can help in defining the properties of a therapeutic antibody that are desired, but certain limitations have to be accepted in interpreting the results. Very few monoclonal antibodies will bind to the equivalent antigen in another species with the same affinity and specificity. The problem becomes more acute the more distantly related are the species. Until recently the only test species for many monoclonal antibodies specific for human antigens were primates, and the use of these species obviously involves serious ethical as well as technical and commercial issues (see Chap. 6) [107,108]. Recent advances in technology have meant that transgenic animals—for example, mice transgenic for human antigens—are becoming more readily available, and these do provide an opportunity to test antihuman monoclonal antibodies in a laboratory setting [109]. An alternative is to make parallel sets of monoclonal antibodies to the equivalent antigens in experimental laboratory animals and then to use the results from studies on these to improve or to select the nearest human-specific equivalent antibodies. It should be obvious from the discussion in section 1.3 (cf. Tables 1.2, 1.3, 1.5) that there are many interactions in addition to the affinity for antigen, such as binding to Fc receptors and to complement, that are likely to be important and that may not show complete conservation between species.

1.4.5. Cloning of Cell Lines

For prolonged stability of the antibody-producing cell lines, it is necessary to clone and then reclone the chosen cells [44,78,85,87]. Cloning consists of subculturing the cells by either limiting dilution at an average of less than one cell in each culture well or by plating out the cells in a thin layer of semisolid agar of methyl cellulose or, alternatively, by single-cell manipulation. The efficiency of cloning varies form cell line to cell line and depends also upon the growth conditions, media, additives, and use of "feeder" cells.

At each stage the cultures must be assayed for production of the appropriate antibody. Some cell lines are intrinsically more stable than others. If the end goal is large-scale production of an antibody, then it is necessary to assess how the production levels are likely to vary with the number of cell generations needed to expand the culture to the required level (see Chap. 5). The stability may also be dependent upon maintaining drug selection on the cells. However, it may be undesirable to use the drug in large quantities or it may be difficult to remove it from the final product.

For large-scale production of therapeutic antibodies, it is necessary to establish a master cell bank and a working cell bank and to monitor the stability of the cell lines derived from these. A complete discussion of these issues is to be found in Chapters 5 and 6. A cell line that produces a large amount of antibody in ascites may not be as good in culture and equally production *in vitro* may be dependent upon not only the particular clone but also on the growth conditions (see Chap. 5). For ethical, regulatory, and commercial reasons, production of antibody by industry and in the research laboratory is increasingly being carried out by scale-up of *in vitro* culture facilities rather than from ascities (see Chaps. 5, 6).

1.5. SECOND-GENERATION ANTIBODIES

As a result of the phenomena of allelic exclusion—i.e., each cell makes only one antibody—conventional antibodies have a single specificity derived from the unique combination of a single pair of a functionally rearranged and expressed heavy chain and light chain. However, if several immunoglobulin genes are artificially introduced into the same cell, then they are all expressed [100,101,110–112]. Multiple expression can be achieved either through the somatic cell fusion of two immunoglobulin-producing cell lines or by transfecting several immunoglobulin genes into a hybridoma cell line. In such circumstances, the individual immunoglobulin chains may pair randomly and give rise to novel combinations. As a crude generalization, it would seem that light chains can pair with any heavy chain class and that the heavy chains can form pairs with a heavy chain of the same class, even of a different subclass or different species.

The unusual pairing of the immunoglobulin chains in such cell lines results in antibody molecules with novel properties (see Fig. 1.5). [100,101]. Functionally, monovalent antibodies arise because the inappropriate pairing of a heavy and a light

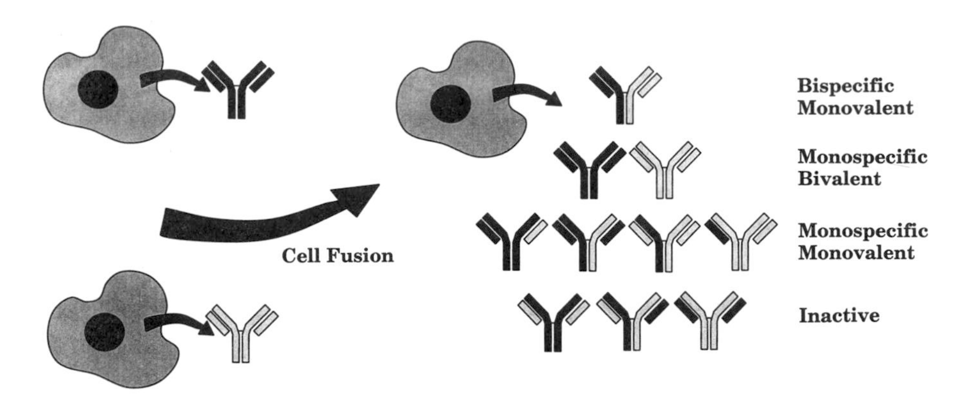

Fig. 1.5. When two antibody-producing cell lines are fused together, the resulting cell line codominanty expresses all of the immunoglobulin chains. These may randomly associate to give rise to the complex mixture of new antibody molecules with different specificities and valencies illustrated.

chain may lead to a nonfunctional binding site. Bispecific antibodies having two different specificities within the same molecule result from the pairing of two different heavy chains, each one associated with its correct light chain. These bispecific and monovalent antibodies have unusual properties and have been found useful in a number of applications, which are discussed below in section 1.8.

1.6. ANTIBODY ENGINEERING

Antibody engineering first started with the manipulation of the immunoglobulin molecule as a protein. Through proteolysis and chemical modifications, antibodies can be fragmented, modified by chemical coupling with other molecules (see Chap. 4), or rendered multispecific by cross-linking different antibodies together [23].

Through the use of recombinant DNA technology (Chap. 3), it is now possible to genetically engineer almost any antibody for the creation and expression at high levels of novel antibodies [113–117]. Recombinant antibodies have the following advantages over conventionally derived monoclonal antibodies.

1. There are no restrictions on the species or isotype from which the recombinant antibodies are derived. Thus through the use of appropriate cloning strategies it should be technically feasible to isolate the genes encoding any antibody made from any immunized species, and so future applications need not be restricted to the derivation of the mouse, rat, and human isotypes listed in Table 1.1

2. Therapeutically useful rodent monoclonal antibodies can be partially "humanized" by making chimeric antibodies with human constant regions, thus introducing the effector mechanisms of the human isotypes (see Table 1.2) while at the same

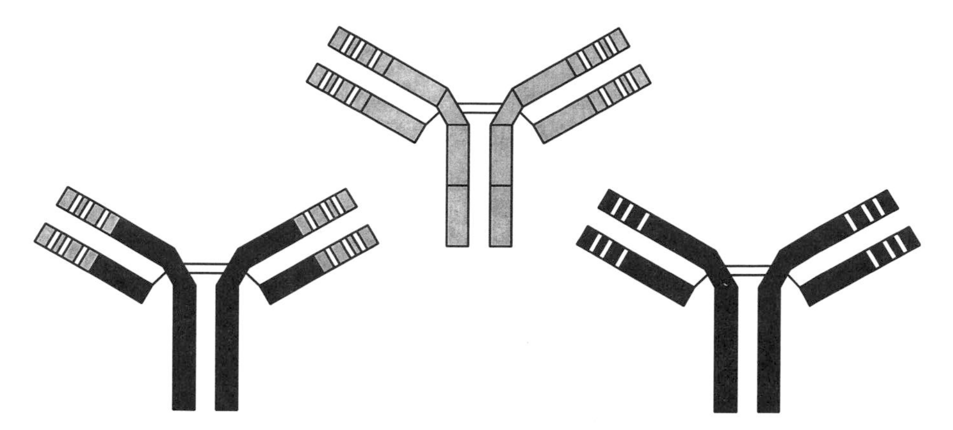

Fig. 1.6. Shown schematically (cf. Figure 1.2) are comparisons by sequence content of a rodent monoclonal antibody (top), a chimeric human/rodent antibody (bottom left), and a fully reshaped or humanized antibody (bottom right). White represents the parts of the structure derived from the rodent sequence and black represents the human content.

time minimizing the number of potential immunogenic epitopes (see Fig. 1.6) [26,29–31,39]. For therapy there are several key features that an antibody should have if it is to be successfully used. Obviously the antibody must possess a desired specificity to bind to a relevant antigen, such as an antigen expressed on a tumor cell surface or, alternatively, a viral antigen or perhaps a bacterial toxin. Once bound to that antigen, the antibody is then normally required to carry out a function such as the targeting of a radioimaging label to or the destruction of the tumor cell or virus, or alternatively the neutralization of the virus or toxin. As discussed in section 1.3, the specificity is a function of the variable region domains, but the effector functions are a product of the constant region domains (Tables 1.2, 1.3). Although some rodent isotypes may be able to activate and exploit human effector systems, as indicated in Table 1.3—and indeed they can show remarkable therapeutic effects, such as in the use of the mouse IgG2a antibody OKT3 for immunosuppression [118] or the use of the rat IgG2b antibody Campath-1G in lymphoma therapy [51,52]—there is still the problem of the antiglobulin response to these "foreign" antibodies, which limits the available time course of therapy to a single treatment of about 7–10 days. The production of a chimeric antibody and the selection of the most appropriate human isotype allow for the retention or addition of desirable functions but at the same time reduce the foreignness of the antibody to the patient (see Chaps. 2, 3) [29,30].

3. As a further step in the reduction of the immunogenicity or foreignness, rodent monoclonal antibodies can also be fully humanized or "reshaped" to produce human antibodies that contain only those key residues from the rodent variable regions responsible for antigen binding [29,53–55,60,102,119] (see Fig. 1.6). The fully humanized or reshaped human IgG1 antibody Campath-1H was derived from the rat

IgG2b antibody Campath-1G, and clinical studies have now shown that the therapeutic benefits of the antibody have been retained and improved by reducing the antiglobulin response [72]. There are many factors to be considered in designing a reshaped antibody, and some of the earliest examples may not have been optimally humanized, as recently discussed in a review by Routledge et al. 1993 [55].

4. As a lead into the therapeutic applications, different isotypes of a given specificity can be constructed, giving rise to panels of antibodies of equivalent specificity and affinity which can then be compared in various assays [26,29–31,39]. As has been outlined above, it is not yet possible to predict with any certainty the properties of a monoclonal antibody that are necessary and an empirical approach, such as comparing a range of isotypes with the same specificity, is likely to be useful in developing the best agent.

5. Specific residues within the constant regions can be modified or replaced so as to generate antibodies with novel effector functions [117,120,121]. Recent studies have indicated that changes in certain key residues within the antibody molecule, particularly within the C_H2 domain and within the hinge, when altered can dramatically affect the binding to different Fc receptors or the activation of complement [63,64,120–130]. At present it is not certain what are the most important effector mechanisms responsible for some of the therapeutic effects seen with antibodies such as Campath-1H, but progress is being made in animal models where panels of different antibodies that exploit different effector functions can all be engineered and then tested for their *in vivo* effects [49,109]. The early results from such studies with mice seem to indicate that for the elimination of CD8-positive T lymphocytes, activation of complement is less important than binding to Fc receptors, although which of the Fc receptors is more important has not yet been determined [49].

It is the modular structure of antibody molecules that are composed of a collection of discrete globular domains, encoded by genes with a similar modular structure whereby each domain is coded in a separate exon (see secs. 1.3.1., 1.3.2; Figs. 1.2–1.4) that makes the manipulation of immunoglobulin genes a relatively straightforward proposition. It is important to note for the reshaping of variable regions that the regions FR1, CDR1, FR2, CDR2, and FR3 are contributed by the V-gene segment, FR4 is contributed by the J-segment, and the most variable protein region CDR3 is contributed by the junction of V_LJ_L for light chains and by the D segment and its junctions V_HDJ_H for the heavy chains (see Figs. 1.1 and 1.2). Thus, although the individual elements of the variable region are inherited in the germline, the precise rearrangements of these elements are unique to a given B cell and, because the possible number of permutations of rearrangements and somatic mutations is so large, every animal even in an inbred population will not be expressing the same repertoire of immunoglobulin genes. This is of relevance when considering therapy with any humanized or even human-derived antibody, because it is possible that the unique sequence and structure of the CDRs of any antibody may eventually be recognized by the patient as foreign and will result in an antiglobulin response. The antigenic site formed by the CDRs is called the "idiotype" and an immune response to this site is called an "antiidiotypic" response. Recent clinical results with the fully humanized antibody Campath-1H indicate that repeated treatment with the antibody does result in an antiidiotypic response, but that the response is delayed in time when compared with the response to the rat antibody Campath-1G, which elicits a strong response following a single course of treatment [72]. However, one possibility is that the human frameworks that were used for the reshaping of Campath-1G are not optimal because they were derived from myeloma-based sequences containing somatic mutations rather than human germline sequences, as discussed in a review by Routledge et al. 1993 [55]. Interestingly, at the time of the reporting of the very first chimaeric antibodies in 1984, it was speculated that CDR grafting might be possible and that it might reduce the immunogenicity further but would probably not avoid the antiglobulin response to the idiotype [131].

Once the recombinant antibody genes have been manipulated, they can be expressed in a number of different systems [54,55,68,69,113,115,132–135], as described in Chapters 3 and 5. Transfection of antibody genes cloned in suitable vectors into myeloma cells may result in expression of the antibody molecules, which are processed and glycosylated in a manner that is characteristic of immunoglobulin produced by hybridoma cells and normal B cells. The level of transcription of the recombinant genes can be controlled in B-lymphoid cells by using immunoglobulin promoters and enhancers. It is also possible to express antibodies in other mammalian cell types, such as CHO cells, but this requires other promoter and enhancer elements that are not B-cell specific [54,55,69]. Antibodies have also been expressed in a number of other eukaryotic and also prokaryotic expression systems, including plant cells, yeast, and bacteria [68,133–135]. As discussed in section 1.3.4, there are certain problems that are encountered in these different expression systems, chiefly with regard to glycosylation and to disulfide bonding, which preclude expression of complete molecules or complicate the purification of the antibody product [68,133]. To reiterate, because antibodies have multiple domains and multiple chains, including intradomain as well as interchain disulfide bonds, the cell must be capable of correctly assembling the molecule. In addition, for most of the antibody effector functions, appropriate N-linked glycosylation is required [62–64]. Limitations in bacteria to the assembly of complex multimeric proteins involving intrachain and interchain disulphides as well as N-linked glycosylation dictate that they are only really suitable for expression of smaller fragments from antibodies, such as Fab or Fv fragments (see sec. 1.3.4; Fig. 1.4; Chap. 3). The plant and yeast expression systems do not secrete the antibody, and to purify the product intact it is necessary to devise protocols to break down the cell walls (see Chap. 5). At present, most commercial large-scale production of recombinant antibody molecules is carried out using either B-lymphoid cell lines or CHO cells, which is described more fully in Chapter 5.

1.7. COMBINATORIAL AND PHAGE DISPLAY LIBRARIES

The derivation of monoclonal antibodies by the Kohler and Milstein technology, as outlined above in section 1.4, is a lengthy and time-consuming process that can take many months to years to follow through and has many other limitations, as mentioned before, and no guarantee of success at the end. Alternatively, recent advances in molecular biology mean that mammalian genes can be cloned and expressed in bacteria over a relatively shorter time scale of weeks to months [53]. The advantage of the Kohler and Milstein technology is that once the immortalized hybridoma cells have been established in culture, the supernatant from each of the cultures can be assayed for the presence of an antibody of interest and the appropriate cultures can then be identified and propagated further (see sec. 1.4). The same procedure needs to be adopted for antibody fragments produced in bacteria. Thus the bacterial colony or phage that encodes an identified antigen-specific Fv of Fab must be isolated and propagated. In the original bacterial expression systems, λ phages were used as vectors that encoded cloned immunoglobulin heavy and light chain genes, which when expressed gave rise to Fab fragments [136]. The bacterial colonies were grown on petri dishes and were replicated using nitrocellulose filters. The bacteria on some of the replica plates were lysed and the "antigen"-specific Fab fragments detected by immunoassay on the filters (a procedure that only works well for some antigens, in particular highly purified proteins). The relevant replica cultures were then identified and further propagated [136,137]. Other expression systems allowed for the expression of the Fab or Fv fragments as fusion proteins with the coat proteins of the bacteriaphage M13 [138–140]. This meant that the phage particles when assembled carried Fab or Fv fragments on their surface and also contained within their phage genomes the cloned DNA that encoded the expressed fragments. These "phage display libraries" can be screened by affinity purification of the phage encoding the specific fragments using antigen immobilized on a solid matrix or column. These expression systems also allow the cloned heavy and light chain genes to be randomly reassorted with each other, giving rise to combinations of immunoglobulin chains that were not necessarily present in the original B cells from which the genes were obtained [53,141]. These libraries are referred to as "combinatorial libraries." It has been argued that the ultimate use for the phage display and combinatorial libraries might be to completely bypass or replace the need for experimental animals in the production of specific antibodies [142–148].

In the first instance, phage display libraries and combinatorial libraries provide the ability to "rescue" the antibody response from almost any immunized animal in the form of cloned genes encoding the individual heavy and light chain variable regions. Combinatorial libraries are generated by cloning a repertoire of immunoglobulin heavy and light chains, usually by using the polymerase chain reaction, from mRNA isolated from tissue containing B cells from an immune donor. Using appropriate expression vectors, the immunoglobulin chains are expressed as pairs of a single heavy chain and a single light chain. However, these combinations represent a random pairing of chains that are not necessarily representative of the original pairings in the B cells from which the mRNA was derived [53]. As discussed in section 1.3, it is the unique pairing of a given heavy chain with a light chain that gives rise to the specificity for antigen. Any reassortment of chains is highly likely to result in the loss of antigen specificity for the original antigen and may result in a specificity for a new antigen. Theoretically, the original pairings will be present in such a library at the square of the frequency of the original B cell [53]. Purely as a theoretical example, if a B cell encoding an antibody of interest was present at a frequency of 1 in 10^6 the same combination of chains would be present in a combinatorial library at a frequency of 1 in 10^{12}. This seems a big disadvantage in that such a low frequency means that a much larger library will need to be generated and screened and this may not be practical. However, in practice it does seem that the combinations of chains for a given specificity are more promiscuously permissive and a higher frequency of positives than expected is found. It should, however, be remembered that the combinations rescued after screening such a library are not necessarily representative of the combinations present in the native B cells. This last point may be of importance, because, as it should be remembered, the *in vivo* B-cell repertoire has been selected through a complex system of random gene rearrangements followed by both positive and negative selection (see sec. 1.2). The reason for this selection, which involves T-cell antigen-specific recognition and help, is to restrict the immune response to "foreign" antigens and to prevent cross-reactions on self antigens. Certain disallowed combinations of heavy and light chains may be generated in combinatorial libraries.

Phage display libraries represent the next step forward in the expressing and screening of combinatorial libraries (see Chap. 3). Genes encoding the surface proteins of a phage are altered so as to allow the insertion of a fragment of an immunoglobulin gene, which is expressed as a fusion protein on the surface of the phage that carries the gene. The phage expressing a Fab of Fv fusion protein can be selectively enriched and isolated by virtue of its avidity for the antigen. The DNA encoding the binding fragment is packaged in the same phage, and this represents a very rapid way of isolating the antigen-specific combinations of genes.

During the course of an animal's immune response, B cells that carry on their surface an antigen-specific receptor in the form of surface antibody can be selected for further differentiation [1–3]. As well as going on to differentiate into antibody-secreting plasma cells, a proportion of B cells allow their expressed antibody V-region genes to undergo somatic mutation followed by further antigen-driven selection. This process of somatic mutation produces a repertoire of antibodies, some with higher affinity than the parental antibodies, called "affinity maturation." Affinity maturation can be mimicked by using phage display libraries in which the genes encoding the antigen-specific chains have been isolated and then by randomly reassorting these chains again or, alternatively, by artificially mutating the CDR sequences using random oligonucleotide primers [143–148]. This process allows for the rescreening and isolation of the phage with altered binding specificity or avidity for antigen. At the extreme, the phage can be used to "mimic" the immune response by generating an artificial randomized library of synthetic genes with random CDR sequences [143,144]. Such libraries have been used to successfully screen for a number of different specificities. However, at present, despite claims that this technology replaces the need for the Kohler and Milstein technology, it can be argued that this is a naive view of how the immune system works and there are many more facets to be mimicked before artificial libraries can wholly replace immunized animals.

A major disadvantage of phage display libraries is that the only antibody function being tested is antigen binding, and this may not be the crucial function for the final application. Although the genes once isolated can be expressed along with any constant regions, the effector function-dependent assays cannot be used for the detection and isolation of the appropriate specificity. Additionally, it is relatively easy to screen phage libraries on purified and homogenous antigen preparations, but it is very difficult for rare antigens or complex mixtures such as cell surface antigens. Also, the animal immune system not only positively selects antibodies for binding to the antigen or pathogen but negatively selects against those specificities that cross-react on the many self antigens (see sec. 1.2). Although it can be argued that this is not necessarily relevant to derivation of antibodies from rodents which are ultimately to be used in therapy in humans, there may be sufficient conservation between species to provide a useful elimination of many undesirable cross-reactions.

1.8. APPLICATIONS OF MONOCLONAL ANTIBODIES

1.8.1. Research Reagents

Kohler and Milstein made their original discovery while pursuing research into the mechanisms for generation of antibody diversity. Although the emphasis of this book is on the problems encountered when applying this technology on a commercial scale, it should be remembered that the origins of the technology and still a major use for monoclonal antibodies is as basic research tools. Monoclonal antibodies are frozen examples from of the B-cell repertoire of an immune animal, at a point in time, and they can be used to investigate the immune response to antigens.

As general research laboratory reagents, they are extremely useful because they can be produced with relatively little effort and in the quantities and purity sufficient for the average research laboratory need. The monoclonality and the ease with which these reagents can be exchanged between laboratories facilitates the precise definition characterization and confirmation of the antibodies and their antigens. A good example of this has been the leucocyte antigen workshops, in which panels of monoclonal antibodies against leucocyte cell surface antigens have been collected and exchanged between a large number of participating laboratories around the world. The clustering of these antibodies into groups related by recognition of the same antigens has allowed the definition of the "cluster of differentiation," or CD nomenclature, for antibodies to human leucocyte cell surface molecules [149].

Once characterized, the monoclonal antibodies are useful for the routine assaying and screening for the presence of antigens [44,50,86,87,149]. They can be used in simple qualitative, semiquantitative, or fully quantitative assays. The binding avidity for antigen can be exploited by using antibodies to affinity-purify antigens from complex mixtures. In a similar fashion, they can be used to remove contaminants from other molecules.

As mentioned earlier (sec. 1.4.4), monoclonal antibodies with specificities for rodent and other experimental animal cell surface antigens are used widely to further medical research. In this context, the antibodies can be used to parallel the potential therapeutic applications in humans. Unfortunately, by their very nature it is very unusual to find a single monoclonal antibody that has an identical binding specificity for a target antigen in several species, and so such model systems in experimental animals only indicate in general terms what types of antibody and which properties will be useful in a given situation (see sec. 1.3.4).

1.8.2. Diagnostics

A major commercial application is the use of monoclonal antibodies in diagnostics, which exploits some of the principles learned from their use in research laboratories [150]. Thus, once an antibody has been fully characterized and an assay that makes use of it established, it may be commercially useful to introduce a diagnostic kit based upon it. Polyclonal antisera have been of use in diagnostics, in particular in radioimmunoassay, for many years. Monoclonal antibodies can easily replace polyclonal antisera in most of these cases. In addition, the ease with which monoclonal antibodies can be purified in quantity means that the assays can be redesigned to exploit this ready availability of the specific antibody (see Chap. 2). Thus, radioimmunoassay conventionally makes use of inhibition of binding by the "unknown" antigen of highly purified radiolabeled antigen combined with relatively specific but not necessarily pure antisera [44,46]. The specificity of the assay is essentially dependent upon the purity of the radiolabeled antigen. The availability of monoclonal antibodies produced in quantity allows for the antibody to be labeled and used in direct capture and binding to antigen, and such systems are less dependent on the purity of the antigen. The use of labeled antibody as opposed to labeled antigen in this assay variation theoretically makes it more sensitive, because a thermodynamically different region of the binding curve is being exploited. As discussed above in section 1.3.4, complications can occur when the antibody cross-reacts on related or other antigen structures. This is in part related to the affinity of the antibody for the antigen, but different monoclonal antibodies will exhibit different cross-reactions. Thus mixing monoclonal antibodies together to blend artificial polyclonal antisera is a solution to some of the complications and has been patented and exploited in a number of "two-site" diagnostic assay systems (see sec. 1.3.4; chap. 2) [46].

Bispecific monoclonal antibodies have been derived that bind to one antigen through one binding site and to a second antigen through the other (see Fig. 1.5), and they thus act as an adapter or bridging molecule. For example, if one specificity is to an antigen such as a neurotransmitter, hormone, or cell-associated antigen and the other is to an enzyme such as horseradish peroxidase, then the antibody will allow the indirect visualization of the localization of the antigen in immunocytochemical screening [110,112,150]. At present it is not clear, at least for diagnostics, what the advantages of bispecific antibodies over chemical conjugates are. Careful consideration needs to be given to such problems as the yield and the purity, particularly taking into account the complex mixture of antibodies with overlapping properties being secreted by these "hybrid-hybridomas" or "quadromas" (see Fig. 1.5, sec. 1.5). The production of two different monoclonal antibodies or their fragments followed by a chemical cross-linking may be considered a commercially simpler procedure, but again the yields and purity of the final product need to be considered (see Chaps. 4, 5). Alternatively, recent efforts have been directed toward the production of single chain bispecific antibodies in bacterial expression systems, although it is too early to comment on how easily these can be produced (see Chap. 3).

One major problem of the use of antibodies as *in vivo* or as *in vitro* diagnostics is that they are antigens in their own right. Thus many people have naturally occurring antibodies in their blood that cross-react on rodent monoclonal antibodies and that often can interfere with diagnostic assays and give rise to misleading results [50]. Thus, for example, in an assay for the presence of a hormone in a plasma sample, the presence of antibody to rodent immunoglobulin could give a falsely high reading by capturing the labeled monoclonal antibody used to detect the hormone. This complication might be reduced by using the minimum fragments that still function in the assay (Fab or Fv; see sec. 1.3.3; Fig. 1.3; Chap. 3) or by reducing the number of "foreign" determinants using recombinant DNA technology (see sec. 1.6, Fig. 1.6, Chap. 3).

Monoclonal antibodies can also be used as *in vivo* diagnostics agents, for example, to allow radioimaging, as discussed in Chapter 2 [50,151–153]. In this use monoclonal antibody is labeled with a radioisotope and is preferentially taken up by tissues carrying a particular antigen, for example, a tumor-associated antigen. The excess concentration of the agent near the tumor allows the specific uptake to be detected, particularly as it is possible to do computer-assisted image enhancement by reference to a control of an isotope-labeled nonspecific antibody. As has been discussed above, and will be discussed in detail in the next section and in Chapter 2, the factors that limit the usefulness of such antibodies are the same as for therapeutic applications and include nonspecific uptake by Fc receptors and carbohydrate receptors and problems with tissue penetration and diffusion characteristics, as well as the serum half-life of the conjugate and the potential for an antiglobulin response. Nonspecific uptake by Fc receptors can be minimized by appropriate choice of antibody fragment or isotype, as described in sections 1.3 and in Tables 1.2, 1.3, and 1.5. Similarly, uptake through carbohydrate receptors is dependent upon the state of glycosylation, which can be influenced by the species and isotype as well as the method of expression of the antibody, as described in section 1.3.4. The diffusion characteristics are very dependent upon fragment size, such that Fab and Fv fragments will penetrate tissues much faster than whole IgG or IgM (sec. 1.3.3; Fig. 1.3). The serum half-life of the conjugate is influenced by many factors, including the isotype (see Table 1.3, for example), the fragment size, the glycosylation of the antibody, as well as the chemical stability of the conjugate. Whole IgG antibody is particularly stable and has a serum half-life that is much longer than most other serum proteins (see Table 1.2), and it appears that this property is a function of the Fc domains [24,34]. The severity and timing of the antiglobulin response is dependent upon the species and isotype and may be minimized through humanization of th antibody, as described in section 1.6, although the modification of the antibody by conjugation with radiolabels may increase its immunogenicity. Antibody engineering can offer a solution here, as it can be used to mutate the residues responsible for Fc receptor binding or to prevent the glycosylation or to express antibody fragments.

1.8.3. Therapeutic Applications

Therapeutic applications of monoclonal antibodies raise a number of key issues (see Chap. 3). Since 1975, considerable effort has been expended in exploring the potential therapeutic benefits of a wide range of different monoclonal antibodies used in different ways. As discussed in Chapter 2, the vast majority of these studies have led to disappointing results, and this in turn has driven researchers in the field to direct their attention to ever more complicated strategies to gain improvements. It might be argued that in some situations much of this effort is wasted if it does not address the key issues, which include the choice of antibody specificity and the required mode of action. In addition, we need to decide how best to overcome limitations such as antibody half-life, lack of effectiveness, and antiglobulin responses, which have all been identified in previous studies. Some changes to the technology—for example, the finding of new ways to make monoclonal antibodies such as through the use of combinatorial libraries—do not in themselves help in answering the question as to why so few antibodies have a demonstrable therapeutic effect.

There are, of course, a large number of different therapeutic uses of monoclonal antibodies (discussed in more detail in Chapter 3), only a few of which can be mentioned here, but they include for example some *in vitro* uses such as the treatment of bone marrow prior to infusion [154]. This can be for the removal of T cells before allografting, which has been shown in animal models to prevent graft-versus-host disease, a major complication in human bone marrow grafts between two individuals (except for identical twins). Alternatively, tumor patients who are receiving a dose of radiotherapy or chemotherapy that would be lethal through the toxic effects on their marrow can have some marrow removed prior to the toxic therapy and then to have it treated with monoclonal antibodies to remove any tumor cells before the marrow is reinfused. Another *in vitro* therapeutic use that is currently being investigated is the removal of "passenger leucocytes" from organ grafts such as kidneys so as to prevent graft-versus-host disease, which is thought to be triggered by the passenger leucocytes that act as efficient antigen-presenting cells (APCs)[106].

There are a large number of *in vivo* uses, including, for example, passive immunization for the neutralization of toxins, venoms, or drug overdoses, or for protection against viruses and other pathogens. The use of antibodies in cancer therapy is an area where there has been considerable interest but only limited success (see Chap. 2). In contrast, the use of monoclonal antibodies in immunosuppression and other more subtle manipulations of the immune system for the treatment of autoimmune disorders, or in treating the complications of organ, tissue, and bone marrow transplants, is an area which has seen considerable progress [71,72,106,151,155,156].

The binding specificity of the antibody is a critical issue, because it raises the questions of what antigens are available as targets, what the possible cross-reactions are, and how they influence the effectiveness of the antibody. For tumor therapy, an ideal specificity would be a tumor-specific cell surface antigen, i.e., an antigen that is unique to tumor cells and that does not show cross-reactions with any other tissue antigen. These are really only readily identifiable in a few situations, such as for the "idiotypes" of immunoglobulin or T-cell antigen receptors on lymphocyte-derived tumor cells [157–159]. More often the antigen to which antibodies are targeted is a "tumor associated" antigen that may be anomalously expressed in an unusual way or at high levels but nevertheless is encountered on some normal cell types at some stage of development. The important issues then become ones of "therapeutic ratio," i.e., the toxicity to the tumor cells versus toxicity to the patient. This can be seen as analogous to the issues faced with many other forms of cytotoxic drug or radiotherapy treatment of malignancies, but the specificity inherent in a monoclonal antibody should produce better targeting.

In the example of using antibodies for immunosuppression, the target antigens are normal cell surface molecules which, depending on the choice of antigen, might be expressed on all lymphocytes (e.g., CDw52, CD45, or MHC Class I) [71,72,106,149], on all T cells (e.g., CD3) [50,101,102,119,149], on only a subset of T cells (e.g., CD4, CD8, or the idiotype of the T-cell antigen receptor) [60,107,108,149,151,154,155], or on activated T cells (e.g., the Interleukin-2 receptor) [154,160–162]. As an example, in *in vitro* systems and in animal models, antibodies directed toward appropriate determinants of major histocompatibility complex (MHC) class I and class II molecules can interfere with antigen presentation and, because these molecules are polymorphic, it is possible to block one allele while leaving the other alleles to carry out their normal function [163]. Similarly, antibodies can be directed at the T-cell antigen receptor either to the invariant chains such as the CD3 ϵ chain [101,102,119], which would affect all T cells, or to V-region family restricted determinants, which would only be expressed on subsets of cells. Other suitable targets include the adhesion and signaling molecules such as CD18/CD11, CD2, ICAMs, CD4, and CD8 [149]. Antibodies to CD4 antigen alone and also combined with antibodies to the CD8 antigen have been shown in animal models to interfere with the immune recognition of antigen and may generate a state of specific immune tolerance, allowing long-term survival of tissue grafts or preventing autoimmunity [154,164].

All of the above uses raise issues, not only of what the relevant binding specificities of the antibodies are, but also of their secondary functions (see sec. 1.3.4). As discussed above, the ability of an antibody to recruit natural effector functions, such as to bind and trigger Fc receptors or to activate the complement cascade, is determined by the antibody isotype (class and subclass; see Tables 1.2, 1.3). In addition, although these functions tend to be conserved between species, there is no absolute relationship between the subclasses and their functions in one species compared to another (see sec. 1.3.4; Tables 1.1–1.5). Thus when mouse or rat monoclonal antibodies are used in humans, it is necessary to determine empirically whether or not they can activate and recruit effector mechanisms (Table 1.3). As mentioned in section 1.6, a possible solution to this is to use engineered "chimeric" antibodies that have human constant regions and rodent variable regions [26,29–

31,39]. These chimeric antibodies should interact with the human host in a more predictable manner. Recent studies have started to analyze the particular structural requirements that dictate the differing efficiencies with which the various isotypes interact with complement and Fc receptors, and this will make it possible to engineer into an antibody only the desired functions for a given application (see sec. 1.6, Chap. 3).

Many of the critical residues that determine the interactions of IgG subclass antibodies with effector functions are located in the C_H2 region of the antibody (see Figs. 1.2, and 1.4; sec. 1.6). It is not clear whether all of the residues that have been identified as controlling the functions are responsible for direct interactions with Fc receptors and complement components or whether they have an indirect effect through determining the tertiary or quarternary structure of the antibody [39,127–130]. A good example is the N-linked carbohydrate attached to a conserved residue, Asparagine 297. Removal of the carbohydrate either through the use of metabolic inhibitors or glycosidases or by mutation of the attachment site seems to abolish most of the functions of IgG, although there is no evidence that Fc receptors or complement interact directly with carbohydrate (see sec. 1.6). Another example is a conserved sequence called the "C1q binding motif," identified by Duncan and Winter for mouse IgG [120]. This sequence motif is present in all four of the human IgG isotypes, even though IgG4 does not bind C1q and it seems that other residues—including a Proline (IgG1, IgG2, and IgG3) to Serine (IgG4) change at position 331—determine the human subclass differences [39,124,127–129].

The isotype is not the only important property of an antibody that influences the potency in killing of human target cells utilizing natural effector mechanisms. Studies indicate that the specificity of the antibody and nature of the target antigen are as important [96]. Most monoclonal antibodies—even when of isotypes such as rat IgG2b or mouse IgG2a (see sec. 1.3.4; Table 1.3), which efficiently bind human C1q—fail to give effective lysis of human cells in the presence of human complement. Sometimes this is a failure to activate the complement, but in other cases it is as a result of the existence of a number of protective mechanisms available to cells that protect them against complement damage. These mechanisms include the shedding of small vesicles from the membranes of cells as well as various membrane-bound inhibitors of the complement cascade, including decay-accelerating factor DAF, membrane cofactor protein MCP, CD59, and homologous restriction factor HRF [99].

Various strategies have been found that enhance cell lysis through monoclonal antibodies. Pairs of monoclonal antibodies, recognizing different determinants on the same antigen, give rise to synergistic complement activation, and this is currently being exploited in the case of a pair of CD45 antibodies, which can purge the "passenger leucocytes" from organ transplants and alleviate the potential for graft rejection [104,106]. Another observation has been that some antibodies are more lytic when they bind antigen monovalently rather than bivalently, although th explanation for this is not clear [101]. The use of blocking antibodies to the complement inhibitory molecules such as CD59 can enhance the lysis caused by other specificities (a different example of synergy to that described above) [99]. Some antigens (or sometimes epitopes of some antigens) seem to be very good targets for antibody-directed cell lysis, and in these cases it is often found that within certain limitations the choice of antibody isotype is important but may not be critical. An example of a good target is the human lymphocyte-associated antigen Campath-1 (CDw52), although again it is still not entirely clear why it is such a good target. A number of therapeutic antibodies have been derived against Campath-1, including a rat IgM (Campath-1M), a rat IgG2b (Campath-1G), and a fully reshaped humanized IgG1 form (Campath-1H), and these have been used successfully in different clinical situations. The rat IgM has been used *in vitro* to lyse lymphocytes present in bone marrow donations using human plasma as a complement source, and the rat IgG2b and human IgG1 have been used *in vivo* for immunosuppression and for treatment of lymphoid malignancies [51,52,71,72].

Given an antibody of suitable target specificity, the main reasons for the engineering of therapeutic antibodies is to improve the effectiveness of the antibody, as well as to reduce the immunogenicity of the antibody and to improve the biological half-life. Chimeric antibodies minimize the risk of the constant regions being a target for specific immune recognition, but there is still the problem of the immunoglobulin allotypes, which occur at different frequencies among different racial groupings and potentially invalidate the applicability of a single recombinant product to all patients [60]. Even assuming the use of human or humanized antibodies, the variable regions pose more problems with regard to potential immunogenicity for two reasons. First, these again show different genetic patterns of inheritance as members of related V-gene-segment families. Second, because of the random rearrangement of these gene segments followed by somatic mutation during B-cell development, the same sequences are unlikely to be represented in most individuals. Thus virtually all monoclonal antibodies are potentially antigenic in their V regions. In practice, it seems that the chance of an antiglobulin response occurring, its magnitude, and the time of its onset are all influenced by the degree of "foreignness" and by the antibody specificity. The immunogenicity of a therapeutic antibody may be reduced in the recipient by generating a state of specific tolerance, but it still seems to be difficult to tolerize for the idiotype (i.e., the unique sequences in the V regions) of a cell-binding antibody.

As discussed in Chapter 2, using antibodies armed with artificial effector mechanisms such as a covalently coupled drug, radioisotope, or toxin has been investigated to deal with the common observation that most monoclonal antibodies have a very poor therapeutic effect [152,153,165,166]. The assumption is that the antibody will specifically target the potent toxic agent. In practice, there are a number of limiting complications to this apparently simple strategy, including the problems of competition for binding by soluble antigen and antigen in the near neighborhood of cells. A major limitation is usually the affinity of the monoclonal antibody for antigen, which for most specificities of interest is found to be such that only a very small proportion of the total antibody will be complexed with antigen at any given point in time. Uptake of the complexes by unwanted tissues is another complication and includes uptake through specific Fc receptors and through carbohydrate receptors. Additionally, some tissues may be exceptionally susceptible to the toxic agent and breakdown products from the complex such as free radioisotopes may accumulate in different organs (e.g., iodine in the thyroid).

Alternative artificial effector mechanisms that have been exploited are the use of bispecific antibodies to target toxic agents and their use to recruit new effector cell types to the target [100]. The bispecific antibodies have been produced either secreted from hybrid-hybridomas (see Fig. 1.3) or, alternatively, as chemically synthesized covalent complexes of two IgGs (Fabs, Fvs) or, more recently, as recombinant single-chain bispecifics (see Chaps. 3, 4). In one simple use of these, they merely replace the need to covalently couple the toxic agent to the antibody by having one specificity for the target and one for the agent. An alternative use has been to exploit the fact that some natural effector cells such as T cells, natural killer (NK) cells, killer (K) cells, and macrophages can be triggered to mediate their cytotoxic effects by triggering them through a small number of cell surface molecules which include CD2, CD3, the T-cell antigen receptor, CD28, and Fc receptors [149]. Thus a bispecific antibody can be used to cross-link a target cell with an effector cell and to bring about "effector cell retargeting," resulting in the destruction of the target by a redirected natural effector mechanism.

In analyzing how natural antibodies mediate their potential, what is often overlooked in therapeutic applications is that natural effector mechanisms have evolved to cleverly exploit the thermodynamics of antibody/antigen interactions. Many antibodies bind antigen with only modest affinity, and thus the amount of bound antibody in the body is very low compared with the total immunoglobulin concentration. Despite this, complement is not unduly activated by this free immunoglobulin but is specifically activated through the higher avidity binding of the first complement component, C1, to immune complexes. The fact that complement activation is an enzyme cascade results in a greatly amplified response to a small local change in amounts of complex generated. Similarly, the activation of cells through low-affinity Fc receptors again allows a response to occur for small immune complexes in the presence of a high concentration of free antibody. Thus the natural immune response seems to function with antibodies of only modest affinity and is not affected by the presence of a massive excess of free antibody. The unbound natural antibody also is not toxic, unlike some of the radioisotope-conjugated complexes, which are quite toxic even at a distance form the cells. Thus in order to generate antibodies that exploit natural effector mechanisms, it may be possible to work with antibodies of low to intermediate affinity; in contrast, for the specific targeting of drugs and radioisotopes, antibodies of very high affinity are desirable.

1.8.4. Abzymes

Generally antibodies are considered as molecules that merely bind to a given ligand, an "antigen." Many of the binding properties of an antibody are shared by other proteins, such as enzymes, for their substrates. Using a simplistic view, enzymes have to bind to one or more substrates, catalyze a reaction involving the making or breaking of chemical bonds, and then release one or more products. Ideally, for an enzyme to be efficient at this process, it must not bind the products with too high an affinity but should assist in the stabilization of the transition states. An antibody that shared these properties would also act as an enzyme. The principle of deriving antibodies that catalyze reactions has been amply demonstrated [167] and has mainly been achieved by predicting the transition state that occurs for a given reaction and then designing and synthesizing a stable chemical analogue of the transition state. The chemical analogues are then used as the immunogen, and monoclonal antibodies are raised using conventional procedures. A significant proportion of such antibodies do catalyze the predicted reaction. Such antibodies are initially interesting because they allow predictions about chemical reactions and transition states to be investigated. Also, the overall structure of an antibody is such that the new abzyme structures can be modeled and determined through a comparison with other antibody structures. The large potential repertoire of immunoglobulins derived by conventional techniques or by recombinant technology may allow for the ready production of abzymes, which will catalyze reactions for which conventional enzymes do not exist. In addition, the structural stability of immunoglobulin Fab fragments may allow such abzymes to be used under a range of conditions. It may even prove possible to derive abzymes that might catalyze reaction *in vivo* such as, for example, the destruction of a toxic compound.

Interestingly, it was as a result of dissatisfaction with the lengthy procedure involved in the production of monoclonal abzymes by the Kohler and Milstein technique that led Lerner's group to investigate the potential for screening combinatorial libraries of Fab fragments from immunized animals, expressed from phage λ in E. coli, in order to speed up the initial screening process [136]. Deriving abzymes is an ideal application for the use of combinatorial libraries and even artificial libraries, because it is only the binding property of the antibody that is required for this function so that Fv or Fab fragments are ideally suited.

1.9. STRATEGIES

From the previous sections of this chapter and also from the other chapters in this book, it should be clear that there are no well-defined and definitive strategies for the development and production of commercially useful monoclonal antibodies. The following brief comments are meant to serve merely as a guide to the possible strategies that can be adopted and, like all generalizations, should not be treated as definitive.

To develop antibodies for *in vitro* use in industrial processes, such as purification of antigens, the species and isotype of the antibody may not be of great importance. The production of monoclonal antibodies from immunized rodents using the conventional Kohler and Milstein technology is currently the most widely used technique. Use of combinatorial libraries and phage expression systems may eventually become widely used, although at present the selection systems need to be improved to make them more applicable to a wider range of antigens beyond the few model systems that have currently been published. Once derived, the antibodies can most conveniently be produced in quantity by scale-up of the hybridoma culture. If fragments of antibodies are useful, then production of antibody from bacterial expression systems might be exploited as a convenient and economical alternative.

Antibodies used in medicine as *in vitro* diagnostics may fall into the above criteria. Again, at present a majority of such reagents are derived from immunized laboratory rodents. Depending on the exact application, the antibody may be used intact or as a fragment. Additionally, the isotype and or species of the antibody may matter for reasons discussed earlier, such as antibody effector functions or reactivity with human antirodent antibodies. The method of large-scale production of the antibody will be dependent upon the nature of the antibody, but again the large-scale tissue culture of a hybridoma is likely to be suitable.

For antibodies used as *in vivo* medical agents, the strategies are less clear at this time. The main problem is that our current state of knowledge of *in vivo* use of monoclonal antibodies does not yet allow the prediction of all of the required properties, including the specificity and functions of the antibody, for any particular use. Without knowing what the end product should be, it is not easy to decide where to start. Current medical use of monoclonal antibodies includes use of conventional rodent-derived monoclonal antibodies, chemically modified antibodies and fragments, and also recombinant antibodies. As discussed above, some success has been achieved using rodent antibodies and these can be suitable for preliminary trials. In the longer term, the prospects of an antiglobulin response is likely to dictate that a majority of therapeutic antibodies be "humanized." Thus these antibodies will eventually be expressed as recombinant products, and so the commercial scale-up will involve decisions on the most suitable expression systems. At the same time, the expression of the antibodies as recombinant products also allows for the protein engineering of the antibody functions to optimize the efficacy. The end point being a human antibody, an alternative strategy might be to start with trying to immortalize human lymphocytes or use a human combinatorial library. On a theoretical basis, there is likely to be very little difference with regard to the immunogenicity of a fully humanized rodent antibody versus a human antibody or a human combinatorial library-derived antibody. The unique nature of the complementarity-determining regions (idiotype) of an antigen-specific antibody may always mean that there is a potential for an antiglobulin response to this region of the molecule, and that this is more likely to be the case for cell-binding antibodies [131,168,169]. At present, the screening systems for conventional monoclonal antibodies specific for complex human antigens may mean that they are easier to work with than combinatorial libraries and phage expression libraries. With regard to the differences in time scale between these different selection systems, they may not have a significant impact on the commercial exploitation of a therapeutic antibody. The clinical testing and licensing of a therapeutic agent is highly complex and is likely to take many years, and so a difference in the early strategies of development of a few weeks or months will not have much of a noticeable effect (see Chap. 6).

1.10. FUTURE PROSPECTS

The previous sections serve as a general introduction to the topics discussed in the rest of this book. Many of the ideas that have only been touched upon briefly are elaborated in much more detail, and in particular there are full discussions of the complications that arise in attempts to commercially exploit monoclonal antibodies. It is to this point that I wish to address this section with some optimism for the future.

It may seem that in the many years since 1975, when Kohler and Milstein first reported their technique for production of monoclonal antibodies, there has been very little evidence of progress with regard to successful commercial applications of antibodies, particularly in the fields of human diagnostics and therapy. This is despite the uncountably large number of scientific papers describing monoclonal antibodies and the antigens they recognize. At first encounter, this might lead the reader who is unfamiliar with this area to adopt a pessimistic attitude for the future prospects. This would be unfortunate, because while it can be argued that the previous history of the development of monoclonal antibodies is littered with false hopes and ambitious projections, the problems encountered have usually been due to underestimating the complexity of the task at hand. Nevertheless, much experience has been gained along the way and the appropriate exploitation of this knowledge in the light of recent advances in monoclonal antibody technology—especially when combined with recombinant DNA technology—holds, I believe, great prospects for the future.

This is nowhere more true than in attempts to use monoclonal antibodies therapeutically, and particularly in the treatment of malignant disorders. Despite the large number of reports, there have been very few in which the antibody has achieved a complete remission, and in those cases where remissions or partial remissions have been reported, there have often been difficulties in demonstrating reproducible clinical effects in a larger trial (see Chap. 2). However, malignant diseases are perhaps the hardest problems to tackle, because a complete kill of the tumor is required for a complete remission. However, in an effort to improve antibody therapy of malignant disorders, much information has been acquired as to which properties of antibodies are important, as outlined above and discussed in more detail in Chapter 2. Progress is inevitably slow because of the ethical considerations in showing that the antibody treatment is itself not harmful, and thus the progression from one antibody to another takes place over a time scale of years. Up until recently, because of this need to show that antibody therapy was safe, the experimental treatments were on the whole confined to life-threatening and end-stage clinical conditions where realistically there was never much chance for dramatic effects. In parallel with the human studies, many research groups have been able to successfully treat animal models of human conditions, such as autoimmune disorders, using various antibody-based strategies and thus raise the prospect for the use of similar strategies in human therapy. Many of these disorders seriously affect large numbers of patients but are rarely life-threatening in the short-term, so consequently the introduction of antibody-based treatments has been rightly slow and cautious. Recently, encouraging results have started to be reported in this area, for example, in the use of Campath-1H in treatment of autoimmune vasculitus and rheumatoid arthritis [71,72].

The use of the Campath-1 (CDw52) series of antibodies serves as a useful example of the slow but significant improvement in antibody-based therapies that have developed over the years since 1975. The use of the Campath-1 antibodies can be charted through the early use of a rat IgM antibody for successful *in vitro* depletion of T lymphocytes from bone marrow, the disappointment in finding that it and a rat IgG2a antibody to the same antigen were ineffective in depleting lymphocytes *in vivo,* but the subsequent appreciation that a rat IgG2b class-switch antibody was much more potent [42,51,52]. The further clinical use of the rat IgG2b highlighted that the limitation to the long-term and repeated use of this antibody was in the antiglobulin response, and so a decision was made to engineer a fully humanized IgG1 form of the antibody Campath-1H [29], which has been used with encouraging results [56,71,72] and is currently in clinical trials with the Wellcome Foundation.

Another excellent example is the use of antibodies to the T-cell CD3 antigen in immunosuppressive treatments for organ transplantation (see chapter 2) [50,101, 102,119,149]. In particular, the antibody Orthoclone OKT3 [50,119] is the only monoclonal antibody to have so far been fully licensed for therapy (see Chaps. 2, 6). OKT3 is a mouse IgG2a antibody and its use is not without complications. Repeated use of the antibody usually provokes an antiglobulin response, which eventually negates the effectiveness of the antibody if repeated therapy proves necessary. As a possible solution to this problem, humanized CD3 antibodies have been constructed [50,102,170]. However, a greater problem is the observation of severe side effects associated with the earlier doses in use of CD3 antibodies and thought to be due to the triggering of cytokine release by T cells [50,171]. The *in vivo* cytokine release is thought to be dependent upon binding of the antibody to accessory cells through Fc receptors. As a response to this problem, humanized forms of CD3 antibodies in which the constant regions have also been engineered to alter Fc receptor binding are now available [50,172,173]. It will be interesting to see if such engineered antibodies are as effective as OKT3 but without many of the problems currently associated with its use.

The future prospects for commercial applications of monoclonal antibodies would seem to be very encouraging. The successful companies are likely to be those that harness the full wealth of experience and expertise available and combine the appropriate technologies in well thought out and highly directed applications of a few very carefully chosen products. The adoption of the new recombinant DNA technologies to choose appropriate specificities, careful selection and manipulation of the antibody isotypes or fragments, and improvements in fermentation systems and protein expression systems will all assist in improving the effectiveness and market potential for the next generation of reagents.

1.11. REFERENCES

1. Abbas, A.K., Lichtman, A.H., Pober, J.S. Cellular and Molecular Immunology. Philadelphia: Saunders, 1991.

2. Clark, W.R. The Experimental Foundations of Modern Immunlogy (4th ed.) New York: Wiley, 1991.
3. Roitt, I.M. Essential Immunology (7th ed.) Oxford: Blackwell, 1991.
4. Kohler, G., Milstein, C. Nature 256, 495 (1975).
5. Burton, D.R., Gregory, L., Jefferis, R. Monogr. Allergy 19, 7 (1986).
6. Stanworth, D.R., Turner, M.W. In: Weir, D.M. (ed.): Handbook of Experimental Immunology, Vol. 1: Immunochemistry. Oxford: Blackwell Scientific, 1986.
7. Pumphrrey, R. Immunol. Today 7, 174 (1986).
8. Burton, D.R. In Metzger, H. (ed.): Fc Receptors and the Action of Antibodies. Washington, DC: American Society for Microbiology, 1990.
9. Jefferis, R. In Shakib, F. (ed.): The Human IgG Subclasses. New York: Pergamon, 1990.
10. Kabat, E.A., Wu, T.T., Reid-Miller, M., Perry, H.M., Gottesman, K. S. Sequences of Proteins of Immunological Interest. Washington, DC: US Department of Health and Human Services, U.S. Government Printing Office, 1987.
11. Hamers Casterman, C., Atarhouch, T., Muyldermans, S., Robinson, G., Hamers, C., Songa, E.B., Bendahman, N., Hamers, R. Nature 363, 446 (1993).
12. Deisenhofer, J. Biochemistry 20, 2361 (1981).
13. Davies, D.R., Padlan, E.A., Sheriff, S. Annu. Rev. Biochem. 59, 439 (1990).
14. Padlan, E.A. In Metzger, H. (ed.): Fc Receptors and the Action of Antibodies. Washington, DC: American Society for Microbiology, 1990.
15. Harris, L.J., Larson, S.B., Hasel, K.W., Day, J., Greenwood, A., McPherson, A. Nature 360, 369 (1992).
16. Esser, C., Radbruch, A. Annu. Rev. Immunol. 8, 717 (1990).
17. Staudt, L.M., Lenardo, M.J. Annu. Rev. Immunol. 9, 373 (1991).
18. Alt, F.W., Oltz, E.M., Young, F., Gorman, J., Taccioli, G. Chen, J. Immunol. Today 13, 306 (1992).
19. Nisonoff, A. Lamoyi, E. J. Immunol. Methods 56, 235 (1983).
20. Parham, P. J. Immunol. 131, 2895 (1983).
21. Rousseaux, J., Rousseaux-Prevost, R., Bazin, H. J. Immunol. Methods 64, 141 (1983).
22. Rousseaux, J., Rousseaux-Prevost, R., Bazin, H. Methods Enzymol. 121, 663 (1986).
23. Stevenson, G.T., Glennie, M.J., Kan, K.S. In Clark, M. (ed.): Protein Engineering of Antibody Molecules for Prophylatic and Therapeutic Applications in Man. Academic Titles, 1993.
24. Morell, A. Terry, W.D., Waldmann, T.A. J. Clin. Invest. 49, 673 (1970).
25. French, M., Monogr. Allergy 19, 100 (1986).
26. Bruggemann, M., Williams, G.T., Bindon, C.I., Clark, M.R., Walker, M.R., Jefferis, R., Waldmann, H., Neuberger, M.S. J. Exp. Med. 166, 1351 (1987).
27. Bindon, C.I., Hale, G., Bruggemann, M., Waldmann, H. J. Exp. Med. 168, 127 (1988).
28. Jefferis, R., Walker, M.R. Monogr. Allergy 23, 73 (1988).
29. Riechmann, L., Clark, M.R., Waldmann, H., Winter, G. Nature 332, 323 (1988).
30. Shaw, D.R., Khazaeli, M.B., LoBuglio, A.F. J. Natl. Cancer Inst. 80, 1553 (1988).
31. Steplewski, Z., Sun, L.K., Shearman, C.W., Ghrayeb, J., Daddona, P., Koprowski, H. Proc.Natl. Acad. Sci. USA 85, 4852 (1988).

32. Unkeless, J.C. Annu. Rev. Immunol. 6, 251 (1988).
33. Lynch, R.G., Sandor, M. In Metzger, H. (ed.): Fc Receptors and the Action of Antibodies. Washington, DC: American Society for Microbiology, 1990.
34. Mariani, G., Strober, W. In Metzger, H. (ed.): FC Receptors and the Action of Antibodies. Washington, DC: American Society for Microbiology, 1990.
35. Ravetch, J.V., Anderson, C.L. In Metzger, H. (ed.): Fc Receptors and the Action of Antibodies. Washington DC: American Society for Microbiology, 1990.
36. Ravetch, J.V., Kinet, J., Annu. Rev. Immunol. 9, 457 (1991).
37. van de Winkel, J.G.J., Capel, P.J.A. Immunol. Today 14, 215 (1993).
38. Sandor, M., Lynch, R.G. Immunol. Today 14, 227 (1993).
39. Greenwood, J., Clark, M. In Clark, M. (ed.): Protein Engineering of Antibody Molecules for Prophylatic and Therapeutic Applications in Man. Academic Titles, 1993.
40. Akiyama Y., Lubeck, M.D., Steplewski, Z., Koprowski, H. Cancer Res. 44, 5127 (1984).
41. Lubeck, M.D., Steplewski, Z., Baglia, F., Klein, M.H., Dorrington, K.J., Koprowski, H. J. Immunol. 135, 1299 (1985).
42. Hale, G., Cobbold, S.P., Waldmann, H., Easter, G., Matejtschuk, P., Coombs, R.R.A. J. Immunol. Methods 108, 59 (1987).
43. Bruggemann, J., Teale, C., Clark, M., Bindon, C., Waldmann, H. J. Immunol. 142, 3145 (1989).
44. Harlow, E., Lane, D., Antibodies—A Laboratory Manual. Cold Spring Harbour, NY: Cold Spring Harbor Laboratories, 1988.
45. Steward, M.W. Immunol. Today 2, 134 (1983).
46. Weir, D.M. (ed.) Handbook of Experimental Immunology, Vol. 1: Immunochemistry. Oxford: Blackwell Scientific, 1986.
47. Milstein, C., Wright, B., Cuello, A.C. Mol. Immunol. 20, 113 (1983).
48. Roberts, S., Cheetham, J.C., Rees, A.R. Nature 328, 731 (1987).
49. Isaacs, J.D., Clark, M.R., Greenwood, J., Waldmann, H. J. Immunol 148, 3062 (1992).
50. Thompson, R.J., Jackson, A.P., Langlois, N. 31, (1985).
51. Dyer, M.J.S., Hale, G., Hayhoe, F.G.J., Waldmann, H. Blood 73, 1431 (1989).
52. Dyer, M.J.S., Hale, G., Marcus, R., Waldmann, H. Leukaemia and Lymphoma (in press).
53. Winter, G., Milstein, C. Nature 349, 293 (1991).
54. Adair J. Immunol. Rev. 130, 5 (1992).
55. Routledge, E.G., Gorman, S.D., Clark, M. In Clark, M. (ed.): Protein Engineering of Antibody Molecules for Prophylatic and Therapeutic Applications in Man. Academic Titles, 1993.
56. Hale, G., Dyer, M.J.S., Clark, M.R., Phillips, J.M., Marcus. R., Riechmann, L., Winter, G., Waldmann, H. Lancet 2, 1394 (1988).
57. WHO. Eur. J. Immunol. 6, 599 (1976).
58. WHO. J. Immunogenet. 3, 357 (1976).
59. Weir, D.M. (ed.). Handbook of Experimental Immunology, Vol. 3: Genetics and Molecular Immunology. Oxford: Blackwell Scientific, 1986.
60. Gorman, S.D, Clark, M.R. Sem. Immunol. 2, 457 (1990).

61. Jefferis, R., Lund, J., Mizutani, H., Nakagawa, H., Kawazoe, Y., Arata, Y., Takahashi, N. Biochem. J. 268, 529 (1990).
62. Leatherbarrow, R.J., Rademacher, T.W., Dwek, R.A., Woof, J.M., Clark, A., Burton, D.R., Richardson, N., Feinstein, A. Mol. Immunol. 22, 407 (1985).
63. Tao, M., Morrison, S.L. J. Immunol. 143, 2595 (1989).
64. Lund, J., Tanaka, T., Takahashi, N., Sarmay, G., Arata, Y., Jefferis, R. Mol. Immunol. 27, 1145 (1990).
65. Goochee, C.F., Monica, T. Bio/Technology 8, 421 (1990).
66. Matsuda, H., Nakamura, S., Ichikawa, Y., Kozai, K., Takano, R., Nose, M., Endo, S., Nishimura Y., Arata, Y. Mol. Immunol. 27, 571 (1990).
67. Patel, T.P., Parekh, R.B., Moellering, B.J., Prior, C.P. Biochem, J. 285, 839 (1992).
68. Pluckthun, A. Bio/Technology 9, 545 (1991).
69. Page, M.J., Sydenham, M.A. Bio/Technology 9, 64 (1991).
70. Crowe, J.S., Hall, V.S., Smith, M.A., Cooper, H.J., Tite, J.P. Clin. Exp. Immunol. 87, 105 (1992).
71. Mathieson, P.W., Cobbold, S.P., Hale, G., Clark, M.R., Oliveira, D.B.G., Lockwood, C.M., Waldmann, H. N. Engl. J. Med. 323, 250 (1990).
72. Isaacs, J.D., Watts, R.A., Hazleman, B.L., Hale, G., Keogan, M.T., Cobbold, S.P., Waldmann, H. Lancet 340, 748 (1992).
73. Parekh, R.B., Dwek, R.A., Sutton, B.J., et al. Nature 316, 452 (1985).
74. Axford, J.S., Sumar, N., Alavi, A., Isenberg, D.A., Young, A., Bodman, K.B., Roitt, I.M. J. Clin. Invest. 89, 1021 (1992).
75. Rook, G.A.W., Stanford, J.L. Immunol. Today 13, 160 (1992).
76. Youinou, P., Pennec, Y.L., Casburnbudd, R., Dueymes, M., Letoux, G., Lamour, A. J. Autoimmunity 5, 393 (1992).
77. Jackson, A.P., Siddle, K., Thompson, R.J. Biochem, J. 215, 505 (1983).
78. Galfre, G., Milstein, C. Methods. Enzymol. 73, 3 (1981).
79. Clark, M.R., Milstein, C. Somatic Cell Genet. 7, (1981).
80. Kohler, G., Howe, S.C., Milstein, C. Eur. J. Immunol. 6, 292 (1976).
81. Schulman, M., Wilde, C.D., Kohler, G. Nature 276, 269 (1978).
82. Galfre, G., Milstein, C., Wright, B. Nature 277, 131 (1979).
83. Kearney, J.F., Radbruch, A., Liesegang, B., Rajewsky, K. J. Immunol. 123, 1548 (1979).
84. Bazin H. In Peters (ed.): Protides of the Biological Fluids. Proceedings Colloquium Oxford. New York: Pergamon, 1982.
85. Clark, M., Waldmann, H. In Beverly, P. (ed.): Methods in Hematology. Churchill Livingstone, 1986.
86. Weir, D.M. (ed.). Handbook of Experimental Immunology, Vol. 4: Applications of Immunological Methods in Biomedical Sciences. Oxford: Blackwell Scientific, 1986.
87. Liddell, J.E., Cryer, A. A Practical Guide to Monoclonal Antibodies. New York: Wiley, 1991.
88. Zimmerman, U., Klock, G., Gessner, P., Sammons, D.W., Neil, G.A. Hum. Antibodies Hybridomas 3, 14 (1992).
89. Clark, M., Cobbold, S., Hale, G., Waldmann, H. Immunol. Today 4, 100 (1983).

90. Foung, S.K.H., Engleman, E.G., Grumet, F.C. Methods Enzymol. 121, 168 (1986).
91. Kozbor, D., Roder, J.C., Sierzega, M.E., Cole, S.P.C., Croce, C.M. Methods Enzymol. 121, 120 (1986).
92. Roder, J.C., Cole, S.P.C., Kozbor, D. Methods Enzymol. 121, 140 (1986).
93. Larrick, J.W., Danielsson, L., Brenner, C.A., Wallace, E.F., Abrahamson, M., Fry, K.E., Borrebaeck, A.K. Bio/Technology 7, 934 (1989).
94. Borrebaeck, C.A.K. Immunol. Today 9, 355 (1988).
95. Boerner, P., Lafond, R., Lu, W., Brams, P., Roysten, I. J. Immunol. 147, 86 (1991).
96. Bindon, C.I., Hale, G., Waldmann, H. Eur. J. Immunol. 18, 1507 (1988).
97. Michaelsen, T.E., Garred, P., Aase, A. Eur.J. Immunol. 21, 11 (1991).
98. Valim, Y.M.L., Lachmann, P.J. Clin. Exp. Immunol. 84, 1 (1991).
99. Lachmann, P.J. Immunol. Today 12, 312 (1991).
100. Clark, M., Gilliland, L., Waldmann, H. Prog. Allergy 45, 31 (1988).
101. Clark, M., Bindon, C., Dyer, M., Friend, P., Hale, G., Cobbold, S., Calne, R., Waldmann, H. Eur. J. Immunol. 19, 381 (1989).
102. Routledege, E.G., Lloyd, I., Gorman, S.D., Clark, M., Waldmann, H. Eur. J. Immunol. 21, 2717 (1991).
103. Hughes-Jones, N.C., Gorick, B.D., Miller, N.G.A., Howard, J.C. Eur.J. Immunol. 14, 974 (1984).
104. Bindon, C.I., Hale, G., Clark, M.R., Waldmann, H. Transplantation 40, 538 (1985).
105. Holmes, N.J., Parham, P. J. Biol. Chem. 258, 1580 (1983).
106. Brewer, Y., Palmer, A., Taube, D., Welsh, K., Bewick, M., Hale, G., Waldmann, H., Dische, F., Parsons, V., Snowden, S. Lancet 2, 935 (1989).
107. Jonker, M., Goldstein, G., Balner, H. Transplantation 35, 521 (1983).
108. Jonker, M., Neuhaus, P., Zurcher, C., Fucello, A., Goldstein, G. Transplantation 39, 247 (1985).
109. Clark, M., Bolt, S., Tunnacliffe, A., Waldmann, H. In Romet-Lemonne, P., Fanger, M., Segal, D.(eds.): Bispecific Antibodies and Targeted Cellular Cytotoxicity: Second International Conference. 1990.
110. Milstcin, C., Cucllo, A.C. Immunol. Today 5, 299 (1984).
111. Milstein, C., Cuello, C. Nature 305, 537 (1983).
112. Suresh, M.R., Cuello, A.C., Milstein, C. Methods Enzymol. 121, 210 (1986).
113. Neuberger, M.S. EMBO J. 2, 1373 (1983).
114. Boulianne, G.L., Hozumi, N., Schulman, M.J. Nature 312, 643 (1984).
115. Morrison, S.L., Oi, V.T. Annu. Rev. Immunol. 2, 239 (1984).
116. Morrison, S.L., Johnson, M.J., Herzenberg, L.A., Oi, V.T. Proc. Natl. Acad. Sci. USA 81, 6851 (1984).
117. Neuberger, M.S., Williams, G.T., Fox, R.O. Nature 312, 604 (1984).
118. Ortho Multicentre Transplant Study Group. N. Engl. J. Med. 313, 337 (1985).
119. Jones, P.T., Dear, P.H., Foote, J., Neuberger, M.S., Winter, G. Nature 321, 522 (1986).
120. Duncan, A.R., Winter, G. Nature 332, 738 (1988).
121. Duncan, A.R., Woof, J.M., Patridge, L.J., Burton, D.R., Winter, G. Nature 332, 563 (1988).

122. Michaelsen, T.E., Aase, A., Westby, C., Sandlie, I. Scand. J. Immunol. 32, 517 (1990).
123. Canfield, S.M., Morrison, S.L. J. Exp. Med. 173, 1483 (1991).
124. Tao, M., Canfield, S.M., Morrison, S.L. J. Exp. Med. 173, 1025 (1991).
125. Michaelsen, T.E., Aase, A., Norderhaug, L., Sandlie, I. Mol. Immunol. 29, 319 (1992).
126. Sarmay, G., Lund, J., Rozsnyay, Z., Gergely, J., Jefferis, R. Mol. Immunol. 29, 633 (1992).
127. Shin, S., Wright, A., Bonagura, V., Morrison, S.L. Immunol. Rev. 130, 87 (1992).
128. Greenwood, J., Clark, M., Waldmann, H. Eur. J. Immunol. 23, 1098 (1993).
129. Morrison, S.L., Canfield, S.M., Tao, M. In Clark, M. (ed.): Protein Engineering of Antibody Molecules for Prophylatic and Therapeutic Applications in Man. Academic Titles, 1993.
130. Jefferis, R., Lund, J. In Clark (ed.): Protein Engineering of Antibody Molecules for Prophylatic and Therapeutic Applications in Man. Academic Titles, 1993.
131. Munro, A. Nature 312, 597 (1984).
132. Weidle, U.H., Borgya, A., Mattes, R., Lenz, H., Buckel, P. Gene 51, 21 (1987).
133. Better, M., Horwitz, A.H. Methods Enzymol. 178, 476 (1989).
134. During, K., Hippe, S., Kreuzaler, F., Schell, J. Plant Mol. Biol. 15, 281 (1990).
135. Swain, W.F. Trends Biotechnol. 9, 107 (1991).
136. Huse, W.D., Sastry, S., Iverson, S.A., Kang, A.S., Alting-Mees, M., Burton, D.R., Benkovic, S.J., Lerner, R.A. Science 246, 1275 (1989).
137. Ward, E.S., Gussow, D., Griffiths, A.D., Jones, P.T., Winter, G. Nature 341, 544 (1989).
138. McCafferty, J., Griffiths, A.D., Winter, G., Chiswell, D.J. Nature 348, 552 (1990).
139. Clackson, T., Hoogenboom, H.R., Griffiths, A.D., Winter, G. Nature 352, 624 (1991).
140. Kang, A.S., Barbas, C.F., Janda, K.D., Benkovic, S.J., Lerner, R.A. Proc. Natl. Acad. Sci. USA 88, 4363 (1991).
141. Kang, A.S., Jones, T.M., Burton, D.R. Proc. Natl. Acad. Sci. USA 88,11120 (1991).
142. Marks, J.D., Hoogenboom, H.R., Bonnert, T.P., McCafferty, J., Griffiths, A.D., Winter, G. J. Mol. Biol. 222, 581 (1991).
143. Barbas, C.F., Bain, J.D., Hoekstra, D.M., Lerner, R.A. Proc. Natl. Acad. USA 89, 4457 (1992).
144. Hawkins, R.E., Russel, S.J., Winter, G. J. Mol. Biol. (1992).
145. Lerner, R.A., Kang, A.S., Bain, J.D., Burton, D.R., Barbas, C.F. Science 258, 1313 (1992).
146. Marks, J.D., Griffiths, A.D., Malmqvist, M., Clackson, T.P., Bye, J.M., Winter, G., Bio/Technology 10, 779 (1992).
147. Griffiths, A., Hoogenboom, H.R. In Clark, M. (ed.): Protein Engineering of Antibody Molecules for Prophylatic and Therapeutic Applications in Man. Academic Titles, 1993.
148. Burton, D.R., Barbas, C.F. In Clark, M. (ed.): Protein Engineering of Antibody Molecules for Prophylatic and Therapeutic Applications in Man. Academic Titles, 1993.

149. Barclay, A.N., Birkeland, M.L., Brown, M.H., Beyers, A.D., Davis, S.J., Somoza C., Williams, A.F. The Leucocyte Antigen Facts Book. Orlando, FL: Academic, 1993.
150. Suresh, M.R., Cuello, A.C., Milstein, C. Proc. Natl. Acad. Sci. USA 83, 789 (1986).
151. Waldmann, T.A. Science 252, 1657 (1991).
152. Abrams, P.G., Carrasquillo, J.A., Schroff, R.W., Eary, J.F., Fritzberg, A.R., Morgan, A.C., Wilbur, D.S., Beaumier, P.L., Larson, S.M., Nelp, W.B. Princ. Cancer Biother. 337 (1987).
153. Goldenberg, D.M., Blumental, R.D., Sharkey, R.M. Semin. Cancer Biol. 1, 217 (1990).
154. Waldmann, H. Annu. Rev. Immunol. 7, 407 (1989).
155. Prinz, J., Braun-Falco, O., Meurer, M. et al. Lancet 338, 320 (1991).
156. Emmrich, J., Seyfarth, M., Fleig, W.E., Emmrich, F. Lancet 338, 570 (1991).
157. Miller, R.A., Maloney, D.G., Warnke, R., Levy, R. N. Engl. J. Med. 306, 517 (1982).
158. Capel, P.J., Preijers, F.W., Allebes, W.A., Haanen, C. Neth. J. Med. 28, 112 (1985).
159. Elliott, T.J., Glennie, M.J., McBride, H.M., Stevenson, G.T. J. Immunol. 138, 981 (1987).
160. Queen, C., Schneider, W.P., Selick, H.E., Payne, P.W., Landolfi, N.F., Duncan, J.F., Avdalovic, N.M., Levitt, M., Junghans, R.P., Waldmann, T.A. Proc. Natl. Acad. Sci. USA 86, 10029 (1989).
161. Queen, C., Schneider, W.P., Waldmann, T.A. In Clark, M. (ed.): Protein Engineering of Antibody Molecules for Prophylatic and Therapeutic Applications in Man. Academic Titles, 1993.
162. Brown, P.S., Parenteau, G.L., Dirbas, F.M., Garsia, R.J., Goldmann, C.K., Bukowski, M.A., Junghans, R.P., Queen, C., Hakima, J., Benjamin, W.R., Clark, R.E., Waldmann, T.A. Proc. Natl. Acad. Sci. USA 88, 263 (1991).
163. Steinman, L. Prog. Allergy 45, 161 (1988).
164. Wofsy, D. Prog. Allergy 45, 106 (1988).
165. Blakey, D.C., Wawrzynczak, E.J., Wallace, P.M., Thorpe, P.E. Prog. Allergy 45, 50 (1988).
166. Wawrznczak, E.J. Br. J. Cancer 64, 624 (1991).
167. Lerner, R.A., Benkovic, S.J., Schultz, P.G. Science 252, 659 (1991).
168. Benjamin, R.J., Cobbold, S.P., Clark, M.R., Waldmann, H. J. Exp. Med. 163, 1539 (1986).
169. Jonker, M., Nooij, F.J.M. Eur. J. Immunol. 17, 1519 (1987).
170. Woodle, E.S., Thistelthwaite, J.R., Jolliffe, L.K., Ziven, R.a., Collins, J.R., Bodmer, M., Athwal, D., Alegre-M.-L., Bluestone, J.A. J. Immunol. 148, 2756 (1992).
171. Chatenoud, L., Ferran, C., Reuter, A., Reuter, C., Legendre, C., Gevaert, Y., Kreis, H., Franchimont, P., Bach, J.F. N. Engl. J. Med. 320, 1420 (1989).
172. Alegre, M.-L., Collins, A.M., Pulito, V.L., Brosius, R.A., Olson, W.C., Zivin, R.A., Knowles, R., Thistlethwaite, J.R., Jolliffe, L.K., Bluestone, J.A. J. Immunol. 148, 3461 (1992).
173. Bolt, S., Routledge, E., Lloyd, I., Chatenoud, L., Pope, H., Gorman, S.D., Clark, M., Waldmann, H. Eur. J. Immunol. 23, 403 (1993).

CHAPTER 2.1

THERAPEUTIC APPLICATIONS OF MONOCLONAL ANTIBODIES FOR HUMAN DISEASE

DAVID A. SCHEINBERG AND PAUL B. CHAPMAN
Leukemia and Clinical Immunology Services, Department of Medicine, Memorial Sloan-Kettering Cancer Center, New York, NY 10021

2.1.1. INTRODUCTION

Since the introduction of techniques for producing monoclonal antibodies (mAb) of defined specificity 15 years ago, there has been steady expansion of their applications into numerous areas of biomedical science. For more than a decade, there have been attempts to use mAb as therapy for diverse human diseases, including coronary artery disease and infarctions, infection control and detection, cardiac and renal allograft failure, graft versus host disease, multiple sclerosis and a host of other autoimmune diseases, and cancer diagnosis and therapy. This chapter will not be a compendium of the tens of thousands of recent papers in the field, nor the hundreds of clinical trials. Rather, it will review the approaches being taken in the field, the strategies to resolve ongoing obstacles to success, and the areas where the most rapid progress is being made. Five approaches to therapy are utilized in the application of mAb in humans *in vivo* (Table 2.1.1). First, mAb can be used to focus an inflammatory response against a target cell. Binding of an mAb to a target cell can result in fixation of complement, which results in lysis, or opsonization, which can target the cell for lysis by various effector cells (a cell of the immune system capable of responding immune-logically, often by killing a target cell) such as natural killer (NK) cells, neutrophils, or monocytes. Second, mAb can be used as carriers simply to deliver another small molecule, atom, radionuclide, peptide, or protein to a specific site *in vivo*. Third, mAb can be directed at critical hormones, growth factors, interleukins, or other regulatory molecules or their receptors in order to control growth or other cell functions. Fourth, anti-idiotypic mAb can be used as vaccines to generate an active immune response. Finally, mAb can be used

Monoclonal Antibodies: Principles and Applications, pages 45–105
© 1995 Wiley-Liss, Inc.

TABLE 2.1.1. Therapeutic Approaches Using Monoclonal Antibodies in Humans

1. To act via intrinsic immunological activity Complement lysis Complement opsonization Neutralization Clearance ADCC by NK cells by neutrophils by monocytes	3. To exert regulatory functions on Hormones Growth factors Interleukins Regulatory molecules Surface Idiotypes (B or T) Receptors (for any of above)
2. To serve as targeting vehicles for Radionuclides Stable atoms Drugs Interleukins, growth factors Toxins Peptides Proteins and enzymes Regulatory molecules Liposomes	4. To act as vaccines Antiidiotype Adjuvants 5. To exert pharmacological intervention To alter pharmacokinetics To alter biodistributions To change properties of agents To clear other mAb

to speed the clearance of organisms or toxins and can fundamentally alter the properties of other therapeutic agents. For example, mAb can be fused to drugs or factors to increase their plasma half-life, change their biodistribution, or render them multivalent. Alternatively, mAb can be used to clear previously infused mAb from the circulation.

Despite the myriad of therapeutic uses of mAb, it is becoming evident, due to the complicated pharmacology and immunogenicity of these agents, that complex strategies will be difficult to achieve with current technology. Intricate schemes of multimolecular interactions of prodrugs and enzymes or of unnatural, genetically engineered constructs with unknown pharmacology may be difficult and time-consuming to implement and validate. New methods for engineering mAb may make these strategies applicable in the future. To date, only one therapeutic mAb and one diagnostic mAb have been approved for use in humans by the FDA in the United States.

In this chapter we will focus on diagnostic uses of mAb in cancer, infection, and cardiovascular disease; therapeutic uses of native, radiolabeled, and conjugated mAb in cancer; bone marrow purging; therapy or shock and viral infections; treatment of autoimmune and inflammatory states; treatment approaches to cardiac, vascular, and renal diseases; and transplantation. We will include discussions of existing problems and strategies for the future. Recent, more exhaustive evaluations in specific areas can be found for the use of mAb in cancer [1–4], hematopoietic cancers [5,6], renal allografts [7], septic shock [8,9], radioimmunodetection of cancer [10–18], conjugation of toxins [19] and isotopes [20,21], immunosuppression [22], cardiovascular disorders [23], and thrombosis [24,25].

2.1.2. DIAGNOSTIC IMAGING

2.1.2.1. General Considerations

The first test of the usefulness of a mAb is its ability to reach its target. This is most often measured by external scintigraphy with a gamma camera or, more recently, by positron emission tomography (PET). The pharmacological issue of delivering a large molecule such as an IgG to a site *in vivo* is not trivial, and failure to selectively image a target has many times led to the abandonment of the mAb. It should be remembered, however, that the criteria for selecting a useful mAb for imaging may be unrelated to its therapeutic value, and vice versa. MAb that are not useful for treatment may be quite valuable as diagnostic tools. One example of this is that smaller molecules, such as Fab and $F(ab')_2$ fragments or genetically engineered single chain antibodies (SCA), may be superior imaging agents compared with larger molecules with more biological effects, such as IgM or IgG3 (Table 2.1.2). In general, it is advisable to use agents with no intrinsic biological activity, since those antibodies with potent ability to fix complement, such as murine IgG3, or to activate cells, such as certain anti-T-cell antibodies, tend to cause significant, and occasionally serious, side effects [26].

Immunogenicity of rodent antibodies is a significant problem, especially if recurrent diagnostic tests or subsequent use of a therapeutic mAb is planned. It is likely that extensively engineered or other non-native Ig structures, even if human-derived, will also be immunogenic on repeated dosing. Despite these problems, few diagnostic studies have been conducted with human or humanized mAb. Those studies that have been tried have used IgM, since human IgG are still difficult to make [27–29].

In designing a diagnostic strategy, both the carrier (the Ig) and the isotope may be selected to be most appropriate for a particular system. Insistence that one isotope

TABLE 2.1.2. Immunoglobulin Structures of Potential Use *In Vivo* for Diagnosis

Structure	Advantages	Disadvantages
IgG1 or 2A (rat) or IgG1 or 2B (mouse)	Economy and availability; lack of biological activity	Large size; immunogenicity
IgG2, IgG4 (human or humanized)	Nonimmunogenic; lack of biologic activity	More difficult to make now; large size
Fab or F(ab')2	Better pharmacokinetics; faster clearance; better penetration	Losses in avidity or stability possible; too rapid clearances
Single chain antibodies	Better pharmacokinetics; faster clearance; better penetration	Losses in avidity likely; unstable; too rapid clearances; immunogenicity possible
Dimeric IgG	Potentially increased binding and retention	Pharmacologic uncertainties; large size

type will solve all the problems of diverse systems will inevitably lead to failure in some of these systems. For example, imaging of metastases of colon carcinoma in the liver may be feasible with an IgG linked to iodine 131, but may fail miserably with the same IgG linked to indium 111, which accumulates in the hepatic parenchyma. By careful combinations of isotopes and Ig, disadvantages of one may be outweighed or overcome by the other. For example, the problem of a human IgG with a long plasma half-life can be compensated by attaching a short-lived emitter. In imaging with radiolabeled mAb, it is critical to maximize the target-to-background ratio. This is in contrast to mAb therapy, in which the longest and largest retention of Ig at the site is critical (largest area under the curve). Depending on tumor accessibility, on modulation and metabolism of the Ig conjugate, and on pharmacokinetics, the optimal imaging window might occur within hours, as in leukemias [30], or after a week [31].

Conventional wisdom has been that hematopoietic cancers are easy, and solid tumors difficult, to image [32]. However, with the appropriate mAb and isotope, many solid tumors image easily, accurately, and with tumor-to-background ratios better than those seen in lymphomas [31].

2.1.2.2. Choice of Isotope

Isotopes can be categorized by the predominant mode of decay: gamma versus positron. The choice of type of decay mode has traditionally been simplified by the lack of available whole-body PET scanners and the abundant wealth of gamma cameras in nuclear medicine facilities. This predilection is likely to change as the potential for highly precise, and most importantly, quantitative imaging becomes possible by use of PET [33]. An alternative to PET is the single photon emission computerized tomographic (SPECT) method, which is already in widespread use and allows a three-dimensional reconstruction of an image from a series (usually 64) of planar images obtained at angular intervals around the patient [34–36]. SPECT can provide improved detectability of lesions, especially deep within the body, but it requires increased isotope dose and more time to obtain the images.

Among gamma emitters, the next decisions usually revolve around radionuclide physical half-life, emission quality, other radioactive emissions (e.g., beta particles) from the nuclide that are potentially cytotoxic, and metabolism; each will be discussed in turn. Half-lives of potential radionuclides for conjugation range from hours to days (Table 2.1.3) and must be considered early in the development of a radioimmunodiagnostic agent. For example, if optimal tumor-to-blood uptake ratios require 5–7 days to achieve, then ^{123}I or another brief-lived nuclide will be inadequate. ^{131}I has been used most often, largely owing to its relatively low cost and ease of conjugation, but its γ photon photoelectric energy peak is too high to allow for best resolution of the targeted sites with available gamma cameras. ^{123}I, ^{111}I, and ^{99m}Tc have better γ emission for gamma camera imaging. ^{131}I also emits a powerful and cytotoxic beta particle, which is usually of no consequence to the patient in the diagnostic doses employed, but can result in partial therapeutic effects in some lymphoma systems that are particularly radiosensitive [37,38].

TABLE 2.1.3. Isotopes of Potential Use in Radioimmunoimaging

Isotope	Advantages	Disadvantages
A. Gamma emitters		
^{131}I	Economical; huge body of experience in humans; metabolism not harmful except to thyroid; simple conjugation.	Suboptimal γ emission characteristics; beta emission as well. Thyroid accumulation; long physical T½ (8 days).
^{123}I	Optimal pure γ emissions; simple conjugation.	Short physical T½ (13 h) thyroid accumulation, very expensive.
^{111}In	Optimal γ emission; reasonable physical T½ (3 days).	Requires chelation; hepatic and marrow metabolism.
^{99m}Tc	Optimal γ emissions for gamma camera imaging.	Short physical T½ (6 h); more difficult chemistry than iodine.
B. Positron		
^{124}I	High resolution and quantitation possible.	Tissue radiation dose comparable to ^{131}I on a mCi/2 mCi basis.

The various nuclides are metabolized in different ways. The metals tend to accumulate in the liver, bone marrow, and reticuloendothelial system. The halides accumulate in the urine, upper gastrointestinal tract via antral secretion, and of course, the thyroid. Thyroid accretion is generally well blocked by Lugol's solution (saturated potassium iodide) given prior to the radioiodine or by perchlorate, which also blocks gastrointestinal secretion. These agents can be associated with minor side effects of their own, including parotitis and nausea, respectively.

Another consideration in choosing a radionuclide is the overall therapeutic plan for the patient. If no therapeutic dose of radiolabeled mAb is planned, then the radionuclide can be chosen for optimizing the diagnostic possibilities. If, however, radioimmunotherapy is to follow radioimmunodiagnosis, it may be prudent to image with a nuclide in the same class as that which will be used for therapy, such as ^{123}I before therapeutic ^{131}I, or ^{111}In before ^{90}Y therapy. Even so, predictions for therapy based on the scan of a nonidentical metal may not be completely relevant.

It seems clear that despite a number of drawbacks, conventional nuclides such as ^{131}I or ^{111}In are adequate for most applications and the development of costly new radionuclide systems is not generally indicated.

2.1.2.3. Imaging Hematopoietic Cancers

Among the cancers derived from hematopoietic cells (acute and chronic leukemias, lymphomas, and myelomas) most effort has been directed toward the development of diagnostic agents for lymphomas. In general, there is little use for diagnostic imaging of myelomas or leukemias, even though the images of the latter are quite dramatic [30], since the sensitivity of imaging is well below that of other simpler diagnostic procedures such as bone marrow aspirate and biopsy or blood samplings. Such imaging does provide useful dosimetry, however [15,30,38–40]. Development of a truly useful diagnostic radioimmunoimaging agent will be difficult, as conventional methods such as computerized tomography (CT) and gallium scanning are both sensitive and specific. Hence, work involving mAb for lymphoma imaging has largely been in preparation for radioimmunotherapy [38,41,42].

Both ^{111}In-labeled and ^{131}I-labeled constructs have been used, and in at least one trial have been compared, for efficacy in a small number of patients [43]. In this study patients received ^{131}I T101 and ^{111}In T101 (anti-CD5) concurrently to image T-cell cutaneous lymphomas. While ^{131}I rapidly cleared from the body and tumor sites, ^{111}In was retained in skin lesions as well as in liver, spleen, and marrow.

The lymphomas are perhaps the best targets available for radioimmunodetection. The cells are commonly found in areas readily accessible to the distribution of mAb, such as in the blood, marrow, spleen, and nodes; the differentiation scheme for lymphoid maturation is well characterized, allowing for the selection of several lineages or stage-specific markers found nowhere else in the body. In addition, since the lymphoid cells are exquisitely radiosensitive, identification of an antibody–antigen system with even moderate tumor-to-nontumor specificity justifies therapeutic trials. CD5, found on T-cell neoplasms and B-CLL, was one early target [43]; CD37 (mAb-1), CD21 (OKB7), an HLA-DR variant (Lym-1), sIgG (anti-idiotypes), and LL2 have also been targeted [37,38,41,42]. In most of these radioimmunodetection trials, imaging was followed immediately, within the same trial, by radioimmunotherapy for the reasons outlined above. Hence, further discussions of these trials in lymphoma will be found in the section on radioimmunotherapy, below.

Because Hodgkin's disease is routinely curable with chemotherapy, it has rarely been studied as a target of radioimmunodetection or radioimmunotherapy. Moreover, the cell surface antigens marking the Reed–Sternberg malignant cells are poorly characterized and often variably expressed. However, curative treatment of Hodgkin's disease depends on accurate staging, so that a complete evaluation of possible nodal sites of involvement is required, often necessitating a diagnostic splenectomy. One important goal of radioimmunodetection in this disease, therefore, is to avoid the need for the splenectomy; another goal is to avoid biopsies of residual masses after presumed curative therapy [44]. In one study, nodal, splenic, and hepatic disease were detected in four or five patients imaged in a pilot trial of radioimmunodetection [44].

2.1.2.4. Radioimmunodiagnosis of Solid Tumors

Tremendous effort has been directed at development of specific targeting agents for solid tumors, particularly colorectal carcinomas, but also cancers of the lung, breast, kidney, brain, ovary, melanoma, and neuroblastoma. The focus is often on occult, residual, or metastatic disease in the pelvis or liver that might avoid conventional detection.

The gastrointestinal carcinomas (colorectal carcinomas), ovarian carcinomas, and breast cancers have been studied in most detail as systems for development of radioimmunodetection or therapy [2,16,17,32,45–48]. These are common diseases worldwide, with several hundred thousand new cases appearing annually in the industrialized countries; they are amenable to the use of a novel diagnostic procedure that might allow detection of residual disease after surgical resection for cure, detection of recurrence after surgery, or determination of stage in advance of surgery. Carcinoembryonic antigen (CEA) is the most frequently used target antigen in colorectal cancers [40] as well as in several of the ovarian and breast trials. Human milk fat globule has been a frequent target in both breast and ovarian carcinomas. A series of mAb against the TAG72 antigen has also seen wide use in ovarian and colorectal carcinomas [50]. An ^{111}In-labeled B72.3 is now licensed for use in the United States. Another mAb, with some of the most specific targeting and highest uptakes into liver metastases as compared to liver tissue, is A33 [31]. As tumor-to-nontumor ratios of 100:1 have been observed, this mAb may be useful in radioimmunotherapy. Radioimmunodetection may find its best use in these diseases in determining whether normal-sized nodes are involved with tumor or in examining the retroperitoneal and pelvic nodes where CT and x-rays are less sensitive and specific. ^{131}I, ^{123}I, ^{131}In, and ^{99m}Tc labels have each been used widely; intact and fragmented mouse mAb as well as chimeric mAb (antibodies derived from two different animal sources) have been tried in models and in humans [51–58]. The radiometals, such as ^{111}In, may be intrinsically inferior for these diseases in which liver metastases are common, since ^{111}In results in a high background in the liver. However, in the pelvis, ^{111}In anti-CEA has compared favorably with CT scans [12,47]. In the case of ovarian cancers, radioimmunodetection may be most useful in the omentum, where CT scans are less useful.

A variety of other tumor types (lung carcinomas, head and neck tumors, neuroblastoma, hepatoma, brain tumors, renal cancers, prostate carcinomas, uterine cancers) also have been studied [59–65]. There are too few data yet to assess whether radiolabeled mAb will be useful in the treatment of these diseases. Melanoma and neuroblastoma have also received considerable attention. In part, this is due to the abundance of tumor-associated differentiation antigens found on these cancers, and in part due to early data that suggested that immunotherapy may be useful in the treatment of these tumors. For reasons that are not clear, neuroblastoma images extremely well and very rapidly with ^{131}I-labeled mAb, and large uptake ratios have resulted in radioimmunotherapy trials with encouraging results [66].

A review of more than 60 human trials of radioimmunodetection has recently been compiled [13].

2.1.2.5. Regional Imaging and Therapy

Due to the myriad of problems associated with the use of radiolabeled mAb *in vivo,* many of which relate to inadequate delivery to target and poor clearance of untargeted mAb, efforts have been made to localize the mAb administration to the region of interest in the patient. For cancers that often involve the peritoneum, such as ovarian and colorectal carcinomas, intraperitoneal (IP) infusions are used [67–70]. For a number of other cancers that spread locally into regional lymph nodes, such as head and neck tumors or breast cancer, and/or cancers that exclusively spread via the lymph nodes, such as lymphomas, radioimmuno–lymphangiography may be useful [71]. Selective injection into the CNS has also been attempted in the imaging and treatment of cancers involving the meninges.

In one study using ^{131}I-B72.3 that looked at both IV and IP administration of mAb into patients with peritoneal spread of a variety of cancers [72], it was found that specific mAb localization to tumors undetectable by conventional imaging methods (CT and x-ray) was possible and that high (70:1) uptake ratios were possible. Direct comparison of IP versus IV in the same patients showed, as expected, that small peritoneal tumors were targeted better with IP injections. As much as one third of the IP dose reached the plasma. It is likely that large implants, with their own blood supply, will still require IV infusions for accessibility. In fact, one comparative study found that mAb targeting to intra-abdominal tumor deposits was superior after IV injection of radiolabeled mAb compared to the IP injections [73].

One application of intralymphatic administration of mAb would be to diagnose tumor-involved nodes that are of normal size and character on CT scan. No studies have yet shown that this is possible or that intralymphatic radioimmunodetection is superior to conventional imaging techniques.

An alternative approach to regional radioimmunodetection is to use a hand-held γ detector to isolate areas of uptake of ^{125}I-labeled mAb during surgery after intravenous infusions of mAb [74]. In one study of 104 patients given ^{125}I-B72.3, 30 sites of otherwise occult disease in 26 patients were identified at surgery. In 18 patients, surgical decisions were changed on the basis of data obtained from the probe: Ten patients were deemed nonresectable while eight patients underwent a more extensive operation. A previous pilot trial had shown similar advantages to this approach [75].

2.1.2.6. Radioimmunodiagnosis of Infections and Inflammatory Sites

Both the infecting organisms and the infiltrating neutrophil response may serve as targets for radioimmunodetection of infectious or inflamed sites. The disadvantages of the use of the organism as a target are the requirement of knowing in advance the presumed infectious agent as well as the small number of organisms that may be found in a given site and thus poor sensitivity. For these reasons, anti-neutrophil mAb have been preferred [76,77]. Although encouraging data for detection of abscesses have been obtained, surprisingly better results have been seen with completely nonspecific polyclonal human IgG [78]. In this latter approach, better than 90% of 56 patients with infections were diagnosed [78]. Imaging of inflammation was not specific, though, since in a second group of 16 cancer patients, most tumors were also visualized [78]. Hence, a variety of inflammatory lesions such as abscesses, cancers, and inflammatory bowel disease might be amenable to such an approach.

2.1.2.7. Cardiac and Vascular Radioimmunoimaging

Radioimmunodiagnostics in the field of cardiovascular disease have largely focused on two clinical problems: myocardial infarction/necrosis and vascular thrombosis. Upon ischemic or inflammatory necrosis of myocardium, cardiac myosin is released from dead and dying cells and this protein becomes a focal new antigen target. Most early work studied the role of radiolabeled mAb in the detection of postinfarction necrosis [23,79,80]. The first use of chelated radiometals (indium 111) for radioimmunodiagnosis was in this field, rather than in cancer imaging [81]. The utility of radioimmunoimaging of cardiac disease may be limited due to the wide number of competing methods that can more economically detect and localize cardiac infarcts and predict cardiac function. Moreover, false negative studies can result from persistent tracer in the blood pool [79]. Cardiomyopathy, a less acute disease that may be more amendable to this form of radioimmunodetection, has also been studied [82,83]. The use of antimyosin imaging may find applications in dilated myopathies, myocarditis, cardiac transplant rejection, and adriamycin toxicity [82,84,85].

An area with great potential for growth because of the lack of definitive noninvasive and nontoxic tests for detection is venous thrombosis [24,86]. Deep venous thrombosis (DVT) is a significant cause of morbidity and, due to the possibility that thrombus will migrate to the lungs, also of mortality via pulmonary embolism. Current diagnosis relies on the clinical exam, Dopplers, impedance plethysmography, and contrast venography, each of which has a significant incidence of false positive and false negative results. Venography also has a high rate of adverse effects, including phlebitis.

Two targets within thrombi have been the focus of considerable work: platelets (including glycoprotein IIIB/IIa, GM 140, or thrombospondin) and fibrin, typically the amino terminus. ^{131}I-, ^{123}I-, ^{99m}Tc-, and ^{111}In-labeled intact or fragmented IgG have been used [24,87]. Because mAb can be made to epitopes that appear during active clot formation, radioimmunodetection has the potential advantage of distinguishing acute from chronic conditions. In addition, both legs and the full extent of each leg are evaluated simultaneously by use of radioimmunodetection. The sensitivity and specificity of these agents have been in the range of 80%–100%, which compare well with conventional methods [86]. Similar approaches are being taken in the diagnosis of pulmonary emboli [88].

A discussion of the features of radioimmunodetection in comparison with other conventional and experimental radiopharmaceuticals has recently been published [25]. Other relevant applications that have been proposed include monitoring thrombolysis during therapy [89] as well as promotion of fibrinolysis by targeting mAb to fibrinolytic enzymes [23,89]. The latter approach might lead to less toxic delivery of urokinase, tissue plasminogen activator (TPA), or streptokinase specifically to vascular, pulmonary, or cardiac clots.

2.1.3. TREATMENT OF CANCER WITH UNCONJUGATED MONOCLONAL ANTIBODIES

In choosing a mAb for therapeutic clinical trial, several factors must be considered regarding the target antigen. The specificity of mAb therapy relies on the assumption that the target antigen is expressed only on the tumor cell. In reality, this is never the case. Virtually all tumor antigens described to date are also expressed, to some degree, on certain normal cells. Thus, the issue becomes the relative degree of expression in normal and tumor cells. Other aspects of the target antigen can affect how well the mAb targets the tumor cell, for example, density, accessibility, heterogeneity, and stability of antigen expression, and whether antigen circulates in the serum.

Aside from issues related to the antigen, characteristics of the mAb have profound effects on the pharmacokinetics, bioavailability, and biological activity of the mAb. It has generally been assumed that mAb with high affinity are preferable for therapy. The pharmacokinetics of the mAb are also affected by the isotype and size of the molecule; $F(ab')_2$ and Fab fragments demonstrate much shorter serum half-lives, although they may be better able to penetrate the endothelium and bind to tumor cells [90]. Unfortunately, mAb fragments often demonstrate decreased avidity for antigen, and since fragments are missing the Fc portion, they do not mediate antibody-dependent cellular cytotoxicity (ADCC) or fix complement (discussed below). On the other hand, the absence of the Fc portion may prevent nonspecific targeting within the reticuloendothelial system.

The characteristics of the tumor itself also have profound effects on the success of mAb therapy. Vascularity, blood volume, and endothelial integrity of the tumor are critical parameters in determining mAb targeting. As a result of these many factors, binding studies *in vitro* often do not predict how well a mAb will target tumor *in vivo* [91], and several investigators have shown that "nonspecific" mAb can target tumor in animal models almost as well as specific mAb [92–94]. This nonspecific targeting is largely due to increased tumor vascularity and leaky tumor capillaries rather than to specific binding by mAb.

2.1.3.1. Mechanisms of mAb-Mediated Tumor Cell Destruction

There are three general mechanisms by which an unconjugated mAb might induce an antitumor effect. Binding to the tumor cell, the mAb can induce a localized inflammatory response, resulting in immune destruction of the tumor cell. Alternatively, the mAb could interfere with the growth or regulation of the tumor cell by binding to a critical growth factor or growth factor receptor. The third mechanism involves anti-idiotypic mAb that can mimic tumor antigens and induce an active antitumor immune response.

Induction of a Localized Inflammatory Response as a Result of mAb Binding to Tumor Cells. Binding of mAb to tumor cells can induce a specific and localized inflammatory response, resulting in tumor cell death. One mechanism is to mediate the activation of complement components, a system of more than 30 proteins that initiates inflammation, neutralization of pathogens, clearance of immune complexes, and disruption of cell membranes. Complement activation leads to the formation of a macromolecular complex known as the membrane attack complex (MAC). The MAC produces cylindrical holes in the membrane of the target cell, leading to cellular damage and usually cell death. While late complement components can directly damage cells, components that are activated earlier in the complement cascade can initiate an array of biologically important functions: (1) Low-molecular-weight complement fragments C3a, C4a, and C5a are anaphylatoxins that can markedly increase vascular permeability and induce smooth muscle contraction; (2) C5a can induce chemotaxis for phagocytic cells, leading to infiltration into sites of inflammation; and (3) cell-bound C3b, C4b, and other components can label target cells and lead to their rapid clearance. Mouse (IgM, IgG2a, IgG3), rat (IgM, IgG2b), and human (IgM, IgG1, IgG3) immunoglobulins are efficient in activating human complement.

There is evidence that the antitumor effects of mouse mAb against human tumors growing in nude mice can be mediated through complement activation [95,96]. Several factors determine whether or not a tumor responds to complement-fixing mAb such as the density of antigen on target tumor cells and whether the tumor produces factors that can make it resistant to complement-mediated killing.

ADCC is another mechanism by which the mAb can induce inflammation at the tumor site. An effector cell expressing an Fc receptor can readily bind to a mAb-coated target cell, inducing the effector cell to destroy the target cell. A wide range of cell types have been shown to be active in ADCC, including macrophages, NK cells, neutrophils, and even non-immune cell types such as platelets. It has been possible to augment ADCC *in vitro* by combining mAb with biological agents that enhance NK cell or macrophage activity such as interleukin 2 (IL-2), α-interferon, β-interferon, γ-interferon, and macrophage colony-stimulating factor [97–102]. Several clinical trials are currently under way in an attempt to exploit this concept.

Not all mAb isotypes are equally active in inducing ADCC; mouse IgG2a and IgG3 mAb and rat IgG2b subclasses have been most effective, while among human IgG isotypes, IgG1 and IgG3 are active. As with complement-mediated cytotoxicity, the number of antibody binding sites may be important in determining whether a mAb is active or inactive in ADCC.

Interference With Growth or Regulation. The mechanisms that underlie neoplastic transformation are incompletely understood, but it is being increasingly appreciated that growth factors and growth factors receptors play an important role. In both *in vitro* and animal models, mAb against growth factors or their receptors can inhibit tumor cell growth. As a result, mAb directed against growth factors (e.g., IL-2, IL-6) and growth factor receptors (e.g., IL-2 receptor, epidermal growth factor (EGF) receptor, transferrin receptor) are being explored for cancer therapy.

Anti-idiotypic mAb. Idiotopes are antigenic markers of the variable region of immunoglobulin molecules, and the set of idiotopes on an antibody molecule constitute and antibody's idiotype. Antibodies that recognize idiotopes are termed anti-idiotypic antibodies and are designated Ab2 antibodies (to distinguish them from the immunizing antibody, Ab1). In clinical trials, Ab2 mAb were first used in the treatment of B-cell malignancies (for review, see [103, 105]) with the goal of targeting the unique idiotope of the transformed B-cell clone. It has been hypothesized that binding of the anti-idiotypic mAb to this tumor-specific marker could induce a specific inflammatory response against the tumor cells by mechanisms discussed above. Alternatively, the anti-idiotypic mAb might down-regulate the malignant B-cell clone via the idiotypic network as proposed by Jerne [106]. *In vitro,* anti-idiotypic mAb can inhibit the proliferation and immunoglobulin synthesis *in vitro* of normal [107] and transformed B cells [108], and can induce apoptosis, or programmed cell death [108].

Anti-idiotypic mAb can also function as vaccines to induce active immunity. Among anti-idiotypic antibodies, some appear to mimic the antigen recognized by the original mAb, or Ab1. These anti-idiotypic mAb are sometimes designated Ab2β mAb and are of special interest for their potential use as vaccines. The molecular basis of antigenic mimicry is not well understood and it is unlikely that the antigen-binding site of the anti-idiotypic antibody "looks like" the antigen. Rather, it is more likely that the three-dimensional conformation of shape and charge mimics that of the antigen, allowing equivalent, but nonidentical interactions with the Ab1. Nevertheless, anti-idiotypic mAb can be used as surrogate immunogens, and in animal models anti-idiotypic mAb have been shown to induce protective immunity against infectious agents [109–111] and against tumors [112,113]. This form of vaccine offers several potential advantages over conventional vaccines: (a) Clinical-grade anti-idiotypic mAb can be easier to obtain in large quantity than purified antigen; (b) anti-idiotypic mAb have no infectious or tumorgenic potential and so may be safer than vaccines that use intact antigen derived from cellular sources; (c) xenogeneic anti-idiotypic mAb may be more immunogenic than conventional antigens and, in some cases, can break immune tolerance [109]; and (d) anti-idiotypic mAb can stimulate both a humoral and a cellular immune response [110,111] in contrast to some classes of antigen (e.g., carbohydrates) that are not effective in inducing cellular immunity. As a result of these observations, several early clinical trials using anti-idiotypic mAb have been carried out or are currently under way.

2.1.3.2. Clinical Trials Using mAb in Cancer Patients

Both rodent and, more recently, chimeric and humanized mAb have been used in clinical trials for cancer. Most trials have been either pilot or Phase I studies that have sought to answer questions about toxicity, pharmacology, tumor targeting, and biological effects of mAb. However, since it is not clear what the relationship is between mAb dose and antitumor effect, many of these early trials have also reported whether any antitumor effects were seen.

Clinical Trials Using Murine mAb for the Treatment of Cancer. Several murine mAb have been used in the treatment of both hematological malignancies and solid tumors. In most cases, mAb have been selected that recognize an antigen specifically or preferentially expressed on the tumor cell. For solid tumors (Table 2.1.4), both protein and glycolipid antigens have been targeted with the intention of inducing an inflammatory response at the site of the tumor. Clinical trials are also under way with mAb designed to block the actions of essential growth factors, such as epidermal growth factor [114] and transferrin.

Patients have been treated with mAb doses of >1 g, and in general toxicity has been mild. Most common toxicities have consisted of urticaria, mild bronchospasm, and low-grade fever. Severe immediate-type hypersensitivity (anaphylactic) reactions to mAb treatment have been rare, but they can occur. In patients receiving high doses of R24, a mAb against the ganglioside G_{D3}, the dose-limiting toxicity was malignant hypertension thought to be due to R24 binding to adrenal medulla, a tissue that expresses some G_{D3}. It has been possible to demonstrate therapeutic mAb targeting to tumor sites, although the efficiency of the targeting varies greatly among antibodies and among tumor types. The majority of solid tumor patients treated with mouse mAb developed human anti-mouse antibody (HAMA), usually within 10 days. Although the presence of HAMA usually does not result in toxicity, it is clear that the serum half-life of mouse mAb is markedly shortened in the presence of HAMA and represents a major limitation of treatment with mouse mAb. Few antitumor responses have been observed in these trials and most were transient, partial responses.

Because of the availability of specific mAb and the ease of obtaining the target cells, many mAb trials have focused on leukemias and lymphomas (Table 2.1.5). While mAb targeting is less problematic than for solid tumors, since the leukemic cells reside within the vascular spaces, identification of an appropriate target antigen has remained an issue. Most of the mAb that have been used have recognized differentiation antigens expressed by normal cells. Despite this, toxicity has generally been mild and has included fevers, chills, rash, and arthralgias; although at high infusions rates, dyspnea has occurred, possibly due to leukoagglutination of target cells. Patients with hematological malignancies are less likely to develop HAMA than patients with solid tumors or patients without malignancies; this is presumably due to a relative immunodeficiency. Short-lived major objective responses were seen in about 50% of patients with cutaneous T-cell lymphomas, but rarely in the leukemias. Two of the trials used mAb that targeted growth factors or growth factor receptors [115,116], and it is of interest that 2 of 11 patients treated with anti-Tac mAb showed a clinical response.

Within the hematological malignancies are a group of diseases that express antigens not found on normal cells. The B-cell malignancies are clonal diseases that express unique surface immunoglobulin, and the immunoglobulin idiotype represents a truly tumor-specific antigen. As a result, anti-idiotypic mAb offer exquisite specificity and have been used in several clinical trials. Binding of an anti-idiotypic mAb to the malignant B cell might trigger an inflammatory response capable of killing the cell. Alternatively, there is evidence that the anti-idiotypic mAb can turn the cell "off" without inducing an inflammatory response. Patients have been treated with anti-idiotypic mAb raised against their own B-cell tumor (anti-private-idiotopes) either alone or in combination with chlorambucil [103,105,117,118] or α-interferon [105,117] (Table 2.1.6, discussed below). Both regimens induced high response rates with a few long-lasting complete responses (CR; complete disappearance of all signs and symptoms of cancer for at least 30 days). It is not clear that either chlorambucil or α-interferon played an important role in these regimens.

In these trials, a "custom-made" anti-idiotypic mAb had to be produced for each patient. Since this process generally requires 9–12 months, this approach has obvi-

TABLE 2.1.4. Clinical Trials With Unconjugated Murine mAb: Solid Tumors

Tumor type	mAb	Antigen[1]	N^2	HAMA, %[3]	Clincal responses[4]	Comments	References
Colorectal carcinoma	17-1A	38-kD protein	54	74	1 CR, 5 PR	CR lasted 10 months.	[295–297]
Melanoma	R24	G_{D3} ganglioside	29	97	4 PR	No dose-response effect seen. Malignant hypertension at high doses (>1 g).	[26,298]
Melanoma	MG21	G_{D3} ganglioside	8	NS	None	No tumor targeting detected at doses up to 100 mg/m^2.	[299]
Melanoma	ME361	G_{D2}/G_{D3} ganglioside	13	100	1 CR	CR lasted >16 months. Mechanism possibly related to induction of anti-iditoypic antibodies.	[300]
Melanoma/neuroblastoma	3F8	G_{D2} ganglioside	17	41	2 PR	One of the PRs lasted >308 weeks.	[301]
Melanoma	14G2a	G_{D2} ganglioside	12	100	1 PR	Pain during infusion (9 patients) and delayed extremity pain (5 patients); 2 patients with Grade 4 neurotoxicity.	[302]

Melanoma	9.2.27	Chondroitin sulfate/HMW-MAA	8	37	None	Tumor targeting to tumor.	[303]
Melanoma	96.5 and 48.7	Chondroitin sulfate/HMW-MAA (96.5) and p97 (48.7)	5	80	None	Four patients received both mAb; one received 96.5 only. Tumor targeting demonstrated.	[304]
Colon/ovarian/breast/lung	L6	Undefined	19	72	1 CR	CR × 14+ weeks in patient with recurrent breast carcinoma who received 400 mg/m^2 of L6.	[305]
Non-small-cell lung cancer	R83852	EGF receptor	11	20	None	Saturation of antigen = 50%–90% at doses of 200 and 400 mg/m^2.	[114]

[1]EGF, epidermal growth factor; HMW-MAA, high-molecular-weight melanoma-associated antigen.

[2]*N*, number of patients.

[3]Percentage of patients developing human anti-mouse antibodies. NS, not stated.

[4]CR, complete response, usually defined as complete disappearance of all tumor for at least 1 month. PR, partial response, defined as a 50% decrease in the sum of the bidimensional diameters of all tumors for at least 1 month.

TABLE 2.1.5. Clinical Trials With Unconjugated Murine mAb: Hematological Malignancies

Tumor type[1]	mAb	Antigen	N^2	HAMA, %[3]	Clincal responses[4]	Comments	References
ALL	J5	CALLA	4	NR	0	Rapid antigen modulation.	[306]
AML	M195	CD33	10	67	0	Good mAb targeting to tumor cells in marrow. Rapid antigen modulation.	[307]
T cell lymphoma; B-CLL;CTCL	T101	CD5	38	24	10	1 CR × 6 weeks in lymphoma patient. 1 PR in CTCL. Transient reduction of circulating CLL cells in 8 patients.	[118, 308–312]
B-cell lymphoma/ leukemia	OKB7	CD21	18	28	0	Antibody delivery related to tumor burden and antigen density.	[40]

Leukemia/ lymphoma	Campath-1M/ Campath-1G	CDw52	29	7	9	Clearance of circulating tumor cells and reduction in splenomegaly with Campath-1G. Response duration ranged from 4 to 32 weeks.	[243, 313]
T-cell leukemia	Anti-Tac	IL-2 receptor	9	11	2	1 CR less than 8 months.	[115]
B-cell lymphoma	Anti-idiotypic mAb	Private idiotopes	14	<10	8	2 CRs lasting more than 5 years.	[103, 105, 117, 118]
B-cell lymphoma	Anti-idiotypic mAb + chlorambucil	Private idiotopes	13	NR	9	1 CR × 31 months.	[105]
B-cell lymphoma	Anti-idiotypic mAb	Shared idiotopes	12	NR	4	1 CR.	[105]
Plasma cell leukemia	Anti-IL-6	IL-6	1	100	0	Transient decrease in serum G protein.	[116]

[1]ALL, acute lymphocytic leukemia; AML, acute myelocytic leukemia, B-CLL, B-cell lymphocytic leukemia; CTCL-cutaneous T-cell lymphoma.
[2]*N*, number of patients.
[3]Percentage of patients developing human anti-mouse antibodies.
[4]CR, complete response, usually defined as complete disappearance of all tumor for at least 1 month. PR, partial response, defined as a 50% decrease in the sum of the bidimensional diameters of all tumors for at least 1 month.

TABLE 2.1.6. Clinical Trials Using mAb in Combination With Cytokines

Tumor type	mAb	Cytokine[1]	N^2	Results[3]	References
Melanoma	R24	IL-2	20	Increased peripheral NK and LAK activity over pretreatment. No *in vitro* cytotoxicity against autologous melanoma in 6 points tested. 1 PR × 6 months.	[123]
Melanoma	R24	IL-2	32	10 PRs. 2 IL-2 related deaths.	[124]
Melanoma	MG22	IL-2	7	Increased NK and LAK activity. No responses.	[125]
Melanoma	XMMME-001	IL-2	9	Eosinophilia in 9/9. Significant skin toxicity and abdominal pain. No clinical responses.	[126]
Breast, colon, lung	L6	IL-2	15	1 PR lasting 12 weeks.	[314]
Melanoma	R24	α-IFN	15	Decrease in $CD8^+$T cells.	[127]
B-cell lymphoma	Private anti-idiotype	α-IFN	11	3 CR lasting more than 5 years.	[105, 117]
Melanoma	R24	M-CSF	19	Mild toxicity. No clinical responses.	[128] and unpublished
Melanoma	R24	GM-CSF	12	Grade 3 hypertension (2 points, and hypotension (1). No clinical responses.	[129]

[1] IL-2, interleukin 2; α-IFN, α-interferon; M-CSF, macrophage colony-stimulating factor; GM-CSF, granulocyte–macrophage–colony-stimulating factor.
[2] *N*, number of patients.
[3] NK, natural killer; LAK, lymphokine-activated killer.

ous limitations. Recently, an attempt has been made to generate a panel of anti-idiotypic mAb that recognize shared idiotypic determinants (anti-SIDs). In this way, it is hoped that typing a patient's lymphoma will identify reactivity with one or more anti-SID mAb already "on the shelf." Using a panel of 51 anti-SID mAb, it has been possible to identify an appropriate anti-idiotypic mAb for almost 50% of B-cell tumors. In an interim report of 12 patients treated with anti-SIDs, 1 CR and 3 PRs (partial response: >50% disappearance of signs and symptoms of disease for at least 30 days) were observed [105].

A major limitation of the use of anti-idiotypic mAb to target B-cell malignancies has been the instability of the B-cell idiotype. Surface immunoglobulin is typically shed from the B cell, resulting in circulating idiotype, which absorbs out the anti-idiotypic. This has required administration of large doses of anti-idiotypic mAb (1–15 g), which is difficult and costly to prepare. In addition, the malignant B-cell clone typically contains subclones of mutated idiotype, or modulates its idiotype. This results in outgrowth of cells that are unreactive with the anti-idiotypic. This is a common mechanism of tumor "escape" and subsequent relapse. It is hoped that with the use of anti-SID mAb, the problem of tumor escape will be minimized, but this remains to be seen. In the future, it will be of interest to compare the clinical responses obtained with anti-SID therapy with the responses obtained with anti-private-idiotype therapy.

Anti-Idiotype Vaccines. Another use of anti-idiotypic mAb is as tumor vaccines. MAb against antigens that are relatively tumor-restricted have been produced, and anti-idiotypic mAb have been raised that presumably mimic these antigens. Early clinical trails have recently been reported in which patients have been injected with anti-idiotypic mAb in an attempt to induce antitumor immunity. Most of the work has been done with melanoma, since melanoma antigens are among the best-characterized tumor antigens. In the first report [119] using mouse anti-idiotypic mAb to immunize patients, 37 patients with metastatic melanoma were given subcutaneous injections of mAb MF11-30, which mimics an epitope on the high-molecular-weight melanoma-associated antigen (HMW-MAA). Antibodies reactive with HMW-MAA were induced in only one patient; however, low-titer anti-HMW-MAA antibodies were detectable in this patient before immunization. Although one objective response was observed (a CR in lymph nodes lasting 43 weeks), this was not in the patient in whom anti-HMW-MAA antibodies were detected. A second anti-idiotypic mAb vaccine that mimics HMW-MAA was constructed with mAb MK2-23 conjugated to kcyhole limpet hemocyanin and mixed with bacillus Calmette-Geurin (BCG) as an immune adjuvant [120]. Twenty-five patients were immunized and 60% developed low-affinity antibodies against HMW-MAA. Three patients experienced partial shrinkage of tumor-infiltrated lymph nodes.

Six patients with Dukes B or C colon cancer were immunized with low-dose human anti-idiotypic mAb mimicking a 72-kD glycoprotein [121]. No antibodies were induced against the antigen but in four of five patients tested, postimmunization peripheral blood mononuclear cells showed an increased proliferation *in vitro* to antigen-positive tumor cells compared with mononuclear cells before treatment. No proliferative response was observed in response to antigen-negative cells.

Clinical trials are also under way in melanoma patients using BEC2, an anti-idiotypic mAb that mimics the ganglioside G_{D3} [122]. In this case a protein (BEC2) is being used to immunize against a nonprotein antigen. If successful, this will greatly expand the range of tumor antigens available for vaccine studies.

Clinical Trials Using Murine mAb in Combination With Cytokines for the Treatment of Cancer. In an attempt to augment the inflammatory response induced by mAb binding to tumor cells, several trials are being conducted with cytokines in combination with mAb (Table 2.1.6). Most of these trials have been carried out in melanoma with mAb R24 against G_{D3} ganglioside. The cytokines used have included interleukin 2 (IL-2) [123–126], α-interferon (α-IFN) [105,117,127], macrophage colony-stimulating factor (M-CSF) [128], granulocyte–macrophage–colony-stimulating factor (GM-CSF) [129], and a trial with R24 + α-TNF (tumor necrosis factor) is under way.

In most cases, toxicities due to the mAb and the cytokine have been nonoverlapping, although when used in combination, the major toxicities are usually attributable to the cytokine. Immunological effects, largely attributable to the cytokine, have been observed in most of the trials. However, there have been few antitumor responses. One exception occurred in a trial in which melanoma patients were treated with high-dose IL-2 and R24 (124). Of the 32 patients treated so far, 10 PRs were observed and responses were seen at all dose levels of R24 (5–100 mg/m^2). In another trial with mAb plus cytokine, 11 B-cell lymphoma patients were treated with anti-idiotypic mAb against private idiotopes plus α-IFN [105,117]. There were 3 long-lasting CRs although the role of α-IFN in this regimen remains undefined.

Trials with mAb plus cytokine are difficult to design for several reasons. The dose–response relationship for either the mAb or the cytokine is usually not established, and combining these two agents further complicates the picture. The timing in administration of the two agents, whether together or one after another, may be critical and adds another variable to these trials. Finally, the mechanism of the antitumor effects is not actually known, so it is unclear which biological effects might reasonably serve as surrogate endpoints. Because of this, many investigators have chosen to look primarily at clinical responses. While this is the ultimate endpoint, in the absence of meaningful clinical responses, other biological endpoints that might reflect important antitumor mechanisms are needed to guide the development of these therapies.

Clinical Trials Using Human or Humanized mAb for the Treatment of Cancer. One of the principal drawbacks to treatment with mAb has been the antimouse response. HAMA has major effects on the pharmacokinetics and tumor targeting of mAb and limits the ability to re-treat patients. Because of this, recent clinical trails with mAb have explored the use of humanized mAb. The potential advantages of a nonimmunogenic mAb include a prolonged serum half-life and better tumor targeting, even after multiple treatments. It is also conceivable that higher doses could be administered safely and that this might result in better therapeutic responses. Another advantage of humanized mAb is that it is possible to convert a mouse mAb with the desired specificity but of an isotype not useful for therapy to a human IgG isotype capable of mediating both complement fixation and ADCC.

Both chimeric mAb (in which the mouse variable domains are transferred to human immunoglobulin constant domains to form an IgG molecule) and CDR-grafted mAb (in which only the mouse complementerity-determining regions are transferred) have been created and clinical trials are under way. To date, a small number of patients have been treated with chimeric mAb (Table 2.1.7). As predicted, the serum half-lives were far longer than seen with mouse mAb (approximately two to three times longer) but significantly shorter than the 21-day half-life of human IgG. Although the incidence of anti-mAb responses varied between the trials, the immunogenicity of the chimeric mAb remains a problem. This is not surprising, since mouse framework regions are retained within the variable domains. In the few patients who received multiple courses of chimeric mAb, serum half-lives have generally been shorter upon re-treatment, which is consistent with the low level immunogenicity.

In order to further decrease the anti-mAb response, CDR-grafted mAb are currently being investigated in clinical trials. The initial experience has been in 2 patients with non-Hodgkin's lymphoma treated with Campath-1H [130]. Neither patient developed detectable anti-Campath-1H antibodies. It will be of extreme interest to expand this experience, to determine whether patients with solid tumors develop antibodies against CDR-grafted mAb, and to study the pharmacokinetics and tumor targeting of these mAb after multiple courses of therapy.

2.1.3.3. Future Directions

In addition to grafting CDRs, other regions of the immunoglobulin molecule can be manipulated. Studies have demonstrated that specific sites within the Fc portions can be altered to manipulate the ability of a mouse mAb to fix complement [131] and to augment binding to high-affinity Fc receptors [132]. This suggests that other sites within the Ig molecule could be engineered to improve mAb efficacy.

As previously noted, antibody fragments may have superior tumor-targeting characteristics due to their smaller size and rapid clearance from the bloodstream. Variable domains alone can be shown to bind antigen, often with reasonably high affinity. These variable regions, which can be cloned and expressed in bacteria, have been termed "single-domain antibodies" (dAb) and might offer advantages over whole immunoglobulin molecules [133].

Two distinct mAb can be combined to form a heterobifunctional mAb possessing one antigen-binding site from each mAb. This has been done either by chemically combining the two mAb molecules or by fusing the two hybridoma cell lines to generate a heterohybridoma. Heterobifunctional mAb have been made in which one arm binds a tumor antigen and the other arm binds a lymphocyte surface antigen. There is evidence that these heterobifunctional mAb can be used to "focus" a cellular immune response against tumor cells, and it has been possible to demonstrate enhanced cell-mediated killing of tumor cells *in vitro* [134–137]. Finally, dimeric and multimeric forms of IgG can be constructed that show enhanced properties of ADCC, complement fixation, and isotope internalization [138,139].

TABLE 2.1.7. Clinical Trials Using Chimeric, Human, or Humanized mAb in Cancer Patients

Tumor type	mAb	Form of mAb[1]	N	Immunogenicity	$T_{1/2}$[2]	Results	References
Non-Hodgkin's lymphoma	Campath-1H	CDR-grafted	2	None	Not formally measured	Clearance of lymphoma cells from blood and bone marrow. Reduction of lymph nodes in 1 patient.	[130]
Colorectal cancer	17-1A	Chimeric	10	1/10 developed HAMA against variable domains.	100.5 h	No antitumor responses.	[315]
Colorectal cancer	17-1A	Chimeric, radioiodinated	6	None	100.9 h	Imaged large lesions (>4 cm).	[316]
Colorectal cancer	B72.3	Chimeric, radioiodinated	12	HAMA in 7/12, 80%–90% of HAMA directed against variable domains.	224 h	HAMA caused rapid clearance when patients re-treated.	[317]
Melanoma	14.18	Chimeric	13	8/13 developed HAMA against variable domains.	181 h	No antitumor effects. Grade 3 abd/pelvic pain in 5 patients.	[318]
Melanoma	L72	Human IgM	8	5/8 developed detectable antibodies against L72.	Not measured	Injected intralesionally. 1 CR, 3 PRs.	[319]
Acute myeloid leukemia	HuG1-M195	CDR-grafted	10	None to date	Variable	Rapid, specific targeting.	[320]

[1]CDR, complementarity-determining region.
[2]$T_{1/2}$ serum half-life.

A series of newly described molecules critical to tumor cell growth may provide novel targets for mAb therapy. MAb against the *neu* oncogene product has been shown to inhibit tumor growth in animals [140,141]. A particularly impressive phenomenon is the programmed cell death, or apoptosis, induced by a mAb against the APO-1 antigen expressed on malignant lymphoid cells [142]. A single injection of the anti-APO-1 mAb has induced regression of established xenografted human B-cell tumors in nude mice. Finally, targeting the p-glycoprotein membrane antigen that mediates multidrug resistance has been reported to potentiate the action of calcium channel antagonists in reversing multidrug resistance [143,144].

New approaches in mAb production promise to expand the availability of useful mAb. Recently it has become possible to isolate and produce mAb by using DNA expression libraries in bacteria [145] instead of the standard technique of creating a splenocyte–myeloma hybridoma cell (discussed further in Chap. 3). One of the major advantages offered by this approach is the ability to screen huge numbers of clones very quickly.

2.1.4. RADIOIMMUNOTHERAPY FOR HEMATOPOIETIC CANCERS

Therapy with radiolabeled mAb typically involves conjugates of ^{131}I, a long-lived β− emitter that also allows gamma camera imaging. However, several other isotopes of therapeutic potential are under active study (Table 2.1.8). As described above for radioimmunodetection and unconjugated mAb therapy, the issues involved in achieving adequate delivery and retention of a radiolabeled mAb at a tumor site are complex and poorly understood, and the obstacles to success are large (Table 2.1.9) [15,32,146–150]. Because of their radiosensitivity and accessibility to mAb, hematopoietic cancers are successful targets of radioimmunotherapy. Often trials combining imaging and therapy are tandemly linked. The best example of the approach of combined radioimmunodetection followed by radioimmunotherapy is at the University of Washington, Seattle, where a series of mAb to CD37 and CD20 or other B-cell non-Hodgkin's lymphoma targets are being evaluated in a compli-

TABLE 2.1.8. Characteristics of Therapeutic Emissions

Emission	Typical Nuclide	Range, mm	Energy, keV	Comment
γ	^{131}I, ^{123}I, ^{111}In	100's	100–2,000	Poorly cytotoxic due to range.
β−	^{131}I, ^{90}Y, ^{186}Rh	0.5–2	100–2,000	Good local toxicity within range.
β+	^{124}I	0.5–2	100–1,000	Similar to β−.
Auger	^{123}I, ^{125}I, ^{111}In	0.0001–0.001	5–20	Highly specific single-cell kill if near or in nucleus.
α	^{212}Bi, ^{213}Bi, ^{211}At	0.05–0.1	1,000–6,000	Extraordinary potency; localized range.

TABLE 2.1.9. Parameters Affecting Efficacy for Radioimmunotherapy

Parameter
Antibody
Antibody avidity
Antibody on-rate
Antibody off-rate
Antibody metabolism
Target
Antigen accessibility
Antigen density
Antigen reexpression rate after modulation
Antigen–antibody complex distribution intracellularly
Circulating antigen levels
Radiosensitivity of target cells; of nontarget tissues
Antigen density on nontarget tissues
Sanctuary sites (central nervous system; testis)
Cell proliferation rate and cycle distribution
Antigen-negative targets (size and number)
Pharmacology
Volume of distribution; plasma $T^{1/2}$, area under the curve
Clearance rates
Clearance routes
Immune complexing and clearance
Tissue and host
Prior therapy
Tumor resistance
Tumor repair
Bone marrow reserve
Tissue oxygenation
Prior antibody treatment; HAMA levels
Tumor blood flow
Binding site barriers
Intralesional back pressures
Isotope characteristics
Dose rate
Emission type
Physical $T^{1/2}$
Metabolism and excretion
Daughter products
Damage to carrier Ig
Stability of linkage to Ig
Short-range β emissions

TABLE 2.1.10. Clinical Trials of Radioconjugated mAb in Lymphoma and Leukemia

Disease[1]	mAb	Antigen	Number of patients	Antitumor activity, %[2]	References
CTCL	^{131}I-T101	CD5	5	100	[152]
B-cell NHL	^{131}I-MB1	CD37	4	100	[38, 39]
	^{131}I-1F5	CD20	1		
B-cell NHL	^{131}I-lym-1	HLA-Dr	18	55	[41]
B-cell NHL	^{131}I-LL2	—	5	80	[37]
B-cell NHL	^{131}I-OKB7	CD21	18	72	[40]
B-cell NHL	^{90}Y-anti-idiotype	sIg	2	100	[321]
Myeloid leukemia	^{131}I-M195	CD33	24	96	[151]

[1]CTCL, cutaneous T-cell lymphoma; NHL, non-Hodgkin's lymphoma.
[2]Responses include complete, partial, mixed, minor, and brief responses.

cated protocol involving serial imaging and pharmacokinetic analysis after multiple escalating doses of radiolabeled mAb [38,39]. In each patient, small, tracer amounts of radiolabeled mAb are injected in order to determine the optimal dose of mAb to achieve maximal binding to tumor with minimal binding to nontarget organs. A large therapeutic dose of ^{131}I-labeled mAb is then administered. Patients who fail to show adequate mAb targeting in the radioimmunodetection phase of study are not treated further. Due to this rigorous selection process and the sensitivity of the lymphomas to the $\beta-$ emissions, complete response rates in this trial are above 80% in the first 20 patients who have received the full course of therapy.

Major responses have also been seen in the pilot trials of ^{131}I-LL2, ^{131}I-lym-1, ^{90}Y-anti-idiotype, and ^{131}I-OKB7, in patients with B-cell lymphomas (Table 2.1.10). As with the Seattle trials, low tumor burden appears to be an important factor in determining the success of adequate delivery of mAb and of therapeutic responses subsequently [40]. Considerable variation was seen in the maximal tolerated doses of these patients, which may be due to changes in volumes of distribution, plasma half-lives, and marrow uptake. Other diseases of interest are acute and chronic myelogenous leukemias; ongoing at Memorial Sloan-Kettering Cancer Center are a series of trials evaluating the use of ^{131}I-labeled M195 (anti-CD33) for leukemia cytoreduction, particularly in preparation for bone marrow transplantation [30,151]. Efficient and specific targeting of cells, followed by killing of more than 99% of the leukemia cells with minimal toxicity, has been achieved.

Less work has been done in T-cell lymphoma, as this is a much less common disease in the United States and Europe. One early and successful trial used ^{131}I-T101 (anti-CD5) in cutaneous T-cell lymphomas [152]. All patients responded to their treatments for up to 3 months with minimal toxicity. CD5 is also a potential target on chronic lymphocytic leukemias.

2.1.5. RADIOIMMUNOTHERAPY FOR SOLID TUMORS

Radioimmunotherapy of solid tumors is complicated by the relative radioresistance of many of these cancers as well as the more difficult pharmacology, as compared with hematopoietic cancers. Thus, not only is it more difficult to saturate these cells with targeted radiation, but once there, the dose is often not effective. In addition, the antigenic profile of solid tumors, particularly as it relates to differentiation of tissue or neoplastic stem cells, is less clearly defined, rendering it difficult to identify antigen targets that may spare normal organ regenerating cells while killing clonogenic cancer stem cells. Because of these inherent difficulties, radioimmunodetection often precedes radioimmunotherapy in order to determine if radioimmunotherapy will be able to deliver 5,000–6,000 rads to the cancer while exposing the marrow to less than 200–300 rads.

Colon carcinomas have received the most attention, with CEA, TAG-72, and A33 antigens predominating [2,16,31,49,153]. Clinical trials show rare remissions and, as expected, myelosuppression as the major adverse effect.

Patients with ovarian carcinoma [67–69] have been treated with intraperitoneally administered mAb with encouraging preliminary results. Pilot trials have also been conducted in glioma [65] and melanoma [57] and hepatoma [64]. These last studies in hepatoma used polyclonal antiferritin antibodies.

Perhaps the most progress has been made in neuroblastoma (20,66) with a ^{131}I-labeled mouse IgG3 mAb against GD2 ganglioside (3F8), which is present in high amounts in neuroblastoma and other neural crest-derived tumors such as melanoma. Large doses of ^{131}I-3F8 result in tumor regressions and severe myelosuppression, which requires the use of bone marrow autotransplantation. Several patients with refractory disease have had prolonged survivals after this approach.

Overall, the use of radioimmunotherapy in solid tumors has seen far fewer successes than similar approaches in hematopoietic cancers; this is largely due to intrinsic differences in these types of cancers in their radiobiology and vascular properties that prevent delivery of adequate doses of radiation. Significant advances in pharmacology and radiation chemistry will be necessary to overcome these obstacles in these types of tumor. For example, smaller agents with better penetration and faster clearance will be needed, as well as isotopes with varied physical half-lives and ranges of emission.

2.1.6. IMMUNOTOXINS

Differential toxicity is a requirement for successful drug therapy for cancer; tumor cells must be killed more readily than all normal tissues. Unfortunately, standard chemotherapy drugs often do not offer a satisfactory therapeutic index and, as a result, the cytotoxicity of normal cells limits the clinical usefulness of these drugs. In an attempt to develop agents with more specificity, immunotoxins have been produced in which a toxin is linked to a mAb specific for tumor antigens in order to deliver toxin specifically to the tumor cell.

To date, the toxins used in immunotoxins have consisted of either bacterial or plant products. These are extraordinarily potent toxins and a single molecule is sufficient to kill a cell. Diphtheria toxin and *Pseudomonas* exotoxin A (PE) are the bacterial toxins most commonly used; they inactivate elongation factor 2, thus inhibiting protein synthesis. Plant products, such as ricin, gelonin, and saponin, block protein synthesis by inactivating the 28S RNA of ribosomes while the toxin calicheamicin degrades double-stranded DNA.

A common feature of most of the bacterial and plant toxins is that one portion of the molecule functions to bind the toxin to the target cell and translocate it into the cytosol, while a separate portion of the molecule, or a separate, covalently linked polypeptide chain contains the enzymatic activity responsible for inhibition of protein synthesis. In the case of ricin, the B chain binds cell surface galactose residues and facilitates transport of the A chain into the cytosol, where the A chain inactivates the ribosomes. Since it is desired that the immunotoxin have specificity only for tumor, it is necessary to neutralize or eliminate the B chain. Most clinical trials have utilized ricin immunotoxins in which the galactose-binding sites are blocked ("blocked" ricin) or conjugated to pure A chain (lacking the B chain). In this way, the specificity of the immunotoxin is defined solely by the mAb to which the toxin in conjugated. Despite this, sugar residues on the ricin A chain allow nonspecific hepatic uptake of the immunotoxin, causing toxicity and limiting the serum half-life. To circumvent this, deglycosylated [154] or recombinant A chain, which is unglycosylated [155], have been used in some clinical trials in the hopes of limiting hepatic uptake.

Other plant toxins, such as the vinca alkaloids [156,157] and methotrexate [158], have been used to construct immunotoxins. These drugs act by disrupting cell division and inhibiting DNA synthesis, respectively. Cell kill with these drugs probably requires delivery of more than one molecule per cell.

2.1.6.1. Clinical Trials With Immunotoxins in Cancer Patients

Several recent reviews have been written on the clinical use of immunotoxins [159–162]. Immunotoxins have been administered to patients with leukemia, lymphoma, melanoma, colorectal cancer, lung cancer, ovarian cancer, and breast cancer (Table 2.1.11). Most of the trials used a ricin-based immunotoxin, although other toxins have also been used [156,158,163,164]. Very few antitumor responses have been reported.

Regional administration of immunotoxins has also been studied. OVB3 conjugated to PE was injected via the intraperitoneal route into patients with ovarian carcinoma [163]. Dose-limiting toxicity was abdominal pain requiring narcotics and encephalopathy that resulted in one death. Encephalopathy was thought to be due to cross-reactivity of the mAb with brain tissue. No antitumor responses were observed.

2.1.6.2. Problems With Immunotoxin Therapy in Cancer Patients

Toxicity. The side effects of immunotoxin therapy have generally been mild. Capillary leak syndrome, characterized by reversible hypoalbuminemia, proteinuria,

TABLE 2.1.11. Clinical Trials Using Mouse mAb Immunotoxins[1]

Disease	Immunotoxin	Antigen	Results	References
Melanoma	Xomazyme–ricin A chain	Chondroitin sulfate, HMW-MAA	1/22 CR seen in Phase I trial. Immunotoxin targeting to tumor demonstrated. 3/43 PR in Phase II trial.	[169, 322]
B-cell lymphoma	Anti-CD22 $F(ab')_2$-deglycosylated ricin A chain	CD22	Transient PR in 38% of patients. 3/15 made HARA; 1/15 made HAMA. Dose-limiting toxicity: pulmonary edema (3), CNS (2), rhabdomyolysis (1).	[154]
NHL, CLL, ALL	Anti-B4-blocked ricin	CD19	1/25 CR, 2/25 PR. Dose-limiting toxicity: transaminitis. 9/15 developed HAMA + HARA.	[166]
Cutaneous T-cell lymphoma	H65-ricin A chain	CD5	4/14 PRs (3–8 months) 7/14 developed neutralizing antibodies. Toxicity related to vascular leak syndrome.	[168]

Refractory Hodgkin's disease	BER-H2-saponin	CD30	Rapid, transient, dramatic tumor shrinkage in 3/4 patients lasting 6–10 weeks. Mild toxicity.	[164]
Colorectal cancer	Xomazyme-791-ricin A chain	72-kD protein	No clinical responses. Vascular leak syndrome.	[323]
Breast cancer	260F9-recombinant ricin A chain	55-kD protein	No clincal responses. Vascular leak syndrome.	[155]
Ovarian cancer	OVB3-Pseudomonas exotoxin	Not stated	Intraperitoneal administration. No clinical responses. Dose-limiting toxicity: CNS (1 death); abdominal pain.	[163]
Small-cell lung cancer	N901-blocked ricin	CD56	0/7 responses. 1 myocardial infarction.	[165]
Non-small-cell lung cancer	KS 1/4-methotrexate	gp40	0/5 responses. Mild toxicity.	[158]
Lung/colorectal cancer	KS 1/4-DAVLB	gp40	Clinical responses not reported. Most patients developed antibodies against DAVLB.	[156, 157]

[1]Abbreviations: HARA, human antiricin antibodies; NHL, non-Hodgkin's lympoma; CLL, chronic lymphocytic leukemia; ALL, acute lymphocytic leukemia; CNS, central nervous system; DAVLB, desacetylvinblastine.

and edema, is commonly, although not invariably, seen with ricin immunotoxins. The pathophysiology of ricin-induced capillary leak syndrome has not been elucidated. Occasionally, serious toxicity has been observed and has been attributed to the expression of the target antigen on normal tissues [163,165], although this has not been formally demonstrated.

Immunotoxin Targeting to Tumor. Several obstacles can prevent delivery of the immunotoxin to the tumor cell cytosol. Many of these relate to tumor targeting by mAb and have already been discussed (Table 2.1.9). Immunotoxins are especially handicapped because of their very short serum half-life (see below). In addition, they are large molecules and, as a result, do not readily escape the intravascular space. The combination of these factors predicts that immunotoxins, as currently constructed, will have limited ability to target tumor. Despite this, several clinical trials have demonstrated immunotoxin bound to tumor cells *in vivo* [158, 164,166].

Once bound to the tumor cell, the immunotoxin must be rapidly internalized and trafficked to the appropriate cellular compartment without being degraded. Not all antigens are efficiently internalized upon being bound by mAb, so it is important to select appropriate antigenic targets. Even the epitope recognized by the mAb may be critical, since subtle changes in mAb specificity can drastically alter the efficiency of internalization and affect the ability of the immunotoxin to kill the cell [167].

Pharmacokinetics of Immunotoxins. Ricin immunotoxins are cleared extremely rapidly from the intravascular space with half-lives generally ranging from 1 h to 8 h [154,155,166,168,169], which is much faster than clearance of unconjugated murine mAb. Ricin immunotoxins lacking sugar moieties did not demonstrate significantly longer serum half-lives [154,155]. One cause of this short half-life may be the instability of the disulfide bond linking the toxin to the mAb. Once dissociated from the mAb, ricin is cleared more rapidly than when it is linked to the mAb. Novel linkers are being developed that are less susceptible to enzymatic degradation and the challenge will be to design these linkers so that biologically-active toxin is successfully delivered into the cytosol.

Immunogenicity of Immunotoxins. Most patients injected with immunotoxin develop anti-mouse antibodies (HAMA) and antitoxin antibodies that can significantly speed the clearance of the immunotoxin from the serum [168] and block binding of the immunotoxin to the target cell [170]. These antibodies usually develop after the first injection, but in the case of diphtheria toxin-based immunotoxins most patients have preexisting high-titer antidiphtheria antibody as a result of previous immunization. Ricin itself is highly immunogenic and human antiricin antibodies (HARA) are routinely induced in patients. Even toxins that are not intrinsically immunogenic can become immunogenic when covalently linked to mouse immunoglobulin [156,157]. It is thought that the mouse mAb functions as a carrier protein and facilitates presentation of the toxin as a hapten. These anti-immunotoxin antibodies severely limit the ability to administer multiple courses of immunotoxin.

2.1.6.3. Future Directions of Immunotoxin Therapy

In order to improve efficacy, immunotoxins with novel specificity elements, toxins, and linkers are being constructed. The immunotoxins used in clinical trials to date have used mouse mAb or mAb fragments [154] as the specificity element. Smaller immunotoxins have been constructed with single chain binding proteins as the specificity element mAb [171]. These immunotoxins are fusion proteins consisting of a toxin and the appropriate complementarity-determining regions of a mAb. While these molecules should more readily exit from the vascular space and may be less immunogenic, they tend to have even shorter serum half-lives and may have less affinity than immunotoxins with intact mAb. To further decrease immunogenicity of immunotoxins, humanized mAb or mAb fragments can be used, although antitoxin antibodies will remain a problem. Use of nonforeign toxins, such as tumor necrosis factor [172], is a novel approach to minimize the antitoxin immune response.

The specificity element of an immunotoxin need not even be an immunoglobulin or an immunoglobulin fragment; immunotoxins have been constructed in which the specificity element is a ligand that recognizes a receptor expressed on the target cell of interest [173–176]. Toxins linked to ligands such as interleukins 2, 4, and 6, insulin-like growth factor, and TGF-α have been constructed and clinical trials are under way in patients with malignancies that overexpress receptors for these ligands. Ligand-based immunotoxins offer several potential advantages. Ligands can be selected that bind to receptors with high affinity and are quickly endocytosed. Also, these ligands are generally smaller than immunoglobulins and would be expected to better exit from the intravascular space. Finally, there should be no host immune response against the ligand. The specificity of these immunotoxins may be a problem, however, since normal cells generally express receptors for these ligands.

The toxins that have been utilized in immunotoxins have been selected for their impressive potency but their limitations have become apparent: They can be difficult to link to mAb, they are immunogenic, and they can be nonspecifically taken up by normal tissue. Since the genes for several toxins have been cloned, it has been possible to molecularly engineer toxin molecules. Much of the work has been done with PE, and variants of PE have been produced that lack domains that are responsible for binding to normal cells [171,174]. This approach offers the possibility of engineering toxins that are more ideally suited for use in immunotoxin constructs.

Ricin is typically linked to mAb using cross-linkers based on disulfide bonds which can be reduced in the intravascular space, liberating the toxin. Novel linkers are being developed that do not release toxin extracellularly [160]—for example, acid-cleavable cross-linkers designed to release the toxin into the cytosol. Another approach has been to utilize the mAb itself as the linker molecule. A bispecific $F(ab')_2$ fragment was constructed in which one arm of the $F(ab')_2$ fragment binds the target tumor cell while the other arm is specific for the toxin saponin [177]. The utility of this approach is likely to be limited, however, by the fact that binding to the target antigen is monovalent, and therefore of lower affinity. Also, toxin binding to $F(ab')_2$ fragments will have an off-rate, which will generate free toxin.

The most novel approach is to construct immunotoxins by molecular engineering techniques. The gene encoding the toxin is ligated to the gene that encodes the specificity element, such as a ligand [173–176] or an Fv mAb fragment [171], and the immunotoxin is expressed as a single protein without the need for a linker molecule. As already discussed, these immunotoxins offer several potential advantages over mAb-based immunotoxins, and the results of initial clinical trials are eagerly awaited.

2.1.7. BONE MARROW PURGING WITH mAb

The use of mAb *ex vivo* to eliminate residual cancer cells from bone marrow harvested from patients who will undergo high-dose chemotherapy and radiotherapy before reinfusion of the marrow has distinct theoretical advantages as a therapeutic approach. It allows for doses of therapy to be given without concern for myelosuppression; it also allows for controlled, mAb-based ablation of the cancer cells *in vitro,* where exact concentrations and times of exposure are determined. However, the use of such an approach requires that the cancer in the patient be sensitive to the proposed high-dose therapy and that the bone marrow be largely free of disease. Therefore, hematopoietic cancers such as acute lymphocytic leukemias and non-Hodgkin's lymphomas have been the primary targets of such a procedure [178–184]. Typically, combinations of mAb are used so as to avoid problems of antigen heterogeneity on target cells and to increase the number of cells killed. MAb with rabbit complement is usually used, but several immunotoxins have also been tried, as well as magnetic immunobeads. More limited work has been done in acute myelogenous leukemia (AML). One difficulty with AML is the elimination of the clonogenic neoplastic cells without eliminating the capacity of the marrow stem cells, which are antigenically similar, to regrow [5,185,186]. Trials with mAb against CD15 in AML have yielded results comparable to results with chemotherapy [183,184,187,188].

Since both mAb-mediated killing and chemotherapy were effective against myeloid leukemia blasts, yet operated by very different mechanisms, another approach for generating additive killing against the leukemias without increased toxicity has been the combination of immunotherapy and chemotherapy [185,186]. For example, a single cycle of mAb M195 (anti-CD33) plus rabbit complement purging was added to the standard combination of VP16 and 4-hydroperoxycyclophosphamide (4-HC) purging in AML. More than 4 logs of leukemia cell lines or more than 3 logs of leukemia cells were killed. Although CFU-GM were also killed, stem cells were spared [186]. In a second study, VP16 purging was combined with cytosine arabinoside (Ara-C), which is one of the most effective agents in therapy for AML, and was then followed by one cycle of purging with MY9 (anti-CD33) plus rabbit complement [185]. These combinations of chemotherapy and mAb were generally able to kill all of the leukemia cells (3–4 logs of cells). Yet at the same time, regrowth of normal progenitor cell colonies, as measured by long-term bone marrow cultures or by CD34 stem cell colony assays on similarly treated marrows, was not eliminated. The use of anti-CD33 purging has been confirmed in a pilot trial of 12 patients, mostly in second remission, by using three cycles of mAb MY9 plus complement *ex vivo* (without chemotherapy) [189]. Survival in this group is comparable to that of similar patients receiving conventionally purged bone marrows.

Immunotoxins against CD33 antigen have also been used to purge residual AML from marrow [190]. One agent consists of MY9 linked to blocked ricin toxin. Studies *in vitro* showed that this immunotoxin can eliminate 2–4 logs of HL60 cells or two logs of fresh leukemia cells, but with substantially less depletion of normal progenitors. Thus the therapeutic index of the immunotoxin was greater than that of MY9 plus rabbit complement. Based on these types of data, a Phase 1 trial of the CD33 immunotoxin was initiated *in vivo*.

Neuroblastoma purging has achieved a measure of success using immunomagnetic bead purging [191,192]; purging of small cell lung cancer is also under investigation [193].

Another use of bone marrow purging is for the removal of normal T cells, rather than neoplastic cells, from donor marrow to prevent graft versus host disease in allogeneic bone marrow transplant [194–198]. This approach is effective and safe, but its ultimate role in allogeneic bone marrow transplant awaits a better understanding of the function of T cells in marrow engraftment and in graft versus leukemia effects, which may be important in prolonging survival.

2.1.8. MONOCLONAL ANTIBODY THERAPY FOR INFECTIONS

Although immunity to infections is multifactorial and depends on cellular functions both to prevent and to eradicate organisms, humoral immunity plays a role in neutralizing viruses, lysing bacteria, and clearing bacterial toxins. MAb-based therapeutics are of interest for infectious diseases for two reasons: First, few effective agents are available for treating viral infections, so mAb may be useful as neutralizing agents or to direct cellular effectors towards infected cells. Second, in the area of bacterial infection, numerous antibiotics are capable of killing bacteria but do not address either the bacterial toxins or soluble mediators of the host immune response that are responsible for many of the complications of infection such as septic shock. For this reason, septic shock still accounts for several hundred thousand deaths annually despite modern antimicrobials.

2.1.8.1. Antivirals

The herpes group of viruses have received the most attention, since they cause a variety of chronic debilitating diseases in humans and can be lethal in immunocompromised hosts. MAb that neutralize Epstein-Barr Virus (EBV) *in vitro* [199] or that neutralize herpes simplex virus (HSV-1) in mouse models [200,201] have been made. Two neutralizing humanized, CDR-grafted mAb to HSV have also been constructed [202]. Progress against cytomegalovirus (CMV) has been most rapid with a number of human mAb developed; two trials in humans have already been conducted [203–205]. Two different Phase 1 trials using human mAb to CMV in allogeneic bone marrow transplant patients at high risk for infection have been published [204,205]. Safety was demonstrated in a total of 33 patients, but an assessment of efficacy awaits further trials. These human mAb demonstrated plasma half-lives (1–2 weeks) that was longer than are usually observed with murine mAb (1–3 days), but far shorter than the expected 21-day plasma half-life reported for natural human IgG [206].

Many groups are attempting to develop mAb to a variety of other pathogenic viruses including human Fab fragments engineered against human immunodeficiency virus (HIV) [207], mouse mAb to hepatitis virus [208], and coxsachievirus [209].

2.1.8.2. Antimicrobial Approaches

There have been attempts to treat bacterial infections in model systems using mAb to *Pseudomonas* [210], to *Vibrio cholerae* [211], to group B streptococci [212], and to *Bordetella pertussis* (using an anti-toxin mAb) [213], as well as attempts to modify fungal infections [214] and protozoal (malarial) infections [215,216], but the largest efforts have been directed not toward the infecting organisms themselves, where cheaper, simpler, and effective antimicrobials are available, but to the host's mediators of the fatal consequences of infections, such as endotoxin, TNF, IL-1, and IL-6. This strategy is advantageous because it would, in principle, be broadly applicable to a variety of infections by attacking a central mediator of the pathogenesis of infection, and would address the major cause of the morbidity and mortality in infections. Inflammatory cytokines induced by endotoxin, such as TNF, can still cause morbidity in the absence of live bacteria [217]. MAb against TNF itself are being tested in models of septic shock [218–220], but most work has focused on antiendotoxin mAb, particularly those that are broadly reactive with the core polysaccharides [221]. These antibodies are postulated to work via the mechanism of neutralization or opsonization (binding of a target, such as bacteria, cells, and viruses, to an immune cell, usually followed by engulfment). IgM mAb appear to be most effective, probably acting via a complement-mediated or multivalent binding process. However, the true reason for the activity of these agents remains elusive [221,222]. Two IgM mAb against *E. coli* endotoxin core polysaccharide have reached clinical trial; one mAb, Xomen E5 [223,224], is murine-derived, while the other, HA-1A [225,226], is largely human with residual mouse sequences (the J chain). Both HA-1A and E5 have been tested in controlled clinical trials in large numbers of patients and each has shown statistically significant activity against subsets of patients with gram-negative sepsis or shock [224,225]. However, because it is difficult to predict which patients have gram-negative sepsis at presentation to the hospital and because the agents are expensive, the Infectious Disease Society of America has recommended guidelines to restrict their use to those patients in whom true gram-negative sepsis is probably [227]. To date, neither agent has been approved by the FDA in the United States, but HA-1A is approved in certain European countries.

2.1.9. THERAPY FOR DIGOXIN INTOXICATION

One of the most successful uses of specific antibodies *in vivo* in humans was in the field of cardiology, where digoxin intoxication was reversed by administration of neutralizing Fab [228]. These were not monoclonal antibodies, but they clearly showed the potential for passive use of specific naturalizing antibodies in both adults and children. The success of this approach in part depends on the use of the Fab fragments, which are quickly cleared by the kidneys, do not persist in the plasma, and do not form immune complexes. In addition, this strategy requires a target drug that is toxic in low quantities but is frequently overdosed either intentionally or by accident. Digoxin, because of its low therapeutic index and its small doses but long plasma half-life, is a good prototype target for this therapeutic approach.

2.1.10. USE OF mAb TO INDUCE IMMUNE SUPPRESSION

In recent years, it has become apparent that autoreactive T cells can induce disease in humans. Autoimmune diseases, including rheumatoid arthritis, multiple sclerosis (MS), myasthenia gravis, and systemic lupus erythematosus (SLE) are thought to be caused by tissue destruction mediated by autoreactive T-cell clones. In the field of transplantation, graft rejection is clearly mediated by host T cells recognizing foreign determinants on the allograft while graft-versus-host disease (GVHD) is mediated by mature T cells within the donor bone marrow. Treatment for autoimmune diseases, graft rejection, and GVHD consists primarily of drugs that nonspecifically suppress immune function. These drugs, such as corticosteroids, methotrexate, azothioprine, and cyclosporin, are each associated with significant toxicity. With the understanding that these diseases are mediated by T cells, several investigators have attempted to use mAb against T cells to suppress the T-cell response in the hopes of inducing a more specific and effective immune suppression without the significant side effects associated with immunosuppressant drugs.

2.1.10.1. Use of mAb to Treat Autoimmune Diseases

Anti-T-cell mAb have been used in several autoimmune diseases; the largest experience, however, has been in rheumatoid arthritis (RA). As shown in Table 2.1.12, mAb against CD4, CD5, CD7, and CD25 (the IL-2 receptor) have been used. The murine mAb were generally well tolerated, although they were, not surprisingly, highly immunogenic; in all but one study [229] the majority of patients developed HAMA. In an attempt to prevent a HAMA response, eight patients with RA were treated with CDR-grafted humanized Campath-1H, which recognizes CDw52 on lymphocytes and on a subset of monocytes. Patients received a 4 mg/day on days 1–5 and 8 mg/day on days 6–10. Although no patient developed antibodies against Campath-1H initially, four patients were re-treated with 40 mg/day $\times$ 5 days and three of these patients developed either anti-idiotypic or anti-allotypic antibodies

TABLE 2.1.12. Clinical Trials Using mAb in Patients With Rheumatoid Arthritis

mAb	Target antigen	Doses	N	Clinical results[1]	HAMA, %[2]	References
Not stated	CD4	10 mg/d × 7	5	Clinical improvement in all patients for 11 weeks. 3 had clinical improvement for 3 months.	0	[229]
VIT4; MT151	CD4	10 mg/d × 7	8	Clinical improvement in all patients lasting 3 weeks to 5 months. Concomitant therapy: steroids, NSAIDs.	75	[232]
MAX.16H5	CD4	0.3 mg/kg/d × 7	10	Clinical improvement in some patients. Concomitant therapy: NSAID (3), NSAID + steroids (7), CTX (1).	60	[231, 234]
BL4	CD4	20–40 mg/d × 10	6	Clinical improvement within 14 days.	67	[324]
cMT412 (chimeric)	CD4	Single dose; 10, 50, 100, 150, 200 mg	15	Decrease in number of tender joints at Day 35 in 8/15 patients. Some with sustained improvement >90 days. Concomitant therapy: Mtx.	Not stated	[233]
Anti-CD5–ricin A chain	CD5	Not stated	79	Sustained clinical responses in 20% of patients.	99	[325]
RFT2	CD7	6 mg/d × 14	16	Transient improvements in 2 patients. No concomitant therapy.	100	[230, 326]
Chimeric RFT2	CD7	Not stated	16	No clinical improvements.	Not stated	[326]
Campath 6	IL2 receptor	25 mg/d × 10	3	Clinical benefit in all 3 patients lasting 1–3 months. Concomitant therapy: NSAID, steroids (2).	Not stated	[327]
Campath-1H (humanized)	CDw52	4 mg/d, days 1–5 8 mg/d, days 6–10	8	Clinical improvement in 7/8 patients lasting up to 8 months. Concomitant therapy: NSAID, steroids (5).	37.5	[235]

[1]NSAID, nonsteroidal antiinflammatory drug; CTX, cyclophosphamide; Mtx, methotrexate.
[2]Percentage of patients developing human anti-mouse antibodies.

against Campath-1H. In most of the studies, it was possible to demonstrate that mAb treatment resulted in a dramatic reduction in circulating T cells [229–235], which is consistent with the mAb targeting to peripheral T cells. In most studies, there were transient clinical improvements in at least a subset of patients treated, although in patients treated with Campath-1H, seven of eight patients experienced clinical improvement for up to 200 days after re-treatment. It is interesting that clinical improvement was not always associated with changes in C-reactive protein, rheumatoid factor, or erythrocyte sedimentation rate (ESR) [229– 235], serum markers that are generally felt to reflect disease activity.

Several factors must be considered in interpreting the significance of the clinical responses observed. Given the fluctuating course of rheumatoid arthritis, the follow-up in these studies is relatively short. The studies each utilized a different mAb, different doses, and different schedules so that the numbers of patients treated in any one way is generally small. In most of the studies, patients were treated concomitantly with other anti-inflammatory drugs, further complicating the interpretation. Despite this, the two consistent observations were that the murine anti-T-cell mAb induced transient clinical improvement in some rheumatoid arthritis patients and the mAb were highly immunogenic in this patient population despite concomitant treatment with immunosuppressive drugs.

Anti-T-cell mAb have also been used in other autoimmune diseases thought to be mediated by autoreactive T cells. MAb against CD2, CD3, CD4, and CD6 have been used to treat patients with MS [236–239]. Biological effects were demonstrated with each mAb but in only one trial could the clinical effects be assessed [236]. In this trial, 16 patients with multiple sclerosis were treated with OKT3 and assessed both clinically and by magnetic resonance imaging (MRI). While there were two patients with clinical improvement at 1 year, the toxicity of the mAb was substantial and the investigators concluded that it is unlikely that OKT3 will be useful in the treatment of MS.

Aplastic anemia, a life-threatening disorder that results from the failure of the bone marrow to produce an adequate quantity of blood cells, is thought to have multiple etiologies. In a subset of cases, autoreactive T-cell clones are thought to be the cause and, because of this, investigators have attempted to use anti-T-cell mAb to treat patients with refractory aplastic anemia [240–242]. Five patients treated with T101 (anti-CD5) developed lymphopenia, but no beneficial hematological responses were observed [242]. Three who had failed treatment with horse anti-lymphocyte globulin with or without high-dose corticosteroids were treated with OKT3 (anti-CD3) [241]. One patient entered a complete response, and after 1 year the patient had normal blood counts and was off all medication; however, this patient had been treated concomitantly with steroids and azothioprine. In a third trial, 25 patients were randomized to receive either horse anti-human thymocyte globulin (ATG) or a mixture of two or three anti-T-cell mAb [240]. Patients were further randomized to either receive or not receive androgens. Not surprisingly, given the small number of patients there were no statistically significant differences between the four treatment groups.

In another series [243], four patients with aplastic anemia were treated with

Campath-1G. All four developed severe lymphopenia and one experienced a slow hematological recovery.

Patients with other autoimmune diseases have been treated with mAb. For example, a patient with systemic vasculitis was treated with four courses of a humanized anti-CDw52 mAb, Campath-1H [244]. Following the fourth course, the patient received a rat anti-CD4 mAb, and peripheral $CD4^+$ T cells were absent for 1 month after treatment. The patient's fever and other symptoms resolved and the patient had a nearly complete response during the subsequent 12 month follow-up period.

The chronic and variable course of the autoimmune diseases, coupled with the fact that treatment usually involved multiple drugs, makes it difficult to assess the impact of mAb therapy. Larger, randomized trials with long-term follow-up will be required in order to determine whether mAb have an important role in the treatment of these disorders, although such trials will be difficult to carry out. It is likely that multiple courses of mAb therapy will be required to treat these diseases, which makes the immunogenicity of murine mAb problematic. Use of humanized anti-T-cell mAb will be the obvious next approach to prevent an anti-immunoglobulin response.

2.1.10.2. Use of mAb to Treat Allograft Rejection

Several mAb against molecules expressed on T cells have been used in the setting of acute allograft rejection or have been used to prevent allograft rejection (Table 2.1.13). Most experience has been with OKT3, an IgG2a against CD3, which is currently approved for clinical use. OKT3 has a role in both the treatment and prevention of solid organ allograft rejection.

TABLE 2.1.13. mAb Used for the Prevention or Treatment of Allograft Rejection or GVHD[1]

mAb	Antigen	Application
OKT3	CD3	Renal, pancreatic, hepatic, cardiac allografts, GVHD
WT32	CD3	Renal allografts
64.1	CD3	GVHD
H65-RTA	CD5	GVHD
T12	CD6	Renal allografts
33B3.1	CD25 (IL-2R)	Renal allografts
Anti-Tac	CD25 (IL-2R)	Renal allografts
2A3	CD25	GVHD
25-3	CD11a (LFA-1)	Renal, bone marrow allografts
CD18 M232	CD18 (LFA-1)	Bone marrow allografts
T10B9.1A-31	TcR	Renal, cardiac allografts
BMA 031	TcR	Renal allografts

[1]GVHD, graft-versus-host disease; TCR, T-cell receptor.

OKT3. OKT3 has been found to be effective in reversing acute graft rejection episodes in patients receiving renal [245–248] or hepatic allografts [249,250]. In renal allograft patients, OKT3 is effective in reversing acute rejection that is unresponsive to conventional steroid therapy [251–253] or associated with acute renal failure occurring within 12 h of transplant [254], although it does not appear to be superior to antilymphocyte globulin (ALG). For the treatment of hepatic allograft rejection, there is evidence that OKT3 may be superior to steroids [255,256]. In a study comparing OKT3 to ALG in the treatment of cardiac allograft rejection, both therapies appeared to have equal efficacy, although OKT3 was associated with more life-threatening infections [257].

OKT3 has also been used as part of immunosuppression regimens to prevent graft rejection. In renal or renal–pancreatic allografts, there was no advantage of OKT3 over ALG or antithymocyte globulin as part of the immunosuppression regimen [258–262]. Two randomized studies reached opposite conclusions regarding the value of OKT3 versus cyclosporin A as part of the immunosuppression regimen; one study found no difference in graft survival [263], while another found improved graft survival in patients treated with OKT3 [264]. In prevention of hepatic allograft rejection, OKT3-containing regiments were not superior to regimens without OKT3 [265,266]. Cardiac allograft patients receiving standard immunosuppression with either OKT3 or ATG showed no differences in graft survival in a nonrandomized trial [267], while a randomized trial showed that the group receiving ATG suffered fewer rejection episodes [268].

Several factors contribute to the difficulty in sorting out these studies. Few studies involve randomized patient groups receiving comparable regimens. Patients always received other immunosuppressive drugs in addition to OKT3, and these drug combinations differed among studies. Finally, there is evidence that the dose [269] and serum concentration [270] of OKT3 is critical for graft protection, which adds another variable to these studies.

There are certain drawbacks associated with OKT3 treatment; the first one is toxicity. Anti-CD3 mAb can induce both inhibition and proliferation of T cells, and the proliferative response is associated with the release of various cytokines. Indeed, OKT3 treatment is associated with high serum levels of TNF and γ-IFN [271]. When OKT3 is injected into patients, the initial dose induces a syndrome of fever, chills, headaches, vomiting, and diarrhea, which is presumably caused by the released cytokines. A second drawback to OKT3 therapy is the induction of HAMA, which can neutralize the therapeutic effect. Virtually all patients develop anti-OKT3 HAMA but the specificity of the humoral response is relatively restricted in that anti-OKT3 antibodies induced in patients generally recognize either isotypic or idiotypic determinants. Of interest, only a subset of IgG antibodies against OKT3 appear to have neutralizing capability *in vitro* and *in vivo* [7,272]. It has been possible to re-treat patients with OKT3 and observe full biological effects, including clearance of $CD3^+$ peripheral lymphocytes and reversal of acute rejection [273]; this is consistent with the notion that not all patients develop anti-OKT3 antibodies with neutralizing activity.

Other mAb Used to Treat Allograft Rejection. Another mAb against CD3, designated WT32, has been used to prevent renal allograft rejection [274]. Compared with nonrandomized control patients receiving similar immunosuppression but without WT32, patients receiving WT32 had fewer rejection episodes, although there was no difference in graft or patient survival. Of interest, side effects of WT32 were similar to those seen with OKT3.

Other mAb against T-cell surface molecules have been used in the setting of renal or bone marrow allograft rejection. MAb T12 (anti-CD6) was used to treat 19 patients undergoing acute renal allograft rejection [275]. Seven patients had complete resolution with T12 therapy; in four additional patients T12 may have contributed to resolution of graft rejection.

Since only activated T cells express the IL-2 receptor, it has been hypothesized that mAb against the IL-2 receptor might prevent or reverse graft rejection. Renal allograft recipients who received mAb 33B3-1 (anti-IL2 receptor) had fewer acute graft rejection episodes than historical controls [276]. A second study compared the ability of mAb 33B3-1 and rabbit ATG to prevent renal allograft rejection in a randomized trial involving 100 patients [277]. MAb 33B3-1 was as effective as ATG in preventing graft rejection, resulted in fewer infectious episodes, and was less toxic. Because of these encouraging results, 33B3-1 was tested for its ability to reverse acute renal allograft rejection [278]. Of 10 patients treated, only two clearly responded to 33B3-1 and the authors concluded that anti-IL2 receptor mAb is not effective in reversing ongoing acute rejection. Another anti-IL2 receptor mAb, anti-Tac, was used to prevent renal allograft rejection [279]. Patients received cyclosporin, azothioprine, and steroids and were randomized to either receive or not receive anti-Tac mAb. The anti-Tac group had fewer rejection episodes but there was no difference between the groups in graft or patient survival.

A mAb against the α/β T-cell receptor (TcR) was able to reverse acute rejection episodes [280], and in a randomized trial against OKT3 it was found to be equally effective and less toxic [281]. Another anti-TcR mAb was used in high doses to prevent renal allograft rejection in high-risk patients [282].

MAb against LFA1, an adhesion molecule expressed on T cells and involved in cell–cell interactions, has been used in the setting of graft rejection. Patients undergoing HLA-mismatched bone marrow transplantation received mAb 25-3 against the alpha chain of LFA1 (CD11a). Most of the patients were children being transplanted because of inherited immunodeficiencies. Seventy-four percent of the patients engrafted, which the authors claimed was better than historical controls [283]. However, when mAb 25-3 was used to prevent graft rejection in adults undergoing bone marrow transplants for treatment of leukemia, only four of nine patients engrafted [284]. The discordant results between these two studies might be due to the fact that these studies involved different patient populations receiving different conditioning regimens. When mAb 25-3 was used to treat acute renal allograft rejection, only one of seven rejection episodes were reversed [285]. Another anti-LFA1 mAb (CD18 M232) recognizes the beta chain of LFA1 (CD18) and was used to prevent bone marrow graft rejection. Of eight leukemia patients undergoing T-cell-depleted allogeneic bone marrow transplantation with CD18 M232, five experienced graft rejection [286]. The authors concluded that this mAb did not enhance bone marrow engraftment.

One of the ligands for LFA1 is ICAM-1, and it has been reasoned that an mAb against ICAM-1 might interfere with T-cell function. In cynomolgus monkeys undergoing heterotopic abdominal cardiac transplantation, mAb R6.5 against ICAM-1 (CD54) prolonged graft survival three times as long as that of untreated control animals [287]. Anti-ICAM-1 mAb have not yet entered human trials.

2.1.10.3. Use of mAb to Treat Graft-Versus-Host Disease

Anti-CD3 mAb have been used to treat GVHD [288,289]. There has only been partial clinical benefit, and in one study [288] four patients developed EBV-associated lymphoproliferative disorders.

MAb against the IL-2 receptor have also been used in clinical trials to treat [290] and to prevent GVHD [291]. In the setting of acute GVHD, anti-IL-2R mAb showed little activity, while the study using anti-IL-2R mAb as GVHD prophylaxis was not designed to assess efficacy.

GVHD has been treated with ricin-based immunotoxins directed against T-cell surface antigens. In patients treated with H65-RTA against CD5, 9 of 16 patients had complete or partial responses [292]. Most responses were in skin, although responses in liver and gut were also observed.

2.1.11 USE OF mAb TO INHIBIT PLATELET FUNCTION

Binding of fibrinogen and von Willebrand factor to platelet glycoprotein IIb/IIIa is a critical step in platelet adhesion and aggregation. MAb 7E3 against glycoprotein IIb/IIIa can inhibit platelet function *in vitro* and in patients with unstable angina [293]. Chimeric 7E3 was successful in partially restoring patency to a coronary artery that had occluded during angioplasty [294].

2.1.12. CONCLUSIONS

The possible applications of mAb in human disease are large and continue to expand. Successful therapeutic interventions in cancers, inflammation, infection, and their consequences have already been demonstrated. A number of radioimmunodiagnostic approaches are becoming competitive with conventional agents in cancer, thrombosis, and inflammation detection. As these are still innovative fields, new advances will continue to improve efficacy in many areas. New genetic engineering methods for constructing human and non-native IgG or variants, perhaps molecules that comprise the minimal binding site necessary for high affinity, new chemistry and radiochemistry for developing immunotoxins and radioconjugates, and new therapeutic strategies for administration of mAb alone or in combination with other drugs and biological agents will likely result in a large array of effective agents in the next decade.

2.1.13. REFERENCES

1. Vaickus, L., Foon, K.A. Overview of monoclonal antibodies in the diagnosis and therapy of cancer. Cancer Invest. 9(2), 195–209 (1991).
2. Mach, J.P., Pelegrin, A., Buchegger, F. Imaging and therapy with monoclonal antibodies in non-hematopoietic cancers. Curr. Opinion Immunol. 3, 685–693 (1991).
3. Mellstedt, H. Monoclonal antibodies in cancer therapy. Curr. Opinion Immunol. 2, 708–713 (1990).
4. Dillman, R.O. Monoclonal antibodies for treating cancer. Ann. Int. Med. 111, 592–603 (1989).
5. Scheinberg, D.A. Current Applications of Monoclonal Antibodies for the Therapy of Hematopoietic Cancers. Curr. Opinion Immunol. 3, 679–684 (1991).
6. Kuzel, T.M., Winter, J.N., Rosen, S.T., Zimmer, A.M. Monoclonal antibody therapy of lymphoproliferative disorders. Oncology 4, 77–93 (1990).
7. Suranyi, M.G., Leenaerts, P.L., Austin, S.M., Hall, B.M. Advances in the immunobiology and clinical practice of transplantation. In Gonick, H.C. (ed.): Current Nephrology, Vol. 14 (pp. 379–431). St. Louis: Mosby Year Book, 1991.
8. Baumgartner, J.D. Immunotherapy with antibodies to core lipopolysaccharide: A critical appraisal. In Moellering, R.C., Young, L.S.G. (eds.): Infectious Disease Clinics of North America (Gram-Negative Speticemia and Septic Shock) (5th ed., pp. 915–928). Philadelphia: W.B. Saunders, 1991.
9. Gorelick, K.J., Chmel, H. The role of monoclonal antibodies in the management of gram-negative sepsis: Experience with the E5 antibody. In Moellering, R.C., Young, L.S.G. (eds.): Infectious Disease Clinics of North America (Gram-Negative Septicemia and Septic Shock) (5th ed., pp. 899–914). Philadelphia: W.B. Saunders, 1991.
10. Waldmann, T.A. Monoclonal antibodies in diagnosis and therapy. Science 252, 1657–1662 (1991).
11. Houghton, A.N., Scheinberg, D.A. Monoclonal antibodies: Potential applications to the treatment of cancer. Semin. Oncol. 13, 165–179 (1986).
12. Goldenberg, D.M., Larson, S.M. Radioimmunodetection in cancer identification. J. Nucl. Med. 33, 803–814 (1992).
13. Larson, S.M. Clinical Radioimmunodetection, 1978–1988: Overview and Suggestions for Standardization of Clinical Trials. Cancer Res. 50(Suppl), 892s–898s (1990).
14. Larson, S.M. Overview of radioimmunodetection: Technical advances. Antibody Immunoconj. radiopharm. 4, 525–530 (1991).
15. Halpern, S.E., Dillman, R.O. Problems associated with radioimmunodetection and possibilities for future solutions. J. Biol. Response Mod. 6, 235–262 (1987).
16. Larson, S.M. Radioimmunology: Imaging and therapy. Cancer 67, 1253–1260 (1991).
17. Mach, J.-P., Buchegger, F., Pèlegrin, A., Bischof-Delaloye, A., Delaloye, B. Progress in radiolabeled monoclonal antibodies for cancer diagnosis and potential for therapy. Accomplishments Cancer Res. 221–252 (1989).
18. Courteny-Luck, N.S., Epenetos, A.A. Targeting of monoclonal antibodies ot tumours. Curr. Opinion Immunol. 2, 880–883 (1990).
19. FitzGerald, D. Pastan, I. Targeted toxin therapy for the treatment of cancer. J. Natl. Cancer Inst. 81, 1455–1463 (1989).

20. Larson, S.M., Cheung, N.-K.V., Leibel, S.A. Section 21.3. Radioisotope conjugates. In DeVita, V.T. Jr., Hellman, S., Rosenberg, S.A. (eds.): Biologic Therapy of Cancer (pp. 496–511). Philadelphia: J.B. Lippincott, 1991.
21. Hnatowich, D.J. Recent developments in the radiolabeling of antibodies with iodine, indium, and technetium. Semin. Nucl. Med. 20, 80–91 (1990).
22. Friend, P.J. Immunosuppression with monoclonal antibodies. Curr. Opinion Immunol. 2, 859–863 (1990).
23. Haber, E. *In vivo* diagnostic and therapeutic uses of monoclonal antibodies in cardiology. Annu. Rev. Med. 37, 249–261 (1986).
24. Koblik, P.D., DeNardo, G.L., Berger, H.J. Current status of immunoscintigraphy in the detection of thromboembolism. Semin. Nucl. Med. 191, 221–237 (1989).
25. Knight, L.C. Radiopharmaceuticals for thrombus detection. Semin. Nucl. Med. 20, 52–67 (1990).
26. Bajorin, D.F., Chapman, P.B. Wong, G.Y., et al. Treatment with high dose mouse monoclonal (anti-GD3) antibody R24 in patients with metastatic melanoma. Melanoma Res. 2, 355–362 (1992).
27. Ryan, K.P., Dillman, R.O., DeNardo, S.J., et al. Breast cancer imaging with In-111 human IgM monoclonal antibodies: Preliminary studies. Radiology 167, 71–75 (1988).
28. Steis, R.G., Carrasquillo, J.A., McCabe, R., et al. Toxicity, immunology, and tumor radioimmunodetecing ability of two human monoclonal antibodies in patients with metastatic colorectal carcinoma. J. Clin. Oncol. 8, 476–490 (1990).
29. Haisma, H.J., Kessel, M.A.P., Silva, C., et al. Human IgM monoclonal antibody 16.88: Pharmacokinetics and distribution in mouse and man. Br. J. Cancer 10, 40–43 (1990).
30. Scheinberg, D.A., Lovett, D., Divgi, C.R., et al. A phase I trial of monoclonal antibody M195 in acute myelogenous leukemia: Specific bone marrow targeting and internalization of radionuclide. J. Clin. Oncol. 9, 478–490 (1991).
31. Welt, S., Divgi, C.R., Real, F.X., et al. Quantitative analysis of antibody localization in human metastatic colon cancer: A phate I study of monoclonal antibody A33. J. Clin. Oncol. 8, 1894–1906 (1990).
32. Epenetos, A.A., Snook, D., Durbin, H., Johnson, P.M., Taylor-Padamitriou, J. Limitations of radiolabeled monoclonal antibodies for localization of human neoplasms. Cancer Res. 46, 3183–3191 (1986).
33. Larson, S.M. Positron emission tomography in oncology and allied diseases. Principles Practice Oncol. 3, 1–12 (1989).
34. DeNardo, G.L., Macey, D.J., DeNardo, S.J., Zhang, C.G., Custer, T.R. Quantitative SPECT of Uptake of Monoclonal Antibodies. Semin. Nucl. Med. 19, 22–32 (1989).
35. Parker, J.A. Quantitative SPECT: Basic theoretical considerations. Semin. Nucl. Med. 19, 3–12 (1989).
36. Kramer, E.L., Noz, M.E., Sanger, J.J.M., Maguire, G.Q. CT-SPECT fusion to correlate radiolabeled monoclonal antibody uptake with abdominal CT findings. Radiology 172, 861–865 (1989).
37. Goldenberg, D.M., Horowitz, J., Sharkey, R.M., et al. Targeting, dosimetry, and radioimmunotherapy of B-cell lymphomas with iodine-131-labeled LL2 monoclonal antibody. J. Clin. Oncol. 9, 548–564 (1991).

38. Press, O., Eary, J.F., Badger, C.C., et al. Treatment of refractory non-Hodgkin's lymphoma with radiolabeled MB-1 (anti-CD37) antibody. J. Clin. Oncol. 7, 1027–1038 (1989).

39. Eary, J.F., Press, O., Badger, C.C., et al. Imaging and treatment of B-cell lymphoma. J. Nucl. Med. 31, 1257–1268 (1990).

40. Scheinberg, D.A., Straus, D.J., Yeh, S.D., et al. A Phase I toxicity, pharmacology, and dosimetry trial of monoclonal antibody OKB7 in patients with non-Hodgkin's lymphoma: Effects of tumor burden and antigen expression. J. Clin. Oncol. 8, 792–803 (1990).

41. DeNardo, G.L., DeNardo, S.J., O'Grady, L.F., Levy, N.B., Adams, G.P., Mills, S.L. Fractionated radioimmunotherapy of B-cell malignancies with ^{131}I-Lym-1[1]. Cancer Res. 50(Suppl), 1014s–1016s (1990).

42. DeNardo, S.J., DeNardo, G.L., O'Grady, L.F., et al. Pilot studies of radioimmunotherapy of B cell lymphoma and leukemia using I-131 Lym-1 monoclonal antibody. Antibody Immunoconj. Radiopharm. 1, 17–33 (1988).

43. Carrasquillo, J.A., Mulshine, J.L., Bunn, P.A., Jr., et al. Indium-111 T 101 monoclonal antibody is superior to iodine-131 T 101 in imaging of cutaneous T-cell lymphoma. J. Nucl. Med. 28, 281–287 (1987).

44. Carde, P., Pfreundschuh, M., da Costa, L., et al. Radiolabeled monoclonal antibodies against Reed-Sternberg cells for *in vivo* imaging of Hodgkin's disease by immunoscintigraphy. Recent Results Cancer Res. 117, 101–111 (1989).

45. Bischof-Delaloye, A., Delaloye, B., Buchegger, F., et al. Clinical value of immunoscintigraphy in colorectal carcinoma patients: A prospective study. J. Nucl. Med. 30, 1646–1656 (1989).

46. Moldofsky, P.J., Sears, H.F., Mulhern, C.B., et al. Detection of metastatic tumor in normal-sized retroperitoneal lymph nodes by monoclonal antibody imaging. N. Engl. J. Med. 311, 106–107 (1984).

47. Vuillez, J.P., Peltier, P., Mayer, J.C., et al. Reproducibility of image interpretation in immunoscintigraphy performed with indium-111- and iodine-131-labeled OC125 F $(ab')_2$ antibody injected into the same patients. J. Nucl. Med. 32, 221–227 (1991).

48. Cohn, K.H., Welt, S., Banner, W.P., et al. Localization of radioiodinated monoclonal antibody in colorectal cancer (abstract). Arch. Surg. 122, 1425–1429 (1987).

49. Mach, J.-P., Buchegger, F., Forni, M., et al. Use of radiolabelled monoclonal anti-CEA antibodies for the detection of human carcinomas by external photoscanning and tomoscintigraphy. Immunol. Today, 2, 239–249 (1981).

50. Carrasquillo, J.A., Sugarbaker, P., Colcher, D., et al. Radioimmunoscintigraphy of colon cancer with iodine-131-labeled B72.3 monoclonal antibody. J. Nucl. Med. 29, 1022–1030 (1988).

51. Hammmond, N.D., Moldofsky, P.J., Beardsley, M.R., Mulhern, C.B., Jr. External imaging techniques for quantitation of distribution of I-131 $F(ab')_2$ fragments of monoclonal antibody in humans. Med. Phys. 11(6), 778 (1984).

52. Buchegger, F., Pèlegrin, A., Delaloye, B., Bischof-Delaloye, A., Mach, J.-P. Iodine-131-labeled mAb $F(ab')_2$ fragments are more efficient and less toxic than intact anti-CEA antibodies in radioimmunotherapy of large human colon carcinoma grafted in nude mice. J. Nucl. Med. 31, 1035–1044 (1990).

53. Khazaeli, M.B., Meredith, R.F., Wheeler, R.H., et al. Pharmacokinetics of single and repeated therapeutic ^{131}I-doses of mouse/human chimeric monoclonal antibody chB72.3 in colorectal carcinoma patients (abstract). Antibody Immunoconj. Radiopharm. 4, 42 (1991).
54. Khazaeli, M., Plott, E., Brezovich, I., Russell, C., Wheeler, R., LoBuglio, A. Phase I trial of ^{131}I-chimeric mouse/human B72.3 (anti-TAG-72) in patients with metastatic colon cancer (abstract). Antibody Immuconj. Radiopharm. 4, 42 (1991).
55. Covell, D.G., Barbet, J., Holton, O.D., Black, C.D.V., Parker, R.J., Weinstein, J.N. Pharmacokinetics of monoclonal immunoglobulin G_1, $F(ab')_2$, and Fab′ in mice. Cancer Res. 46, 3969–3978 (1986).
56. Ballou, B., Reiland, J., Levine, G., Knowles, B., Hakala, T.R. Tumor location using $F(ab')_{2\mu}$ from a monoclonal IgM antibody; Pharmacokinetics. J. Nucl. Med. 26, 283–292 (1985).
57. Carrasquillo, J.A. Krohn, K.A., Beaumier, P.L., et al. Diagnosis and treatment of solid tumors with radiolabeled antibodies and immune fragments. Cancer Treat. Rep. 68 (1), 317–328 (1984).
58. Goldenberg, D.M., Goldenberg, H., Sharkey, R.M., et al. Clinical studies of cancer radioimmunodetection with carcinoembryonic antigen monoclonal antibody fragments labeled with ^{123}I or ^{99m}Tc. Cancer Res. 50(Suppl.), 909s–921s (1990).
59. Rosenblum, M.G., Murray, J.L., Haynie, T.P., et al. Pharmacokinetics of ^{111}In-labeled anti-p97 monoclonal antibody in patients with metastatic malignant melanoma (abstract). Cancer Res. 45, 2382–2386 (1985).
60. Zimmer, A.M., Rosen, S.T., Spies, S.M., et al. Radioimmunoimaging of human small cell lung carcinoma with I-131 tumor specific monoclonal antibody. Hybridoma, 4, 1–11 (1985).
61. Yeh, S.D., Larson, S.M., Burch, L., et al. Radioimmunodetection of neuroblastoma with iodine-131-3F8: Correlation with biopsy, iodine-131-metaiodobenzylguanidine and standard diagnostic modalities. J. Nucl. Med. 32, 769–776 (1991).
62. Divgi, C.R., Welt, S., Kris, M., et al. Phase I and imaging trial of indium 111-labeled anti-epidermal growth factor receptor monoclonal antibody 225 in patients with squamous cell lung carcinoma. J. Natl. Cancer. Inst. 83, 97–104.
63. Hayes, D.F., Zalutsky, M.R., Kaplan, W., et al. Pharmacokinetics of radiolabeled monoclonal antibody B6.2 in patients with metastatic breast cancer (abstract). Cancer Res. 46, 3157–3163 (1986).
64. Order, S.E., Stillwagon, G.B., Klein, J.L., et al. Iodine 131 antiferritin, a new treatment modality in hepatoma: A radiation therapy oncology group study. J. Clin. Oncol. 3, 1573–1582 (1985).
65. Brady, L.W., Woo, D.V., Markoe, A.M., et al. ^{123}I-labeled monoclonal antibody against epidermal growth factor receptor. Antibody Immunoconj. Radiopharm. 3, 169–179 (1990).
66. Cheung, N.-K.V., Miraldi, F.D. Iodine 131 labeled G_{D2} monoclonal antibody in the diagnosis and therapy of human neuroblastoma. In Evan, A.E., D'Angio D.J., Seeger, R.C., Knudson, R. (eds.): Advances in Neuroblastoma Research 2 (pp. 595–604). New York: Alan R. Liss, 1988.
67. Stewart, J.S.W., Hird, V., Snook, D., et al. Intraperitoneal radioimmunotherapy for ovarian cancer: Pharmacokinetics, toxicity, and efficacy of I-131 labeled monoclonal antibodies. Int. J. Radiat. Oncol. Biol. Phys. 3, 169–179 (1990).

68. Epenetos, A.A., Munro, A.J., Stewart, S., et al. Antibody-guided irradiation of advanced ovarian cancer with intraperitoneally administered radiolabeled monoclonal antibodies. J. Clin. Oncol. 5, 1890–1899 (1987).
69. Finkler, N.J., Muto, M.G., Kassis, A.I., et al. Intraperitoneal radiolabeled OC 125 in patients with advanced ovarian cancer. Gynecol. Oncol. 34, 339–344 (1989).
70. Hird, V., Stewart, J.S., Snook, D., et al. Intraperitoneally administered ^{90}Y-labelled monoclonal antibodies as a third line of treatment in ovarian cancer. A phase 1-2 trial: Problems encountered and possible solutions. Br. J. Cancer 10(Suppl.), 48–51 (1990).
71. Mulshine, J.L., Carrasquillo, J.A., Weinstein, J.N., et al. Direct intralymphatic injection of radiolabeled ^{111}In-T101 in patients with cutaneous T-cell lymphoma. Cancer Res. 51, 688–695 (1991).
72. Colcher, D., Esteban, J., Carrasquillo, J.A., et al. Complementation of intracavitary and intravenous administration of a monoclonal antibody (B72.3) in patients with carcinoma. Cancer Res. 47, 4218–4224 (1987).
73. Ward, B.G., Mather, S.J., Hawkins, L.R., et al. Localization of radioiodine conjugated to the monoclonal antibody HMFG2 in human ovarian carcinoma: Assessment of intravenous and intraperitoneal routes of administration. Cancer Res 47, 4719–4723 (1987).
74. Cohen, J., Martin, E.W., Jr., Lavery, I., et al. Radioimmunoguided surgery using iodine 124 B72.3 in patients with colorectal cancer. Arch. Surg. 126, 349–352 (1991).
75. Martin, D.T., Hinkle, G.H., Tuttle, S. et al. Intraoperative radioimmunodetection of colorectal tumor with hand-held radiation detector. Am. J. Surg. 150, 672–675 (1985).
76. Thakur, M.L. Immunoscintigraphic imaging of inflammatory lesions: Preliminary findings and future possibilities. Semin. Nucl. Med. 20, 92–98 (1990).
77. Strauss, H.W., Fischman, A.J., Khaw, B.A., Rubin, R.H. Non-tumor applications of radioimmune imaging. Int. J. Appl. Instrum. 18, 127–134 (1991).
78. Rubin, R.H., Fischman, A.J., Callahan, R.J. et al. In-labeled nonspecific immunoglobulin scanning in the detection of focal infection. N. Engl. J. Med. 321(14), 935–940 (1989).
79. Johnson, L.L., Seldin, D.W. The role of antimyosin antibodies in acute myocardial infarction. Semin. Nucl. Med. 19, 238–246 (1989).
80. Khaw, B.A., Gold, H.K., Yasuda, T., Leinbach, R.C., Strauss, H.W., et al. Acute myocardial infarct imaging with technetium-99m-DPTA-antimyosin Fab (abstract). Circulation 66, 272 (1982).
81. Khaw, B.A., Fallon, J.T., Strauss, H.W., Haber, E. Myocardial infarction imaging of antibodies to canine cardiac myosin with indium-11-diethylenetriamine pentaacetic acid. Science 209, 295–297 (1980).
82. Yamada, T., Matsumori, A., Tamaki, N., et al. Detection of adriamycin cardiotoxicity with indium-111 labeled antimyosin monoclonal antibody imaging. Jpn. Circ. J. 55(4), 337–383 (1991).
83. Khaw, B.A., Fallon, J.T., Russell, P.S., Rosseel, J., Ferguson, P. et al. New approach to determination of heart transplant rejection by 125I-antimyosin antibody and 99m-Tc-fibrinogen (abstract). Clin. Res. 29, 496A (1981).
84. Frist, W., Yasuda, T., Segall, G. et al. Noninvasive detection of human cardiac transplant rejection with indium-111 antimyosin (Fab) imaging. Circulation 76(5, Pt. 2), V81–V85 (1987).

85. Yasuda, T., Palacios, I.F., Dec, G.W., et al. Indium 111-monoclonal antimyosin antibody imaging in the diagnosis of acute myocarditis. Circulation 76(2), 306–311 (1987).
86. Schaible, T.F., Alavi, A. Antifibrin scintigraphy in the diagnostic evaluation of acute deep venous thrombosis. Semin. Nucl. Med. 21, 313–324 (1991).
87. Som, P., Oster, Z.H., Zamora, P.O. et al. Radioimmunoimaging of experimental thrombi in dogs using technetium-99m-labeled monoclonal antibody fragments reactive with human platelets. J. Nucl. Med. 27, 1315–1320 (1986).
88. Kanke, M., Matsueda, G.R., Strauss, H.W., Yasuda, T., Liau, C.S., Khaw, B.A. Localization and visualization of pulmonary emboli with radiolabeled fibrin-specific monoclonal antibody. Int. J. Radiat. Appl. Instrum. 18 (1), 127–134 (1991).
89. Wang, G.-J., Oster, Z.H., Som, P., Zamora, P.O. A monoclonal antibody reacting with platelets for monitoring thrombolysis. Nucl. Med. Biol. 18, 275–280 (1991).
90. Colapinto, E.V., Humphrey, P.A., Zalutsky, M.R. et al. Comparative localization of murine monoclonal antibody Mel-14 $(Fab')_2$ fragment and whole IgG2a in human glioma xenografts. Cancer Res. 48, 5701–5707 (1988).
91. McCready, D.R., Balch, C.M., Fidler, I.J., et al. Lack of comparability between binding of monoclonal antibodies to melanoma cells in vitro and vivo. J. Natl. Cancer Inst. 81, 682–687 (1989).
92. Hagan, P.L., Halpern, S.E., Dillman, R.O. et al. Tumor size: Effect on monoclonal antibody uptake in tumor models. J. Nucl. Med. 27, 422–427 (1986).
93. Mann, B.D., Cohen, M.B., Saxton, R.E., et al. Imaging of human tumor xenografts in nude mice with radiolabeled monoclonal antibodies. Cancer 54, 1318–1327 (1984).
94. Sands, H., Jones, P.L., Shah, S.A., et al. Correlation of vascular permeability and blood flow with monoclonal antibody uptake by human Clouser and renal cell xenografts. Cancer Res. 48, 188–193 (1988).
95. Capone, P.M., Papsidero, L.D., Chu, T.M. Relationship between antigen density and immunotherapeutic response elicited by monoclonal antibodies against solid tumors. J. Natl. Cancer. Inst. 72, 673–677 (1984).
96. Chapman, P.B., Lonberg, M., Houghton, A.N. Light chain variants of an IgG3 anti-GD3 monoclonal antibody and the relationship between aviditiy, effector functions, tumor targeting, and antitumor activity. Cancer Res. 50, 1503–1509 (1990).
97. Basham, T.Y., Kaminski, M., Kitamura, K., et al. Synergistic antitumor effect of interferon and anti-idiotype monoclonal antibody in murine lymphoma. J. Immunol. 137, 3019–3024 (1986).
98. Berinstein, N., Levy, R. Treatment of a murine B-cell lymphoma with monoclonal antibodies and IL-2. J. Immunol. 139, 971 (1987).
99. Latzova, E., Savary, C.A., Herberman, R.B., et al. Augmentation of antileukemia lytic activity by OKT3 monoclonal antibody: Synergism of OKT3 and interleukin-2. Natl. Immun. Cell Growth Regul. 6, 219–223 (1987).
100. Munn, D.H., Cheung, N.K.V. Interleukin-2 enhancement of monoclonal antibody-mediated cellular cytoxicity against human melanoma. Cancer Res. 42, 6600–6605 (1987).
101. Weiner, L.M., Moldofsky, P.J., Gatenby, R.A., et al. Antibody delivery and effector cell activation in a phase II trial of recombinant gamma-interferon and the murine monoclonal antibody C017-1A in advanced colorectal carcinoma. Cancer Res. 48, 2568–2573 (1988).

102. Weiner, L.M., Steplewski, Z., Koprowksi, H., et al. Divergent dose-related effects of gamma-interferon therapy on in vitro antibody-dependent cellular cytoxicity by human peripheral blood monocytes. Cancer Res. 48, 1042–1046 (1988).

103. Brown, S.L., Miller, R.A., Levy, R. Antiidiotypic antibody therapy of B-cell lymphoma. Semin. Oncol. 16, 199–210 (1989).

105. Maloney, D.G., Levy, R., Miller, R.A. Monoclonal anti-idiotype therapy of B cell lymphoma. Biol. Ther. Cancer Updates 2 (6), 1–10 (1992).

106. Jerne, N.K. Towards a network theory of the immune system. Ann. Immunol. (Inst. Pasteur) 125C, 373–389 (1974).

107. Koide, J., Takeuchi, T., Hosono, O., et al. Suppression of in vitro production of anti-U1-ribonucleoprotein antibody by monoclonal anti-idiotypic antibody to anti-U1-ribonucleoprotein antibody. Scand. J. Immunol. 28, 687–696 (1988).

108. Hasbold, J., Klaus, G.G.B. Anti-immunoglobulin antibodies induce apoptosis in immature B cell lymphomas. Eur. J. Immunol. 20, 1685–1690 (1990).

109. Stein, K.E., Soderstrom, T. Neonatal administration of idiotype or antiidiotype primes for protection against Eschericia coli K13 infection in mice. J. Exp. Med. 160, 1001–1011 (1984).

110. Ertl, H.C.J., Finberg, R.W. Sendai virus-specific T-cell clones: Induction of cytolytic T cells by an anti-idiotypic antibody directed against a helper T-cell clone. Proc. Natl. Acad. Sci. USA 81, 2850–2854 (1984).

111. Sharpe, A.H., Gaulton, G.N., McDade, K.K., Fields, B.N., Greene, M.I. Syngeneic monoclonal antiidiotype can induce cellular immunity to reovirus. J. Exp. Med. 160, 1195–1205 (1984).

112. Dunn, P.L., Johnson, C.A., Styles, J.M., Pease, S.S., Dean, C.J. Vaccination with syngeneic monoclonal anti-idiotype protects against tumor challenge. Immunology 60, 181–186 (1987).

113. Raychaudhuri, S., Saeki, Y., Fuji, H., Kohler, H. Tumor-specific idiotype vaccines. I. Generation and characterization of internal image tumor antigen. J. Immunol. 137, 1743–1749 (1986).

114. Perez-Soler, R., Donato, N.J., Zhang, HZ, et al. Phase I study of anti-epidermal growth factor receptor (EGFR) monoclonal antibody (MoAb) RG838 in patients (PTS) with non-small cell lung cancer (abstract). Proc. Am. Soc. Clin. Oncol. 11, 254 (1992).

115. Waldmann, T.A., Goldman, C.K., Bongiovanni, K.F., et al. Therapy of patients with human T-cell lymphotropic virus I-induced adult T-cell leukemia with anti-Tac, a monoclonal antibody to the receptor for interleukin-2. Blood 72, 1805–1816 (1988).

116. Klein, B., Wijdenes, J., Zhang, X.G., et al. Murine anti-interleukin-6 monoclonal antibody therapy for a patient with plasma cell leukemia. Blood 78, 1198–1204 (1991).

117. Brown, S.L., Miller, R.A., Horning, S.S., et al. Treatment of B-cell lymphomas with antiidiotype antibodies alone and in combination with alpha interferon. Blood 73, 651–661 (1989).

118. Miller, R.A., Oseroff, A.R., Stratte, P.T., et al. Monoclonal antibody therapeutic trials in seven patients with T cell lymphoma. Blood 62, 988–995 (1983).

119. Mittelman, A., Chen, Z.J., Kageshita, T., et al. Active specific immunotherapy in patients with melanoma. A clinical trial with mouse antiidiotypic monoclonal antibodies elicited with syngeneic anti-high-molecular-weight melanoma-associated antigen monoclonal antibodies. J. Clin. Invest. 86, 2136–2144 (1990).

120. Mittelman, A., Chen, Z.J., Yang, H., Wong, G.Y., Ferrone, S. Human high molecular weight melanoma-associated antigen (HMW-MAA) mimicry by mouse anti-idiotypic monoclonal antibody MK2-23: Induction of humoral anti-HMW-MAA immunity and prolongation of survival in patients with stage IV melanoma. Proc. Natl. Acad. Sci. USA 89, 466–470 (1992).

121. Robins, R.A., Denton, G.W.L., Hardcastle, J.D., Austin, E.B., Baldwin, R.W., Durrant, L.G. Antitumor immune response and interleukin-2 production induced in colorectal cancer patients by immunization with human monoclonal anti-idiotypic antibody. Cancer Res. 51, 5425–5429 (1991).

122. Chapman, P.B., Livingston, P.O., Steffens, T.A., Oettgen, H.F., Houghton, A.N. Pilot trial of anti-idiotypic monoclonal antibody BEC2 in patients with metastatic melanoma (abstract). Proc. Am. Assoc. Cancer Res. 33, 208 (1992).

123. Bajorin, D.F., Chapman, P.B., Dimaggio, J., et al. Phase I evaluation of a combination of monoclonal antibody R24 and interleukin 2 in patients with metastatic melanoma. Cancer Res. 50, 7490–7495 (1990).

124. Creekmore, S., Urba, W., Koop, W., et al. Phase IB/II trial of R24 antibody and interleukin-2 (IL2) in melanoma (abstract). Proc. Am. Soc. Clin. Oncol. 1186, 345 (1992).

125. Goodman, G.E., Hellstrom, I., Stevenson, U., Steen, K., Hellstrom, K.E. Phase I trial of murine monoclonal antibody MG-22 and IL-2 in patients with disseminated melanoma (abstract). Proc. Am. Soc. Clin. Oncol. 1190, 346 (1992).

126. Zukiwski, A.A., Itoh, K., Benjamin, R., et al. Pilot study of rIL-2 administered with a murine anti-melanoma antibody in patients with metastatic melanoma (abstract). Proc. Am. Assoc. Cancer Res. 1448, 365 (1989).

127. Caulfield, J., Barna, B., Murthy, S., et al. Phase Ia-Ib trial of an anti-GD3 monoclonal antibody in combination with interferon-α in patients with malignant melanoma. J. Biol. Response Mod. 9, 319–328 (1990).

128. Steffens, T.A., Bajorin, D.F., Williams, L.J., et al. A phase I trial of R24 monoclonal antibody (Mab) and recombinant human macrophage colony stimulating factor (rhM-CSF) in patients (pts) with advanced melanoma (abstract). Proc. Am. Soc. Clin. Oncol. 1182, 344 (1992).

129. Felice, A.J., Chachoua, A., Oratz, R., et al. A phase IB trial of GM-CSF with murine monoclonal antibody R24 in patients with metastatic melanoma (abstract). Proc. Am. Soc. Clin. Oncol. 1188, 346 (1992).

130. Hale, G., Clark, M.R., Marcus, R., et al. Remission induction in non-Hodgkins lymphoma with reshaped human monoclonal antibody Campath-1H. Lancet 1394–1399 (1988).

131. Duncan, A.R., Winter, G. The binding site for Clq on IgG. Nature 332, 738–740 (1988).

132. Duncan, A.R., Woof, J.M., Partridge, L.J., Burton, D.R., Winter, G. Localization of the binding site for the human high affinitiy Fc receptor on IgG. Nature 332, 563–546 (1988).

133. Ward, E.S., Gussow, D., Griffiths, A.D., Jones, P.T., Winter, G. Binding activities of a repertoire of single immunoglobulin variable domains secreted from Escherichia coli. Nature 341, 544–546 (1989).

134. Jung, G., Ledbetter, J.A., Muller-Eberhard. Induction of cytotoxicity in resting human

T lymphocytes bound to tumor cells by heteroconjugates. Proc. Natl. Acad. Sci. USA 84, 4611–4615 (1987).

135. Perez, P., Hoffman, R.W., Titus, J.A., et al. Specific targeting of human peripheral blood T cells by heteroaggregates containing anti-T3 cross-linked to anti-target cell antibodies. J. Exp. Med. 163, 166–178 (1986).
136. Perez, P., Titus, J.A., Lotze, M.T., et al. Specific lysis of human tumor cells by T cells coated with anti-T3 cross linked to anti-tumor antibody. J. Immunol. 137, 2069 (1986).
137. Staerz, U.D., Bevin, M.J. Hybrid hybridoma producing a bispecific monoclonal antibody that can focus effector T-cell activity. Proc. Natl. Acad. Sci. USA 83, 1453–1457 (1986).
138. Caron, P.C., Laird, W., Co, M.S., Avdalovic, N.M., Queen, C., Scheinerg, D.A. Engineered humanized dimeric forms of IgG are more effective antibodies. J. Exp. Med. 176, 950–953 (1992).
139. Shopes, B. A genetically engineered human IgG mutant with enhanced cytolytic activity. J. Immunol. 148, 2918 (1992).
140. Drebin, J.A., Link, V.C., Greene, M.I. Monoclonal antibodies specific for the neu oncogene product directly mediate antitumor effects in vivo. Oncogene 2, 387–394 (1988).
141. Drebin, J.A., Link, V.C., Greene, M.I. Monoclonal antibodies reactive with distinct domains of the neu oncogene-encoded p185 molecule exert synergistic antitumor effects in vivo. Oncogene 2, 273–277 (1988).
142. Trauth, B.C., Klas, C., Peters, A.M.J., et al. Monoclonal antibody-mediated tumor regression by induction of apoptosis. Science 245, 301–305 (1989).
143. Pearson, J.W., Fogler, W.E., Volker, K., et al. Reversal of drug resistance in a human colon cancer xenograft expressing MDR1 complementary DNA by in vivo administration of MRK-16 monoclonal antibody. J. Natl. Cancer Inst. 83, 1386–1391 (1991).
144. Broxterman, H.J., Kuiper, C.M., Shuurhus, G.J., et al. Increase of daunorubicin and vincristine accumulation in multidrug resistant human ovarian carcinoma cells by a monoclonal antibody reacting with P-glycoprotein. Biochem. Pharmacol. 37, 2389–2393 (1988).
145. Huse, W.D., Sastry, L., Iverson, S.A., et al. Generation of a large combinatorial library of the immunoglobulin repertoire in phage lambda. Science 246, 1275–1281 (1989).
146. Langmuir, V.K., Sutherland, R.M. Radiobiology of radioimmunotherapy: Current status. Antibody Immunoconj. Radiopharm. 1, 195–211 (1988).
147. Kyriakos, R.J., Shih, L.B., Ong, G.L., Patel, K., Goldenberg, D.M., Mattes, M.J. The fate of antibodies bound to the surface of tumor cells *in vitro*. Cancer Res. 52, 601–608 (1992).
148. Makrigiorgos, G.M., Adelstein, S.J., Kassis, A.I. Limitations of conventional internal dosimetry at the cellular level. J. Nucl. Med. 30, 1856–1864 (1989).
149. Scheinberg, D.A., Strand, M. Kinetic and catabolic considerations of monoclonal antibody targeting in erythroleukemic mice. Cancer Res. 43, 265–272 (1983).
150. Pimm, M.V., Perkins, A.C., Armitage, N.C., Baldwin, R.W. The characteristics of blood-borne radiolabels and the effect of anti-mouse IgG antibodies on localization of radiolabeled monoclonal antibody in cancer patients. J. Nucl. Med. 26, 1011–1023 (1985).

151. Schwartz, M.A., Lovett, D.R., Redner, A., et al. Leukemia cytoreduction and marrow ablation after therapy with ^{131}I labeled monoclonal antibody M195 for acute myelogenous leukemia (AML) (abstract). Proc. Am. Soc. Clin. Oncol. 10, 230 (1991).

152. Rosen, S.T., Zimmer, A.M., Goldman-Leikin, R., et al. Radioimmunodetection and radioimmunotherapy of cutaneous T cell lymphomas using an ^{131}I-labeled monoclonal antibody: An Illinois Cancer Council Study. J. Clin. Oncol. 5, 562–573 (1987).

153. Schroff, R.W., Fer, M.F., Weiden, P.L., et al. Rhenium 186 labelled antibody in patients with cancer; report of a Phase 1 study. Antibody Immunoconj. Radiopharm 3, 99–111 (1990).

154. Vitteta, E.S., Stone, M., Amlot, P., et al. Phase I immunotoxin trial in patients with B-cell lymphoma. Cancer Res. 51, 4052–4058 (1991).

155. Weiner, L.M., O'Dwyer, J., Kitson, J., et al. Phase I evaluation of an anti-breast carcinoma monoclonal antibody 260F9-recombinant ricin A chain immunoconjugate. Cancer Res. 49, 4062–4067 (1989).

156. Schneck, D., Butler, F., Dugan, W., et al. Disposition of a murine monoclonal antibody vinca conjugate (KS1/4-DAVLB) in patients with adenocarcinomas. Clin. Pharmacol. Ther. 47, 36–41 (1990).

157. Petersen, B.H., DeHerdt, S.V., Schneck, D.W., Bumol, T.F. The human immune response to KS1/4-desacetylvinblastine (LY256787) and KS1/4-desaceytlvinblastine hydrazide (LY203728) in single and multiple dose clinical studies. Cancer Res. 51, 2286–2290 (1991).

158. Elias, D.J., Hirschowitz, L., Kline, L.E., et al. Phase I clinical comparative study of monoclonal antibody KS1/4 and KS1/4-methotrexate immunoconjugate in patients with non-small cell lung carcinoma. Cancer Res. 50, 4154–4159 (1990).

159. Hertler, A.A., Frankel, A.E. Immunotoxins; A clinical review of their use in the treatment of malignancies. J. Clin. Oncol. 7, 1932–1942 (1989).

160. Press, O.W. Immunotoxins. Biotherapy 3:65–76 (1991).

161. Rybak, S.M., Youle, R.J. Clinical use of immunotoxins: Monoclonal antibodies conjugated to protein toxins. In Oettgen, H.F., (ed.): Immunology and Allergy Clinics of North America. Human Cancer Immunology II (pp. 359–380). Philadelphia: W.B. Saunders, 1991.

162. Vitetta, E.S., Thorpe, P.E. Immunotoxins. In DeVita Jr., V.T., Hellman, S., Rosenberg, S.A. (eds.): Biologic Therapy of Cancer (pp. 482–495). Philadelphia: J.B. Lippincott, 1991.

163. Pai, L.H., Bookman, M.A., Ozols, R.F., et al. Clinical evaluation of intraperitoneal Pseudomonas exotoxin immunoconjugate OVB3-PE in patients with ovarian cancer. J. Clin. Oncol. 9, 2095–2103 (1991).

164. Falini, B.B. Response of refractory Hodgkin's disease to monoclonal anti-CD30 immunotoxin. Lancet 339, 1195–1196 (1992).

165. Lynch, T.J., Shefner, J., Wen, P., et al. Immunotoxin therapy with N901-blocked ricin (N901-BR) for small cell lung cancer (SCLC): A phase I study (abstract). Proc. Am. Soc. Clin. Oncol. 11, 314 (1992).

166. Grossbard, M.L., Freedman, A.S., Ritz, J., et al. Serotherapy of B-cell neoplasms with anti-B4-blocked ricin: A phase I trial of daily bolus infusion. Blood 79, 576–585 (1992).

167. May, R.D., Finkelman, F.D., Wheeler, H.T., Uhr, J.W., Vitetta, E.S. Evaluation of

ricin A chain-containing immunotoxins directed against different epitopes on the δ-chain of cell surface-associated IgD on murine cells. J. Immunol. 144, 3637–3642 (1990).

168. LeMaistre, C.F., Rosen, S., Frankel, A., et al., Phase I trial of H65-RTA immunoconjugate in patients with cutaneous T-cell lymphoma. Blood 78, 1173–1182 (1991).

169. LoBuglio, A.F., Khaeli, M.B., Lee, J., et al. Pharmacokinetics and immune response to Xomazyme-Mel in melanoma patients. Antibody Immunoconj. Radiopharm. 1, 305–310 (1988).

170. Durrant, L.G., Byers, V.S., Scannon, P.J., et al. Humoral immune responses to XMMCO-791 a immunotoxin in colorectal cancer patients. Clin. Exp. Immunol. 75, 258–264 (1989).

171. Kreitman, R.J., Chaudhary, V.K., Waldmann, T., Willingham, M.C., FitzGerald, D.J., Pastan, I. The recombinant immunotoxin anti-Tac(Fv)-Pseudomonas exotoxin 40 is cytotoxic toward peripheral blood malignant cells from patients with adult T-cell leukemia. Proc. Natl. Acad. Sci. USA 87, 8291–8295 (1990).

172. Rosenblum, M.G., Cheung, L., Murray, J.L., Bartholomew, R. Antibody-mediated delivery of tumor necrosis factor (TNF-α): Improvement of cytotoxicity and reduction of cellular resistance. Cancer Commun. 3, 21–27 (1991).

173. Prior, T.I., Helman, L.J., FitzGerald, D.J., Pastan, I. Cytotoxic activity of a recombinant fusion protein between insulin-like growth factor I and Pseudomonas exotoxin. Cancer Res. 51, 174–180 (1991).

174. Pai, L.H., Gallo, M.G., FitzGerald, D.J., Pastan, I. Antitumor activity of a transforming growth factor 6-Pseudomonas exotoxin fusion protein (TGF-α-PE40). Cancer Res. 51, 2808–2812 (1991).

175. Siegall, C.B., Kreitman, R.J., FitzGerald, D.J., Pastan, I. Antitumor effects of interleukin 6-Pseudomonas exotoxin chimeric molecules against the human hepatocellular carcinoma, PLC-PRF-5 in mice. Cancer Res. 51, 2831–2836 (1991).

176. Puri, R.K., Ogata, M., Leland, P., Feldman, G.M., FitzGerald, D.J., Pastan, I. Expression of high-affinity interleukin 4 receptors on murine sacroma cells and receptor-mediated cytotoxicitiy of tumor cells to chimeric protein between interleukin 4 and Pseudomonas exotoxin. Cancer Res. 51, 3011–3017 (1991).

177. Flavell, D.J., Cooper, S., Morland, B., Flavell, S.U. Characteristics and performance of a bispecific $F(ab')_2$ antibody for delivering saporin to a CD7+ human acute T-cell leukaemia cell line. Br. J. Cancer 64, 274–280 (1991).

178. Ramsay, N.C., LeBien, T., Nesbit, M., et al. Autologous bone marrow transplantation for patients with acute lymphoblastic leukemia in second or subsequent remission: Results of bone marrow tested with monoclonal antibodies BA-1, BA-2, BA-3 plus complement. Blood 66(3), 508–513 (1985).

179. Ramsay, N.K.C., Kersey, J.H. Bone marrow purging using monoclonal antibodies. J. Clin. Immunol. 8, 81–88 (1988).

180. Uckun, F.M., Gajl-Peczalska, K., Meyers, D.E., et al. Marrow purging in autologous bone marrow transplantation for T-lineage acute lymphoblastic leukemia: Efficacy of *ex vivo* treatment with immunotoxins and 4-hydroperoxycyclophosphamide against fresh leukemic marrow progenitor cells. Blood 69, 361–366 (1987).

181. Trickett, A.E., Ford, D.J., Lam-Po-Tang, P.R.L., Vowels, M.R. Immunomagnetic bone marrow purging of common acute lymphoblastic leukemia cells: suitability of BioMag particles. Bone Marrow Transplant. 7, 199–203 (1991).

182. Kvalheim, G., Sorensen, O., Fodstad, O., et al. Immunomagnetic removal of B-lymphoma cells from human bone marrow: A procedure for clinical use. Bone Marrow Transplant. 3, 31–41 (1988).

183. Ball, E.D., Vredenburgh, J.J., Mills, L.E., et al. Autologous bone marrow transplantation for acute myeloid leukemia following *in vitro* treatment with neuraminidase and monoclonal antibodies. Bone Marrow Transplant. 6, 277–280 (1990).

184. Ball, E.D. *In vitro* purging of bone marrow for autologous marrow transplantation in acute myelogenous leukemia using myeloid-specific monoclonal antibodies. Bone Marrow Transplant. 3, 387–392 (1988).

185. Stiff, P.J., Schultz, W.C., Bishop, M., et al. Anti-CD33 monoclonal antibody and etoposide/cytosine arabinoside combinations for the *ex vivo* purification of bone marrow in acute nonlymphocytic leukemia. Blood 77, 355–362 (1991).

186. Lemoli, R.M., Gasparetto, C., Scheinberg, D.A. Autologous bone marrow transplantation in acute myelogenous leukemia; *In vitro* treatment with myeloid-specific monoclonal antibodies and drugs in combination. Blood 77, 1829–1836 (1991).

187. Howell, A.L., Fogg-Leach, M., Davis, B.H., Ball, E.D. Continuous infusion of complement by an automated cell processor enhances cytotoxicity of monoclonal antibody sensitized leukemia cells. Bone Marrow Transplant. 4, 317–322 (1989).

188. DeFabritiis, F., Ferro, D., Sandrelli, A., et al. Monoclonal antibody purging and autologous bone marrow transplantation in acute myelogenous leukemia in complete remission. Bone Marrow Transplant. 4, 669–674 (1989).

189. Robertson, M.J., Soiffer, R.J., Freedman, A.S., et al. Human bone marrow depleted of CD33-positive cells mediates delayed but durable reconstitution of hematopoiesis: clinical trial of MY9 monoclonal antibody-purged autografts for the treatment of acute myeloid leukemia. Blood 79, 2229–2236 (1992).

190. Roy, R.C., Griffin, J.D., Belvin, M., et al. Anti-MY9-blocked ricin: An immunotoxin for selective targeting of acute myeloid leukemia cells. Blood 77, 204–2412 (1991).

191. Pole, J.G., Gee, A., Janssen, W., Lee, C., Gross, S. Immunomagnetic purging of bone marrow: A model for negative cell selection. Am. J. Pediatr. Hematol. Oncol. 12 (3), 257–261 (1990).

192. Pole, J.G., Casper, J., Elfenbein, G., et al. High-dose chemoradiotherapy supported by marrow infusions for advanced neuroblastoma: A Pediatric Oncology Group Study. J. Clin. Oncol. 9, 152–158 (1991).

193. Ball, E.D., Powers, F.J., Vredenburgh, J.J., Heath, C.A., Converse, A.O. Purging of small cell lung cancer cells from bone marrow using immunomagnetic beads and a flow-through device. Bone Marrow Transplant. 8, 35–40 (1991).

194. Martin, P.J., Hansen, J.A., Torok-Storb, B., et al. Effects of treating marrow with a CD3-specific immunotoxin for prevention of acute graft-versus-host disease. Bone Marrow Transplant. 3, 437–444 (1988).

195. Kogler, G., Capdeville, A.B., Hauch, M., et al. High efficiency of a new immunological magnetic cell sorting method for T cell depletion of human bone marrow. Bone Marrow Transplant. 6, 163–168 (1990).

196. Knobloch, C., Spadinger, U., Rueber, E., Friedrich, W. T cell depletion from human bone marrow using magnetic beads. Bone Marrow Transplant. 6, 21–24 (1990).

197. Filipovich, A.H., Vallera, D.A., Youle, R.J., Quinones, R.R., Neville, D.M., Jr., Kersey, J.H. *Ex vivo* treatment of donor bone marrow with anti-T-cell immunotoxins for prevention of graft-versus-host disease. Lancet 1, 469–471 (1984).

198. Waldmann, H., Polliak, A., Hale, G., et al. Elimination of graft-versus-host disease by *in vitro* depletion of alloreactive lymphocytes with a monoclonal rat and anti-human lymphocyte antibody (Campath-1). Lancet 2, 483–486 (1984).

199. Thorley-Lawson, D.A., Geilinger, K. Monoclonal antibodies against the major glycoprotein (gp350/220) of Epstein-Barr virus neutralize infectivity. Proc. Natl. Acad. Sci. USA 77, 5307–5311 (1980).

200. Staats, H.F., Oakes, J.E., Lausch, R.N. Anti-glycoprotein D monoclonal antibody protects against herpes simplex virus type 1-induced diseases in mice functionally depleted of selected T-cell subsets of asiolo GM1+ cells. J. Virol. 65, 6008–6014 (1991).

201. Mester, J.C., Glorioso, J.C., Rouse, B.T. Protection against zosteriform spread of herpes simplex virus by monoclonal antibody. J. Infect. Dis. 163, 263–269 (1991).

202. Co, M.S., Deschamps, M., Whitley, R.J., Queen, C. Humanized antibodies for antiviral therapy. Proc. Natl. Acad. Sci. USA 88, 2869–2873 (1991).

203. Foung, S.K.H., Perkins, S., Bradshaw, P., et al. Human monoclonal antibodies to human cytomegalovirus. J. Infect. Dis. 159, 436 (1989).

204. Drobyski, W.R., Gottlieb, M., Carrigan, D., et al. Phase I study of safety and pharmacokinetics of a human anticytomegalovirus monoclonal antibody in allogeneic bone marrow transplant recipients. Transplantation 51, 1190–1196 (1991).

205. Aulitzky, W.E., Schulz, T.F., Tilg, H., et al. Human monoclonal antibodies neutralizing cytomegalovirus (CMV) for prophylaxis of CMV disease: Report of a phase I trial in bone marrow transplant recipients. J. Infect. Dis. 163, 1344–1347 (1991).

206. Waldmann, T.A., Strober, W. Metabolism of immunoglobulins. Progr. Allergy 13, 1–110 (1969).

207. Burton, D.R., Barbas, C.F. III, Persson, M.A.A., Koenig, S., Chanock, R.M., Lerner, R.A. A large array of human monoclonal antibodies to type 1 human immunodeficiency virus from combinatorial libraries of asymptomatic seropositive individuals. Proc. Natl. Acad. Sci. USA 88, 10134–10137 (1991).

208. Shouval, D., Shafritz, D.A., Zurawski, V.R., Jr., Wands, J.R. Immunotherapy of nude mice of human hepatoma using monoclonal antibodies against hepatitis B virus. Nature 298, 567–569 (1982).

209. Kishimoto, C., Abelmann, W.H. Monoclonal antibody therapy for prevention of acute coxsackievirus B3 myocarditis in mice. Circulation 79, 1300–1308 (1989).

210. Hector, R.F., Collins, M.S., Pennington, J.E. Treatment of experimental Pseudomonas aeruginosa pneumonia with a human IgM monoclonal antibody. J. Infect. Dis. 160, 483 (1989).

211. Sciortino, C.V. Protection against infection with Vibrio cholerae by passive transfer of monoclonal antibodies to outer membrane antigens. J. Infect. Dis. 160, 248 (1989).

212. Raff, H.V., Bradley, C., Brady, W., et al. Comparison of functional activities between IgG1 and IgM class-switched human monoclonal antibodies reactive with group B streptococci or Escherichia coli K1. J. Infect. Dis. 163, 346–354 (1991).

213. Halperin, S.A., Issekutz, T.B., Kasina, A. Modulation of Bordetella pertussis infection with monoclonal antibodies to pertussis toxin. J. Infect. Dis. 163, 355–361 (1991).

214. Dromer, F., Charreire, J. Improved amphotericin B activity by a monoclonal anti-Cryptococcus neoformans antibody: Study during murine cryptococcosis and mechanisms of action. J. Infect. Dis. 163, 1114–1120 (1991).

215. Potocnjak, P., Yoshida, N., Nussenzweig, R.S., Nussenzweig, V. Monovalent fragments (Fab) of monoclonal antibodies to a sporozoite surface antigen (Pb44) protect mice against malarial infection. J. Exp. Med. 151, 1504–1513 (1980).
216. Perrin, L.H., Ramirez, E., Lambert, P.H., Miescher, P.A. Inhibition of P. falciparum growth in human erythrocytes by monoclonal antibodies. Nature 289, 301–303 (1981).
217. Young, L.S., Proctor, R.A., Beutler, B., McCabe, W.R., Sheagren, J.N. University of California/Davis Interdepartmental Conference on Gram-Negative Septicemia. Rev. Infect. Dis. 13, 666–687 (1991).
218. Silva, A.T., Bayston, K.F., Cohen, J. Prophylactic and therapeutic effects of a monoclonal antibody to tumor necrosis factor-α in experimental gram-negative shock. J. Infect. Dis. 162, 421–427 (1990).
219. Beutler, B., Milsark, I.W., Cerami, A.C. Passive immunization against cachectin/tumor necrosis factor protects mice from lethal effect of endotoxin. Science 229, 869–871 (1985).
220. Opal, S.M., Cross, A.S., Kelly, N.M., et al. Efficacy of a monoclonal antibody directed against tumor necrosis factor in protecting neutropenic rats from lethal infections with Pseudomonas aeruginosa. J. Inf. Dis. 161, 1148–1152 (1990).
221. Young, L.S., Gascon, R., Alam, S., Bermudez, L.E.M. Monoclonal antibodies for treatment of gram-negative infections. Rev. Infect. Dis. 2(Suppl. 7), S1546–S1571 (1989).
222. Overbeek, B.P., Veringa, E.M. Role of antibodies and antibiotics in aerobic gram-negative septicemia: Possible synergism between antimicrobial treatment and immunotherapy. Rev. Infect. Dis. 13, 751–760 (1991).
223. Harkonen, S., Scannon, P., Mischak, R.P., et al. Phase I study of a murine monoclonal anti-lipid A antibody in bacteremic and nonbacteremic patients. Antimicrob. Agents Chemother. 32, 710–716 (1988).
224. Greenman, R.L., Schein, R.M.H., Martin, M.A., et al. A controlled clinical trial of E5 murine monoclonal IgM antibody to endotoxin in the treatment of gram negative sepsis. The Xoma Sepsis Study Group. J. Am. Med. Assoc. 266, 1097–1102 (1991).
225. Zeigler, E.J., Fisher, C.J., Jr., Sprung, C.L., et al. Treatment of gram-negative bacteremia and septic shock with Ha-1A human monclonal antibody against endotoxin. N. Engl. J. Med. 324 429–436 (1991).
226. Smith, C.R., Straube, R.C., Ziegler, E.J. HA-1A – A human monoclonal antibody for the treatment of gram-negative sepsis. In Moellering, R.C., Young, L.S.G. (eds.): Infectious Disease Clinics of North America (5th ed., pp. 253–267). Philadelphia: W.B. Saunders, 1991.
227. Wenzel, R.P., Andriole, V.T., Bartlett, J.G., et al. Antiendotoxin monoclonal antibodies for gram-negative sepsis: Guidelines from the IDSA. Clin. Infect. Dis. 14, 973–976 (1992).
228. Smith, T.W., Butler, V.P., Jr., Haber, E., Fozzard, H., Marcus, F.I., et al. Treatment of life-threatening digitalis intoxication with digoxin-specific Fab antibody fragments: Experience in 26 cases. N. Engl. J. Med. 307, 1357–1362 (1982).
229. Herzog, C., Walker, C., Pichler, W., et al. Monoclonal anti-CD4 in arthritis. Lancet 2, 1461–1462 (1987).
230. Kirkham, B., Chikanza, I., Pitzalis, C., et al. Response to monoclonal CD7 antibody in rheumatoid arthritis. Lancet 1, 589 (1988).

231. Emmrich, F., Horneff, G., Becher, W., et al. An anti-CD4 antibody for treatment of chronic inflammatory arthritis. Agents Actions 32 (Suppl.), 165–170 (1991).

232. Herzog, C., Walker, C., Muller, W., et al. Anti-CD4 antibody treatment of patients with rheumatoid arthritis: I. Effect on clinical course and circulating T cells. J. Autoimmun. 2, 627–642 (1989).

233. Moreland, L.W., Bucy, R.P., Pratt, P.W., et al. Treatment of refractory rheumatoid arthritis (RA) with a chimeric anti-CD4 monoclonal antibody (abstract). J. Cell. Biochem. Suppl. 15E, 183 (1991).

234. Horneff, G., Burmester, G.R., Emmrich, F., Kalden, J.R. Treatment of rheumatoid arthritis with an anti-CD4 monoclonal antibody. Arthritis Rheum. 34, 129–140 (1991).

235. Isaacs, J.D., Watts, R.A., Hazleman, B.L., et al. Humanised monoclonal antibody therapy for rheumatoid arthritis. Lancet 340, 748–752 (1992).

236. Weinshenker, B.G., Bass, B., Karlik, S., Ebers, G.C., Rice, G.P.A. An open trial of OKT3 in patients with multiple sclerosis. Neurology 41, 1047–1052 (1991).

237. Hafler, D.A.R., Ritz, J., Schlossman, S.F., Weiner, H.L. Anti-CD4 and anti-CD2 monoclonal antibody infusions in subjects with multiple sclerosis. J. Immunol. 141, 131–138 (1988).

238. Hafler, D.A.R., Fallis, R.J., Dawson, D.M., Schlossman, S.F., Reinherz, E.L., Weiner, H.L. Immunologic response of progressive multiple sclerosis patients treated with an anti-T-cell monoclonal antibody, anti-T12. Neurology 36, 777–784 (1986).

239. Hafler, D.A., Weiner, H.L. Immunosuppression with monoclonal antibodies in multiple sclerosis. Neurology 38(Suppl. 2), 42–47 (1988).

240. Doney, K., Martin, P., Storb, R., et al. A randomized trial of antihuman thymocyte globulin versus murine monoclonal antihuman T-cell antibodies as immunosuppressive therapy for aplastic anemia. Exp. Hematol. 13, 520–524 (1985).

241. Jansen, J., Gratama, J.W., Zwaan, F.E., Simonis, R.F.A. Therapy with monoclonal antibody OKT3 in severe aplastic anemia. Exp. Hematol. 12(Suppl. 15), 46–47 (1984).

242. Champlin, R., Ho, W., Bayever, E., et al. Treatment of aplastic anemia: Results with bone marrow transplantation, antithymocyte globulin, and a monoclonal anti-T cell antibody. In Young, N.S., Levine, A.S., Humphries, R.K. (eds.): Aplastic Anemia: Stem Cell Biology and Advances in Treatment (pp. 227–238). New York: Alan R. Liss, 1984.

243. Dyer, M.J.S. Hale, G., Marcus, R., Waldmann, H. Remission induction in patients with lymphoid malignancies using unconjugated CAMPATH-1 monoclonal antibodies. Leuk. Lymph. 2, 179–193 (1990).

244. Mathieson, P.W., Cobbold, S.P., Hale, G., et al. Monoclonal-antibody therapy in systemic vasculitis. N. Engl. J. Med. 323, 250–254 (1990).

245. Ortho Multicenter Transplant Study Group. A randomized clinical trial of OKT3 monoclonal antibody for acute rejection of cadaveric renal transplants. N. Engl. J. Med. 313, 337 (1985).

246. Ponticelli, C., Rivolta, E., Tarantino, A., et al. Clinical experience with Orthoclone OKT3 in renal transplantation. Transplant. Proc. 18, 942–948 (1986).

247. Deierhoi, M.H., Barber, W.H., Curtis, J.J., et al. A comparison of OKT3 monoclonal antibody and corticosteroids in the treatment of acute renal allograft rejection. Am. J. Kidney Dis. 11, 86–89 (1988).

248. Schroeder, T.J., Weiss, M.A., Smith, R.D., Stephens, G.W., Carey, M., First, M.R.

The use of OKT3 in the treatment of acute vascular rejection. Transplant. Proc. 23, 1043–1045 (1991).

249. Starzl, T.E., Fung, J.J. Orthoclone OKT3 in treatment of allografts rejected under cyclosporine-steroid therapy. Transplant. Proc. 18, 937–941 (1986).

250. Woodle, E.S., Thistlewaite, J.R., Emond, J.C., et al. OKT3 therapy for hepatic allograft rejection: Comparison of results in adults and children. Transplant. Proc. 22, 1765–1766 (1990).

251. Cosimi, A.B., Jenkins, R.L., Rohrer, R.J., Delmonico, F.L., Hoffman, M., Monaco, A.P. A randomized clinical trial of prophylactic OKT3 monoclonal antibody in liver allograft recipients. Arch Surg. 125, 781–784 (1990).

252. Haberal, M., Sert, S., Gulay, H., Arslan, G., Bilgin, N. The treatment of steroid-resistant renal allograft rejection with OKT3 and plasmapheresis. Transplant. Proc. 22, 1761–1763 (1990).

253. Hesse, U.J., Wienand, P., Baldamus, C., Arns, W. Preliminary results of a prospectively randomized trial of ALG vs OKT3 for steroid-resistant rejection after renal transplantation in the early postoperative period. Transplant. Proc. 22, 2273–2274 (1990).

254. Steinmuller, D.R., Hayes, J.M., Novick, A.C., et al. Comparison of OKT3 with ALG for prophylaxis for patients with acute renal failure after cadaveric renal transplantation. Transplantation 52, 67–71 (1991).

255. Samuel, D., Gugenheim, J., Canon, C., Saliba, F., Bismuth, H. Use of OKT3 for late acute rejection in liver transplantation. Transplant. Proc. 22, 1767–1768 (1990).

256. Rohrer, R.J., Jenkins, R.L., Khettry, U., et al. Histologic response to OKT3 therapy for hepatic allograft rejection. Transplant. Proc. 19, 2459–2461 (1987).

257. Deeb, G.M., Bollng, S.F., Steimle, C.N., Daws, J.E., McKay, A.L., Richardson, A.M. A randomized prospective comparison of MALG with OKT3 for rescue therapy of acute myocardial rejection. Transplantation 51, 180–183 (1991).

258. Grino, J.M., Castelao, A.M., Seron, D. et al., Prophylactic OKT3, CyA, and steroids versus antilymphoblast globulin, CyA, and steroids in cadaveric kidney transplantation. Transplant. Proc. 24, 39–41 (1992).

259. Hanto, D.W., Jendrisak, M.D., McCullough, C.S., et al. A prospective randomized comparison of prophylactic ALG and OKT3 in cadaver kidney allograft recipients. Transplant. Proc. 23, 1050–1051 (1991).

260. Frey, D.J., Matas, A.J., Gillingham, K.J., et al. MALG vs OKT3 following renal transplantation: A randomized prospective trial. Transplant. Proc. 23, 1048–1049 (1991).

261. Illner, W.D., Theodorakis, J., Abendroth, D., et al. Quadruple-drug induction therapy in combined renal and pancratic transplantation—OKT3 versus ATG. Transplant. Proc. 22, 1586–1587 (1990).

262. Melzer, J.S., D'Alessandro, A.M., Kalayoglu, M., Pirsch, J.D., Belzer, F.O., Sollinger, H.W. The use of OKT3 in combined pancreas-kidney allotransplantation. Transplant. Proc. 22, 634–635 (1990).

263. Pauw, L.D., Abramowicz, D., Goldman, M., Vereerstraeten, P., Kinnaert, P., Toussaint, C. Comparison between prophylactic use of OKT3 and cyclosporine in cadaveric renal transplantation. Transplant. Proc. 22, 1759–1760 (1990).

264. Goldman, M., Abramowicz, D., De Pauw, L., et al. Benefical effects of prophylactic

OKT3 in cadaver kidney transplantation: Comparison with cyclosporin A in a single-center prospective randomized study. Transplant. Proc. 23, 1046–1047 (1991).

265. Cosimin, A.B., Cho, S.I., Delmonico, F.L., Kaplan, M.M., Roher, R.J., Jenkins, R.L. A randomized clinical trial comparing OKT3 and steroids for treatment of hepatic allograft rejection. Transplant. Proc. 19, 2431–2433 (1987).

266. McDiarmid, S.V., Busuttil, R.W, Levy, P., Millis, M.J., Terasaki, P.I., Ament, M.E. The long-term outcome of OKT3 compared with cyclosporine prophylaxis after liver transplantation. Transplantation 52, 91–97 (1991).

267. Pulpon, L.A., Dominguez, P., Chafer, M., et al. Induction immunosuppression with OKT3 monoclonal antibody in cardiac transplant recipients. Transplant. Proc. 22, 2319 (1990).

268. Ippoliti, G., Negri, M., Abelli, P., et al. Preoperative prophylactic OKT3 vs RATG. A randomized clinical study in heart transplant patients. Transplant. Proc. 23, 2272–2274. (1991).

269. Norman, D.J., Barry, J.M., Bennett, W.M., et al. OKT3 for induction immunosuppression in renal transplantation: A comparative study of high versus low doses. Transplant. Proc. 23, 1052–1054 (1991).

270. McDiarmid, S.V., Millis, M., Terashita, G., Ament, M.E., Busuttil, R., Terasaki, P. Low serum OKT3 levels correlate with failure to prevent rejection in orthotopic liver transplant patients. Transplant. Proc. 22, 1774–1776 (1990).

271. Chatenoud, L., Legendre, C., Ferran, C., Bach, J.-F., Kreis, H. Corticosteroid inhibition of the OKT3-induced cytokine-related syndrome—Dosage and kinetics prerequisites. Transplantation 51, 334–338 (1991).

272. Hesse, C.J., Heyse, P., Stolk, B.J.M., et al. The incidence and quality of antiidiotypic antibody formation after OKT3 monoclonal therapy in heart-transplant recipients. Transplant. Proc. 22, 1772–1773 (1990).

273. Colledan, M., Gridelli, B., Rossi, G., et al. Prolonged and repeated courses of OKT3 after liver transplantation. Transplant. Proc. 22, 1769–1771 (1990).

274. Frenken, L.A.M., Hoitsma, A.J., Koene, R.A.P. Prophylactic use of anti-CD3 monoclonal antibody WT32 in kidney transplation. Transplant. Proc. 23, 1072–1073 (1991).

275. Kirkman, R.L., Araujo, J.L., Busch, G.J., et al. Treatment of acute renal allograft rejection with monoclonal anti-T12 antibody. Transplantation 36, 620–626 (1983).

276. Soulillou, J.P., Le Mauff, B., Olive, D., et al. Prevention of rejection of kidney transplants by monoclonal antibody directed against interleukin 2. Lancet 1, 1339–1342 (1987).

277. Soulillou, J.-P., Cantarovich, D., Le Mauff, B., et al. Randomized controlled trial of a monoclonal antibody against the interleukin-2 receptor (33B3.1) as compared with rabbit antithymocyte globulin for prophylaxis against rejection of renal allografts. N. Engl. J. Med. 322, 1175–1182 (1990).

278. Cantarovich, D., Le Mauff, B., Hourmant, M., et al. Anti-interleukin 2 receptor monoclonal antibody in the treatment of ongoing acute rejection episodes of human kidney graft—A pilot study. Transplantation 47, 454–457 (1989).

279. Kirkman, R.L., Shapiro, M.E., Carpenter, C.B., et al. A randomized prospective trial of anti-Tac monoclonal antibody in human renal transplantation. Transplantation 51, 107–113 (1991).

280. Waid, T.H., Lucas, B.A., Amlot, P., et al. T10B9.1A-31 anti-T-cell monoclonal

antibody: Preclinical studies and clinical treatment of solid organ allograft rejection. Am. J. Kidney Dis. 14(5), Suppl. 2, 61–70 (1989).

281. Waid, T.H., Lucas, B.A., Thompson, J.S., et al. Treatment of acute cellular rejection with T10B9.1A-31 or OKT3 in renal allograft recipients. Transplantation 53, 80–86 (1992).
282. Dendorfer, U., Hillebrand, G., Kasper C., et al. Effective prevention of interstitial rejection crises in immunological high risk patients following renal transplantation: Use of high doses of the new monoclonal antibody BMA 031. Transplant. Proc. 22, 1789–1790 (1990).
283. Fisher, A., Blanche, S., Le Deist, F., et al. Prevention of graft failure by an anti-LFA-1 monoclonal antibody in HLA incompatible marrow transplantation. Bone Marrow Transplant. 3 (Suppl.), 204–205 (1988).
284. Maraninchi, D., Mawas, C., Reiffers, J., et al. Anti-LFA1 monoclonal antibody and bone marrow graft rejection in adults. Lancet 2, 579–580 (1988).
285. LeMauff, B., Hourmant, M., Rougier, J.-P., et al. Effect of anti-LFA1 (CD11a) monoclonal antibodies in acute rejection in human kidney transplantation. Transplantation 52, 291–296 (1991).
286. Baume, D.K., Kuentz, M., Pico J.-L., et al. Failure of a CD18/anti-LFA1 monoclonal antibody infusion to prevent graft rejection in leukemic patients receiving T-depleted allogeneic bone marrow transplantation. Transplantation 47, 472–474 (1989).
287. Flavin, T., Ivens, K., Rothlein, R., et al. Monoclonal antibodies against intercellular adhesion molecule 1 prolong cardiac allograft survival in cynomolgus monkeys. Transplant. Proc. 23, 533–534 (1991).
288. Martin, P.J., Hansen, J.A., Anasetti, C., et al. Treatment of acute graft-versus-host disease with anti-CD3 monoclonal antibodies. Am. J. Kidney Dis. 11, 149–152 (1988).
289. Gleixner, B., Kolb, H.J., Holler, E., et al. Treatment of aGVHD with OKT3; Clinical outcome and side-effects associated with release of TNF alpha. Bone Marrow Transplant 8, 93–98 (1991).
290. Anasetti, C., Martin, P.J., Hansen, J.A., et al. A Phase I-II study evaluating the murine anti-IL-2 receptor antibody 2A3 for treatment of acute graft-versus-host disease. Transplantation 50, 49–54 (1990).
291. Blaise, D., Olive, D., Hirn, M., et al. Prevention of acute GVHD by in vivo use of anti-interleukin-2 receptor monoclonal antibody (33B3.1): A feasibility trial in 15 patients. Bone Marrow Transplant. 8, 105–111 (1991).
292. Byers, V.S., Henslee, P.J., Kernan, N.A., et al. Use of an anti-pan T-lymphocyte ricin in a chain immunotoxin in steroid-resistant acute graft-versus-host disease. Blood 75, 1426–1432 (1990).
293. Gold, H.K., Gimple, L.W., Yasuda, T., et al. Pharmacodynamic study of $F(ab')_2$ fragments of murine monoclonal antibody 7E3 directed against human platelet glycoprotein IIb/IIa in patients with unstable angina pectoris. J. Clin. Invest. 86, 651–659 (1990).
294. Anderson, H.V., Revana, M., Rosales, O., et al. Intravenous administration of monoclonal antibody to the platelet gp IIb/IIa receptor to treat abrupt closure during coronary angioplasty. Am. J. Cardiol. 69, 1373–1376 (1992).
295. Sears, H.F., Atkinson, B., Mattis, J., et al. The use of monoclonal antibody in phase I clinical trial of human gastrointestinal tumors. Lancet 1, 762–765 (1982).

296. Sears, H.F., Herlyn, D., Steplewski, Z., et al. Phase II clinical trial of a murine antibody cytotoxic for gastrointestinal adenocarcinoma. Cancer Res. 45, 5910–5913 (1985).
297. Sears, H.G., Herlyn, D., Steplewski, Z., et al. Effects of monoclonal antibody immunotherapy in patients with gastrointestinal adenocarcinoma. J. Biol. Response Mod. 3, 138–150 (1984).
298. Vadhan-Raj, S., Cordon-Cardo, C., Carswell, E.A., et al. Phase I trial of a mouse monoclonal antibody against GD3 ganglioside in patients with melanoma: Induction of inflammatory responses at tumor sites. J. Clin. Oncol. 6, 1636–1648 (1988).
299. Goodman, G.E., Hellström, I., Hummel, D., Brodzinsky, L., Yeh, M.Y., Hellström, K.E. Phase I trial of monoclonal antibody MG-21 directed against a melanoma associated GD3 ganglioside antigen (abstract). Proc. Am. Soc. Clin. Oncol. 6, 209 (1987).
300. Lichtin, A., Iliopoulos, D., Guerry, D., Elder, D., Herlyn, D., Steplewski, Z. Therapy of melanoma with an anti-melanoma ganglioside monoclonal antibody; A possible mechanism of a complete response (abstract). Proc. Am. Soc. Clin. Oncol. 7, 247 (1988).
301. Cheung, N.V., Lazarus, H., Miraldi, F.D., et al. Ganglioside GD2 specific monoclonal antibody 3F8: A phase I study in patients with neuroblastoma and malignant melanoma. J. Clin. Oncol. 5, 1430–1440 (1987).
302. Saleh, M.N., Khazaeli, M.B., Wheeler, R.H., et al. phase I trial of murine monoclonal anti-GD2 antibody 14G2a in metastatic melanoma. Cancer Res. 52, 4342–4347 (1992).
303. Oldham, R.K., Foon, K.A., Morgan, A.C., et al. Monoclonal antibody therapy of malignant melanoma: In vivo localization in cutaneous metastasis after intravenous administration. J. Clin. Oncol. 2, 1235–1244 (1984).
304. Goodman, G.E., Beaumier, P., Hellstrom, I., Fernyhough, B., Hellstrom, K.E. Pilot trial of murine monoclonal antibodies in patients with advanced melanoma. J. Clin. Oncol. 3, 340 (1985).
305. Goodman, G.E., Hellstrom, I., Brodzinsky, L., et al. Phase I trial of murine monoclonal antibody L6 in breast, colon, ovarian, and lung cancer. J. Clin. Oncol. 8, 1083–1092 (1990).
306. Ritz, J., Pesando, J.M., Sallan, S.E., et al. Serotherapy of acute lymphoblastic leukemia with monoclonal antibody. Blood 58, 141–152 (1981).
307. Scheinberg, D.A., Tanimoto, M., McKenzie, S., Strife, A., Old, L.J., Clarkson, B.D. Monoclonal antibody M195: A diagnostic marker for acute myelogenous leukemia. Leukemia 3, 440–445 (1989).
308. Schroff, R.W., Farrell, M.M., Klein, R.A., et al. T65 antigen modulation in a phase I monoclonal antibody trial with chronic lymphocytic leukemia patients. J. Immunol. 133, 1641–1648 (1984).
309. Betram, J.H., Gill, P.S., Levine, A.M., et al. Monoclonal antibody T101 in T cell malignancies: A clinical, pharmacokinetic, and immunologic correlation. Blood 68, 752–761 (1986).
310. Dillman, R.O., Shawler, D.L., Dillmann, J.B., et al. Therapy of chronic lymphocytic leukemia and cutaneous T-cell lymphoma with T101 monoclonal antibody. J. Clin. Oncol. 2, 881–891 (1984).
311. Dillman, R.O., Shawler, D.L., Sobel, R.E., et al. Murine monoclonal antibody therapy in two patients with chronic lymphocytic leukemia. Blood 59, 1036–1045 (1982).

312. Foon, K.A., Schroff, R.W., Bunn, P.A., et al. Effects of monoclonal antibody therapy in patients with chronic lymphocytic leukemia. Blood 64, 1085–1093 (1984).

313. Dyer, M.J., Hale, G., Hayhoe, F.G., et al. Effects of CAMPATH-1 antibodies in vivo in patients with lymphoid malignancies. Influence of antibody isotype. Blood 73, 1431–1439 (1989).

314. Ziegler, L.D., Palazzolo, P., Cunningham, J., et al. Phase I trial of murine monoclonal antibody L6 in combination with subcutaneous interleukin-2 in patients with advanced carcinoma of the breast, colorectum, and lung. J. Clin. Oncol. 10, 1470–1478 (1992).

315. LoBuglio, A.F., Wheeler, R.H., Trang, J., et al. Mouse/human chimeric monoclonal antibody in man: Kinetics and immune response. Proc. Natl. Acad. Sci. USA 86, 4220–4224 (1989).

316. Meredith, R.F., LoBuglio, A.F., Plott, W.E., et al. Pharmacokinetic, immune response, and biodistribution of iodine-131-labeled chimeric mouse/human IgG1,κ 17-1A monoclonal antibody. J. Nucl. Med. 32, 1162–1168 (1991).

317. Khazaeli, M.D., Saleh, M.N., Liu, T.P., et al. Pharmacokinetics and immune response of 131I-chimeric mouse/human B72.3 (human $\gamma 4$) monoclonal antibody in humans. Cancer Res. 51, 5461–5466 (1991).

318. Saleh, M.N., Khazaeli, M.D., Wheeler, R.H., et al. Phase I trial of the chimeric anti-GD2 monoclonal antibody ch14.18 in patients with malignant melanoma. Hum. Antibody Hybridomas 3, 19–23 (1992).

319. Irie, R.F., Morton, D.L., Regression of cutaneous metastatic melanoma by intralesional injection with human monoclonal antibody to ganglioside GD2. Proc. Natl. Acad. Sci. USA 83, 8694–8698 (1986).

320. Caron, P.C., Scott, A., Graham, M., et al. Specific targeting of humanized anti-CD33 monoclonal antibody M195. A phase I trial for acute myelogenous leukemia (abstract). Proc. Am. Soc. Clin. Oncol. (in press).

321. Halpern, S.E., Parker, B.A., Vassos, A., Frincke, J.M. ^{90}Yttrium (^{90}Y) anti-idiotype monoclonal antibody therapy of non-Hodgkin's lymphoma (abstract). J. Nucl. Med. 39, 778 (1989).

322. Spitler, L.E., del Rio, M., Khentigan, A., et al. Therapy of patients with malignant melanoma using a monoclonal antimelanoma antibody-ricin A chain immunotoxin. Cancer Res. 47, 1717–1723 (1987).

323. Byers, V.S., Rodvien, R., Grant, K., et al. Phase I study of monclonal antibody-ricin A chain immunotoxin XomaZyme-791 in patients with metastatic colon cancer. Cancer Res. 49, 6153–6160 (1989).

324. Goldberg, D., Morel, P., Chatenoud, L., et al. Immunological effects of high dose administration of anti-CD4 antibody in rheumatoid arthritis patients. J. Autoimmun. 4, 617–630 (1991).

325. Strand, V. The emerging role of biologics in rheumatic disease. J. Rheum. 19(Suppl. 33), 49–45 (1992).

326. Panayi, G.S. Anti-CD7 monoclonal antibodies—Additional remarks. In Strand V. (ed.) Proceedings: Early decisions in DMARD development II (p. 94). Atalanta: Arthritis Foundation, 1991.

327. Kyle, V., Coughlan, R.J., Tighe, H., Waldmann, H., Hazleman, B.L. Beneficial effect of monoclonal antibody to interleukin 2 receptor on activated T cells in rheumatoid arthritis. Ann. Rheum. Dis. 48,428–429 (1989).

CHAPTER 2.2

THE ROLE OF MONOCLONAL ANTIBODIES IN THE ADVANCEMENT OF IMMUNOASSAY TECHNOLOGY

M.J. PERRY
Celltech Ltd., Slough, Berkshire, SL1 4EN, UK

2.2.1. INTRODUCTION

The invention of monoclonal antibodies [1] occurred at a time when essentially two immunoassay technologies were available. One was radioimmunoassay (RIA), introduced by Yalow and Berson [2] in 1960 and well established as a routine analytical method; the other was the two-site, or sandwich, immunoradiometric assay (IRMA), first reported in 1971 by Addison and Hales [3] but struggling to find significant usage. However, in the space of 10 years monoclonal antibodies revolutionized immunoassay, established the IRMA as the predominant technology, and spawned a generation of conceptually novel methods. To understand why monoclonals had such an impact it is helpful to review the two methods discussed above (see Figs. 2.2.1 and 2.2.2).

RIAs are competitive assays; that is, the sample analyte is determined as a result of its ability to compete with ^{125}I-labeled analyte for a limited concentration of specific antibody. However, from both practical and theoretical considerations [4] it was apparent that the sensitivity of such assays was dictated, and hence limited, by the affinity of the antibody, since for maximal sensitivity the optimal amount of antibody tends towards zero, whence its affinity must be high. With the IRMA, however, a large excess of antibody is employed, and hence by the law of mass action the sensitivity of the assay is not limited by the affinity of the antibodies. Thus, for maximal sensitivity the optimal amount of antibody tends to infinity [5]. In the two-site IRMA format two antibodies are employed that recognize spatially separated epitopes on the analyte. The first antibody is coupled to a solid phase, such as agarose or plastic, and is present in great excess concentration over the analyte. It thus effectively extracts the analyte from solution. A second antibody,

Monoclonal Antibodies: Principles and Applications, pages 107–120
© 1995 Wiley-Liss, Inc.

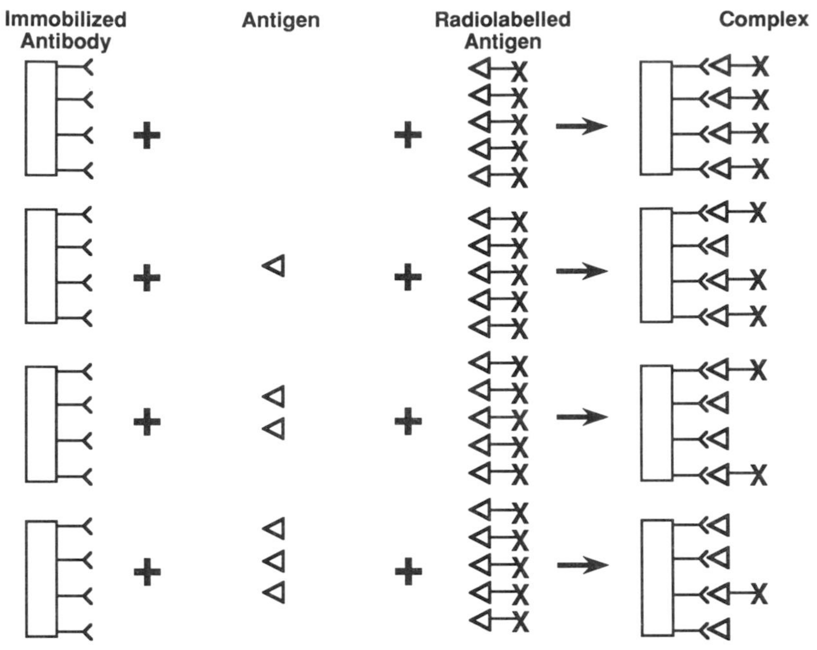

Fig. 2.2.1. Schematic representation of the RIA, showing from top to bottom the sequence of events with increasing amounts of antigen. Increased antigen results in a reduced amount of bound label. The immobilization of the antibody can be achieved by direct attachment as shown here or more commonly by reaction with a solid phase reagent that binds to the Fc region of the antibody.

carrying the label, is then added and binds to the solid phase in direct proportion to the concentration of the analyte. The remaining free label can be easily removed, since the "bound" complex is attached to a solid phase. In practice, the principle of the two-site IRMA has been modified to produce either an "inclusive" assay, in which the two antibodies are added together, or a "reverse" assay, in which the analyte is reacted with the labeled antibody before the addition of the solid-phase antibody. Unlike RIA, the IRMA uses excess antibody; thus, while the affinity of the antibodies is still of importance, the principle limitation of sensitivity is the specific activity of the label together with the ability to minimize the nonspecific binding (NSB) (i.e., non-analyte-mediated binding) of the label to the solid phase [5].

Not only did the IRMA offer potentially greater sensitivity but it was also a much faster assay, since at the low antibody concentrations necessary for maximum sensitivity the RIA frequently required 2–3 days to establish equilibration of binding, whereas the higher concentrations of antibody employed in the IRMA permitted assay times of 2–4 h. The use of excess reagents in the IRMA also had the advantage that acceptable assay performance could be achieved by using relatively low affinity antibodies, which failed to perform in the corresponding RIA systems. With so much going for it, why had the IRMA not been an instant success upon its

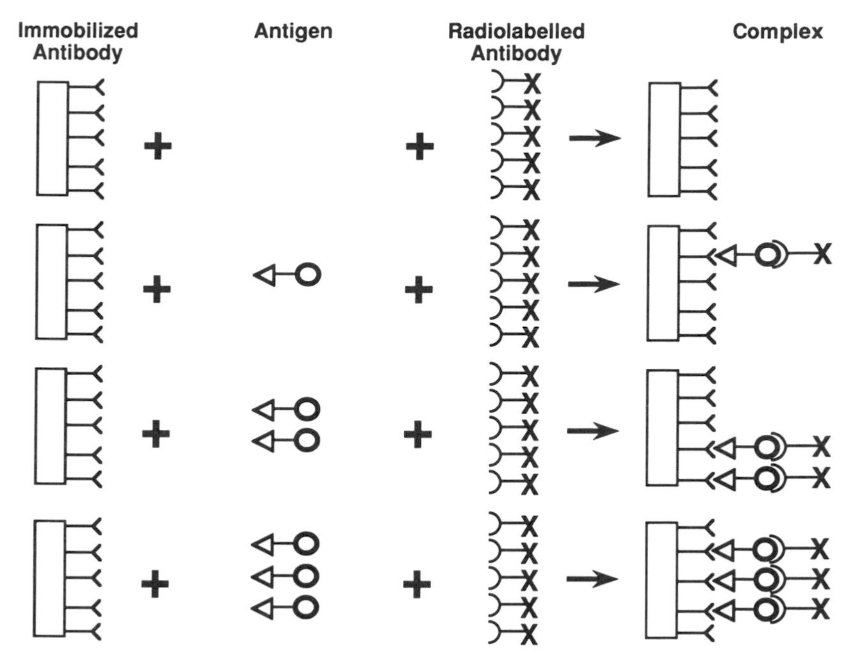

Fig. 2.2.2. Schematic representation of the two-site immunoradiometric assay (IRMA), showing from top to bottom the sequence of events with increasing amounts of antigen. Increased antigen results in an increased amount of bound radiolabel.

first report in 1971 [3]? The progress of the IRMA at that time was largely constrained by the relatively high amounts of pure antibody required. With polyclonal antisera as the sole source of antibody this demand could not be met, since affinity purification of the specific antibody gave low yields and was not applicable to high-affinity antibodies whose tightness of binding to the affinity column required elution with denaturing buffers.

Thus it was against this technology backdrop of a well-established but limited RIA and a potentially exciting but reagent-hamstrung IRMA that monoclonals appeared.

As would be predicted from their properties (Table 2.2.1), monoclonals had little to offer the RIA, since in terms of affinity they were if anything generally found to be inferior to polyclonals. This may in part reflect the fact that only a small proportion of the polyclonal antibodies need to be high affinity for the whole serum to apparently have high affinity. In addition the apparent high affinity (or more correctly avidity if studied under conditions other than a simple biomolecular reaction) of polyclonal antisera relative to monoclonal antibodies may reflect cooperativity. This phenomenon depends on antibody bivalency, which leads to the formation of multimeric complexes. It is envisaged that the high avidity of antibody binding in such complexes reflects the fact that antigen is attached simultaneously to more than one antibody rather than any change in the affinity of interaction at any

TABLE 2.2.1. Properties of Monoclonal Antibodies

Advantage

1. Unlimited supply of antibody with defined characteristics.
2. Easily purified by chromatography under mild conditions.
3. High purity often gives low background binding.
4. Production of fragments is easier and more controlled.
5. Antibodies can be generated to discrete epitopes on the same antigen.
6. Rapid equilibration in binding to antigen.
7. Monospecific antibody can be obtained even from an impure immunogen.

Disadvantages

1. Frequently of lower affinity than equivalent polyclonal antisera.
2. Specificity gives susceptibility to polymorphic variation of antigen and labile epitopes
3. Usually do not form good precipitation complexes with most antigens.

one epitope. Alternative mechanisms in which conformational changes influence one antibody so as to favor binding of the other to a modified epitope have been reported [6]. This "false" overestimate of polyclonal affinity is supported by two observations: (1) Mixtures of two monoclonals for different epitopes on a given antigen demonstrate higher avidity than that of the individual monoclonals; and (2) a relatively long time (often days at 4°C) is required for polyclonal sera to reach binding equilibrium with antigen at low concentration compared with monoclonals, which achieve equilibrium much faster (typically hours), presumably reflecting a simpler bimolecular reaction. In the future it may become a realistic option to use protein engineering techniques to increase the affinity of monoclonal antibodies. It is already possible [7] to produce Fv fragments by recombinant technology and to link these to give multimeric binding entities with increased affinities compared with the parent immunoglobulin.

However, with respect to the two-site assay monoclonals were exactly what the researcher had been waiting for to exploit the potential of the format. For the first time an unlimited supply of the base raw material was available. Milligram (ascitic fluid) to gram (large-scale fermenters) quantities of antibody were available. Moreover, they were easily purified (ion exchange; protein A or G) under mild conditions with good yield. The availability of large amounts of pure homogeneous protein meant that researchers could explore the replacement of ^{125}I with labels of greater activity. New chemical strategies to covalently link these molecules to the antibody could more easily be optimized to maintain maximum activity of both label and antibody. The production of antibody fragments such as Fab_2 from monoclonals was much easier to optimize than from polyclonals; in production from polyclonals different susceptibilities of the antibodies to the protease resulted in variable and heterogeneous yields. The use of antibody fragments enabled NBS to be reduced and interference due to substances in the test sample that bind to Fc (e.g., rheumatoid factor) to be negated. This reduction of NSB allowed the full benefit of the increased specific activities of the new labels to be translated fully into increased assay sensitivity. It is now also possible to produce antibody fragments directly by recombinant DNA technology.

With unlimited amounts of homogeneous antibody the researchers' imagination was given a free rein to devise assay formats that were simpler and quicker to use. As a result of the experimental advances made during the 1980s, immunoassays today can deliver sensitivities down to 10^{-17} mole/L, compared with 10^{-12} mole/L for RIA with assay times determined in minutes as opposed to days. The fact that monospecific antibody can be obtained from impure immunogen has meant that this improved assay performance has been achieved against an ever increasing number of analytes, many of which would be difficult to measure using polyclonal serum. Improvements in ease of use have been equally dramatic, immunoassays no longer being the domain of skilled technicians in the laboratory but susceptible to accurate determinations of patients in the privacy of their own homes.

These improvements, which have all been driven by the availability of monoclonals, are discussed with examples in the following sections.

2.2.2. ASSAY SENSITIVITY

Given monoclonals of good avidity and a solid phase of low NSB, the detection limit of the two-site assay is largely determined by the specific activity of the labeled antibody [5]. The search to replace ^{125}I led to a investigation of multitude of compounds, some of which are shown in Table 2.2.2. These can be divided into three categories; (1) enzymes, (2) fluorescent molecules, and (3) chemiluminescent molecules. Enzymes were an obvious choice due to their ability to generate many molecules of product from a single molecule of label, in contrast to ^{125}I, with typically one detectable event per second per 7.5×10^6 labeled molecules. Horseradish peroxidase (HRP) [8] and alkaline phosphatase [9] have proved the favorite choice due to their ease of conjugation, robustness, and availability of suitable colorimetric substrates. While convenient, colorimetry does impose limitations due to the low absorption maximum of many chromogens and the limited dynamic range of colorimeters (i.e., maximum optical density of 2.0). Ishikawa [10] reported a detection limit of 1,000 amol of HRP by colorimetry compared with 10 amol with a fluorogenic substrate. A similar increase in sensitivity with fluorimetry has been

TABLE 2.2.2. Frequently Used Labels

Enzyme	Fluorescent
Alkaline phosphatase	Coumarin
β-D-galactosidase	Fluorescein
Horseradish peroxidase	Europium
Urease	Terbium
β-Lactamase	Samarium
	Phycoerythrin
Luminescent	Indirect
Isoluminol	Streptavidin
Acridinium	Avidia
	Biotin

reported for alkaline phosphatase [11]. The finding that the detection of HRP using the chemiluminescent substrate luminol can be enhanced some 500-fold by phenolic compounds [12] has resulted in the commercial availability of enhanced chemiluminescence enzyme immunoassay and reagents under the trade name Amerlite, from Amersham International.

A particularly novel and sensitive approach utilizes enzyme cycling in the detection of alkaline phosphatase [13]. In this system alkaline phosphatase converts NADP to NAD, which is the trigger for the redox cycle. The two enzymes that complete the cycle are alcohol dehydrogenase and diaphorase, which convert NAD to NADH and back to NAD. The net reaction is the reduction by ethanol of a tetrazolinum dye to the intensely colored formazan. Thus, since each phosphatase molecule produces many hundred NAD molecules and because each NAD coenzyme can be responsible for the conversion of many hundreds of secondary substrate molecules, the original phosphatase activity is amplified greatly, giving clinical detection limits of 0.01 amol for alkaline phosphatase and 0.46 amol for antigen. A similar concept for enzyme cascade, again using alkaline phosphatase, has been reported [14]. Alkaline phosphatase triggers an amplification cascade based on hydrolysis of synthetic flavin adenine dinucleotide phosphate that yields peroxide, which is detected with an HRP-catalyzed colorimetric assay. (Fig. 2.2.3).

Fluorescent labels were recognized as capable of yielding very high specific activities, since each labeled molecule can be excited to yield many photons in response to exposure to a high-energy light source. However, the principle problem with fluorescence is the background fluorescence generated by many biological substances and plastics. This interference can be negated by recourse to time-solve

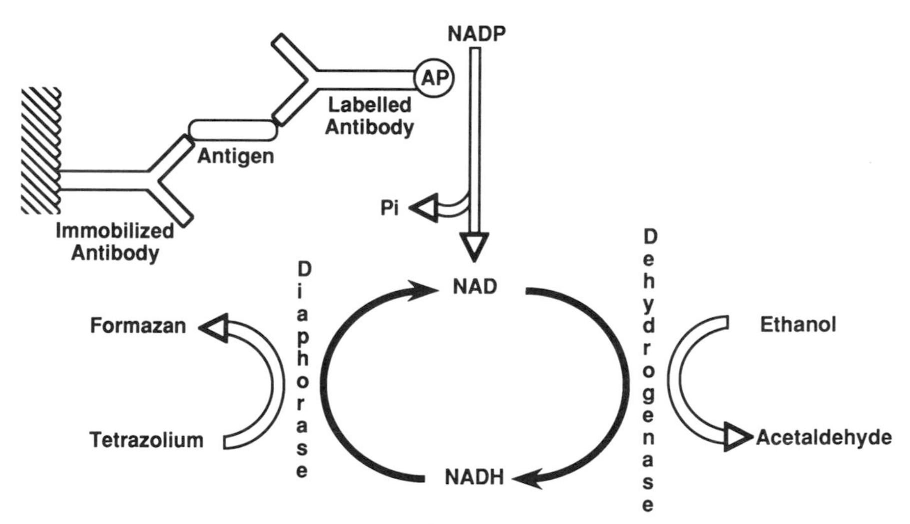

Fig. 2.2.3. Detection of alkaline phosphatase (AP) by enzyme cycling. Alkaline phosphatase, shown here as the label in a two-site (sandwich) immunoassay, converts NADP to NAD. The latter product acts as the trigger for the redox cycle in which alcohol dehydrogenase and diphorase act to oxidize a tetrazolium dye to a colored formazan by the reduction of ethanol.

techniques [15]. The fluorescence associated with serum proteins and other background components is characterized by lifetimes of approximately 10 ns, whereas the fluorescent lifetimes of the lanthanide chelates, such as europium, are 10–1,000 μs (i.e., some 3–4 orders of magnitude greater). This, coupled with their large Stokes shift (i.e., difference between the wavelengths for emitting and exciting radiations), means that using a time resolved fluorimeter with appropriate delay, counting, and cycle times, europium chelates can be detected at concentrations as low as 10^{-15} mole/L with immunoassay detecting analytes such as Thyrotropin (TSH) down to 10^{-13} mole/L. Indeed, the concept of time-resolved fluorimetry has been successfully commercialized and has gained widespread acceptance in the diagnostic market. Further increases in sensitivity by using this technology have recently been reported [16], in which alkaline phosphatase is used as the label to produce a fluorescent ternary complex containing the lanthanide terbium.

One of the first alternative labels to ^{125}I was the chemiluminescent acridinium ester [17,18], which gave specific activities of 10^6 photon counts per nanogram of labeled antibody, compared with approximately 10^4 radioactive counts per minute per nanogram with ^{125}I. Such a label was used in an assay for thyroid-stimulating hormone (TSH) with an antigen detection limit of approximately 5×10^{-13} mole/L, compared with 6×10^{-12} mole/L for a typical RIA, which for the first time allowed a clear distinction between TSH levels in thyrotoxic and euthyroid subjects.

A recurring approach to maximizing the final signal obtained in an immunoassay is to attach multiple molecules of the label to the component to be detected. The use of the high affinity of biotin ($K_D = 10^{15}$ L/mole) for avidin or preferably streptavidin, which due to its lower pI and lack of glcosylation exhibits lower nonspecific binding, has been a widely adopted approach to this problem [19,20]. The tag antibody can be easily conjugated with multiple biotin molecules (typically at least 10 : 1 (M : M) without a detrimental effect on antibody affinity. The bound biotiny lated antibody can then be revealed by reaction with streptavidin that is itself multiply labeled with the signal generator, be it enzyme or chemiluminophore/fluorophore. Thus, each molecule of biotinylated antibody bound yields many more molecules of label, providing a significant amplification. Alternatively the quadrivalency of streptavidin for biotin can be used to tag the biotinylated antibody with a complex formed with streptavidin and biotinylated enzyme. Antistreptavidin monoclonals can be used to further enhance sensitivity by selectively enlarging the streptavidin–biotin–label complex by bridging to a second layer of this complex [19].

However, the most impressive sensitivity has recently been obtained by marrying immunoassay to the polymerase chain reaction (PCR) [21]. PCR has become a powerful tool in molecular biology based on its ability to generate enormous amounts of a specific DNA segment flanked by a set of primers [22]. In the immuno-PCR technique [21] a streptavidin–protein A chimera is used to tag the immobilized antibody via the protein A–Fc interaction. A specific DNA fragment labeled with biotin is then bound through its interaction with streptavidin. Then a segment of the attached DNA is amplified by PCR and the PCR products are analyzed by electrophoresis with ethidium bromide staining. The authors claim this technique allows as few as 580 antigen molecules (9.6×10^{-22} moles) to be reproducibly detected. Given that the sensitivity of the system can be controlled by the number of PCR amplification cycles, future advances may permit the researcher to obtain the ultimate goal of detecting a single antigen molecules.

2.2.3. DEVELOPMENT OF NOVEL ASSAY FORMATS

While improvement in sensitivity is important to many areas such as viral diagnostics, for many routine analytes such as reproductive hormones it is an irrelevance, since adequate sensitivity has been achieved long ago. Increasing emphasis has been placed on developing assays that are easier to use, in the clinic, in the home, or to satisfy novel requirements in the research laboratory.

During the 1980s there was a realization that clinical diagnostic services could be made less expensive if it were possible to move some of the routine analysis, certainly of non-life-threatening cases such as pregnancy and fertility, out of the centralized laboratory and into the doctor's office or even the patient's home. To do this required radically different assay formats that were essentially homogeneous assays—that is, once the sample was added no further manipulation, such as washing to remove unbound label, was required. Fluorescent labels have proved popular with developers of homogenous systems, due to the ability of antibody–antigen reaction to modify the emission of light from the label. Fluorescence polarization assays were developed by Perkin Elmer in the early 1980s [23], while others adopted the approach of fluorescence energy transfer [24]. A novel europium chelate was the basis for another assay [25], in which the binding of an antihapten antibody quenches fluorescence of the conjugated hapten. In a similar vein Amersham International has commercialized a scintillation proximity immunoassay (SPA) [26]. In SPA, antibody is immobilized on fluorescent microspheres such that bound (but not free) label (either ^{125}I or ^{3}H) can efficiently elicit scintillations. A particular advantage of this method is that it is equally applicable to both haptens or large-molecular-weight antigens.

Despite the undoubted sensitivity and reliability of the above methods they do require dedicated instrumentation to detect the label, which limits the application to the larger doctor's office. For this reason enzyme labels with their ability to generate colored endpoints that can be determined by eye or, for increased precision, by a simple inexpensive colorimeter have proved most popular. Recombinant DNA technology [27] was exploited to produce new large, inactive fragments and small, inactive fragments of β-D-galactosidase that when combined gave a fully active enzyme. These fragments were used to establish a combined enzyme donor immunoassay (CEDIA) such that hapten-linked small fragments could combine with the larger enzyme fragment but not if it was bound by antihapten antibody. Thus, competition of sample analyte with the hapten enzyme fragments modulated total enzyme activity.

Another development that is targeted at assays for use in the home or the doctor's office is the immunochromatographic test strip, which requires only one operation (application of the sample), is self-timed, and is complete within 10 min [28,29]. The immunostrip is an integral device that contains all the necessary reagents for carrying out automatically any one of a variety of formats of heterogeneous enzyme immunoassay. The essential elements of the immunostrip are a small bibulous paper strip (15.5 × 1 × 0.1 cm) on which immunochemicals have been deposited in a stable dry chemistry form; and, at one end of the paper strip, a reservoir containing approximately 2 ml of a developing solution, which is physically separated from the dry chemistry component by a rupturable membrane. The immunochemicals are laid down as discrete transverse zones along the length of the strip. Near the reservoir end is a sample location zone for receiving an aliquot of the sample to be tested. In the midregion of the strip is a zone, called the measuring location, at which the test result is observed. Near the end of the strip distal to the reservoir is an indicator zone, which acts as a procedural control and end-of-test signal. The immunostrip is operated by applying an aliquot of sample to the sample location, and the test is initiated by application of finger pressure to the reservoir. This action ruptures the reservoir membrane and allows developing solution to contact the paper strip. Under capillary action, developing solution is drawn along the paper strip from one end to the other. As it does so, sample and resolubilized immunochemicals mix and migrate along the strip, reacting as they do so in a predetermined manner. Some 10 min later a blue color appears at the indicator zone, signaling that the test can be read at the measuring location. The presence or absence of a blue color or the intensity of the blue color at the measuring location indicates the test result. (Fig. 2.2.4).

Availability of submicron colored latex particles has allowed a somewhat simpler form of this assay to be commercialized by Unipath for pregnancy testing, that is, measurement of human chorionic gonadotrophin (hCG) [30]. The essential features of the test are an absorbent wick that is held in contact with a porous membrane material. The membrane has on it three separate zones of antibody. When the sampler wick, which is held in contact with and at the bottom of a porous membrane material, is saturated with urine, this passes along the membrane material until it reaches the mobile zone. This zone consists of colored particles sensitized with monoclonal antibody to the alpha subunit of hCG. The urine picks up the colored particles and carries it to the second zone, which is located directly below the large window. The second zone is composed of monoclonal antibody to the beta subunit of hCG immobilized in the membrane. If the urine contain hCG, the colored particles will bind, causing the formation of a blue line in the large window. If no hCG is present, all of the color flows past the large window. The urine continues to move along the strip and comes into contact with the third zone (below the small window). Excess or unreacted particles passing the first zone will be trapped, leading to the appearance of a blue line in the small window every time, whether hCG is present or not. Thus a positive result is indicated by a line in the large (test) and small (control) windows. A negative result is indicated by a line in the small window only.

An alternative method of homogeneous immunoassays that is proving useful in the research laboratory is based on monitoring the antibody–antigen reaction at a

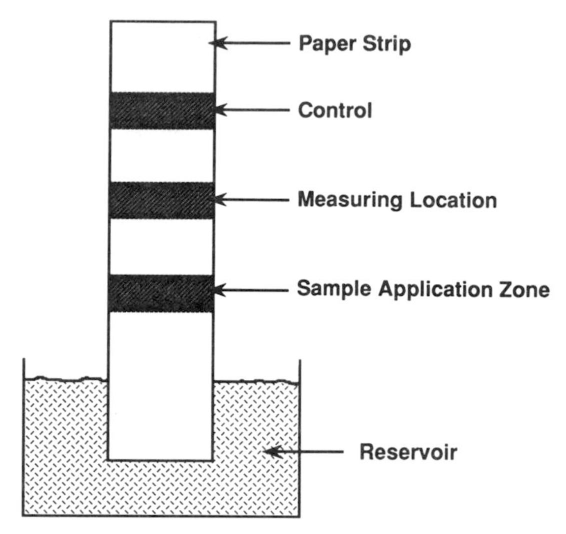

Fig. 2.2.4. Immunochromatographic strip assay. Schematic representation of the immunochromatographic test strip set up for a two-site immunoassay. Sample is applied to the sample application zone, which also contains an enzyme-labeled antibody. Rupture of a membrane allows the developing solution containing a substrate for the enzyme to flow along the strip. In doing so the sample and enzyme antibody conjugate are transported to the measuring location, where a second antibody is immobilized on the paper. The immune complex becomes bound at the measuring location, while excess enzyme conjugate is carried to the control zone, where it is immobilized by an anti-Fc interaction. The substrate in the developing solution is then converted to a colored product at both the control and measuring zones. The amount of color formed at the measuring location is proportional to the amount of antigen present in the sample.

continuous surface that is an integral part of a sensing mechanism. The basic concept of these immunosensors involves the immobilization of either the antibody or antigen to the sensor surface, which is constructed in such a manner that some property of the surface changes when an antigen–antibody reaction occurs there. The antibody–antigen interaction at the sensor surface can be monitored by one of the following methods: (1) Change in refractive index or layer thickness. (2) Attenuation of light due to the introduction of specific absorbing groups at the surface. (3) Increase in fluorescence due to binding of a specific fluorophore. (4) Quenching of fluorescence at the surface.

By far the most successful method to date has been evanescent wave immunoassays that utilize the internal reflection systems of (1) and (2) [31,32]. In particular two systems, BIAcore from Pharmacia and IAsys from Fisons, are now commercially available. One partner of the immunological binding reaction is immobilized onto the surface of a solid support that has certain optical characteristics. These enable the specific antibody–ligand interaction that occurs at the surface to be measured by internal reflection spectroscopy (IRS). This technique can be used to monitor surface reactions that occur within a distance from the surface of the order of a wavelength of light. The solid support is optically denser than the aqueous medium containing the binding partner. Thus, when a light beam is totally internally reflected within the optically denser medium, a unique waveform known as the evanescent wave is generated in the optically less dense medium. This evanescent wave penetrates a fraction of a light wavelength into the optically less dense medium and in doing so interacts only with compounds close to or at the surface. Thus the evanescent wave is the sensing component, being able to determine changes occurring at the sensor surface due to the specific antibody–ligand interaction. The two most common optical techniques employed to measure the interaction of the evanescent wave with molecules close to the surface are attenuated total reflection (ATR) and total internal reflectance fluorescence (TIRF). In ATR the antibody–ligand interaction forms a new optically absorbing film at the surface. This film absorbs energy from the evanescent wave, which can be measured as an attenuation of the internally reflected light beam. With TIRF the soluble binding partner carries a fluorescent label. As with ATR, binding to the immobilized partner results in absorption of energy from the evanescent wave, but in this case the energy is reemitted at a longer wavelength as fluorescence. The disadvantage of TIRF compared with ATR is that it requires a fluorescent analogue of one of the components of the binding reaction and is thus generally limited to competitive type assays.

An additional advantage of this technology apart from ease of use is that it allows the molecular interaction of the antibody–antigen to be monitored in real time. It thus permits accurate determination of association, dissociation, and equilibrium constants with the molecules in their native state, without the need for labels or prior purification of samples. This application is proving increasingly important in the research laboratory.

A key feature of these optical systems is the waveguide, the component at whose surface the interactions take place. Waveguides vary from simple reflection prisms to the more sophisticated optical fibers. Greater sensitivity can be achieved with thiner and longer waveguides, and for this reason much attention has been focused on the use of optical fibers. However, despite the advances made with optical fiber technology in the communications industry, successful optical fiber-based immunosensors have yet to be developed. If the problems associated with the manufacture of such devices can be overcome, then smaller, more compact equipment relative to current systems may facilitate the movement of immunosensors into areas such as the doctor's office and even permit *in vivo* analysis to continuously monitor selected analytes.

2.2.4. EXAMPLES OF SPECIFIC ANALYTE MEASUREMENTS MADE POSSIBLE WITH MONOCLONAL ANTIBODIES

One of the earliest advantages recognized for monoclonals was the fact that they could be obtained for unique epitopes on a given antigen. This situation was much harder to achieve with polyclonal antibodies, which occur as a complex mixture of multiple epitopes and were difficult to resolve except by repeated affinity purification, with all the associated problems discussed previously. This specificity of monoclonals is particularly valuable when the target antigen is found in the presence of potentially cross-reacting compounds, whether these are naturally occurring homologues or degradation products. The assay of human chorionic gonadotrophin was one of the first examples of the value of monoclonal specificity. This hormone has common α subunits with thyrotropin (TSH) and lutropin (LH) and ~80% homology of the β subunits with LH. Immunization with intact hCG enabled a panel of monoclonals to be established [33]. Some were specific for hCG, and others cross-reacted strongly with LH, while specificity for intact α β structure as opposed to the free subunits was also obtained.

In a similar vein, monoclonals were used to discriminate between various forms of proinsulin [34]. This hormone is synthesized as a single polypeptide chain and proteolytically cleaved at two sites to remove c-peptide and leave the disulfide-linked A and B chains of insulin. Monoclonals specific to intact proinsulin and the partially proteolyzed forms enabled the authors to investigate the *in vivo* processing of proinsulin and the physiochemical nature of the circulating proinsulin.

The ability to select antibodies to defined spatially distinct epitopes has enabled much smaller proteins and peptides to be assayed by the two-site immunoassay than was initially believed feasible. Brown [35] generated a panel of monoclonals to the 37 amino acid calcitonin-gene-related peptide (CGRP). Monoclonals were obtained to the N terminus, midsection, and C terminus of thc peptide, permitting two-site assays to be set up that were specific for the intact molecule, and the N-terminus and C-terminus fragments. These assays were subsequently used to establish the pharmokinetic profile of CGRP in more detail than would have been possible with RIA and polyclonal sera [36].

Certain proteins are poor immunogens, frequently because they are closely conserved between different species, and with these targets polyclonal antibodies have not been successfully produced. However, certain computer programs [37] appear to be able to identify potential antigenic sequences within a protein, and subsequent immunization with peptides containing these sequences have yielded monoclonals to the intact protein [38]. This approach has recently [39] been put to good effect in obtaining monoclonals specific for inhibins and activins, proteins of considerable current interest as hormones, growth factors, and potential tumor markers [40,41].

2.2.5. SUMMARY

The availability in bulk of monoclonal antibodies as highly standardized reagents of exquisite specificity has made possible the development of a new generation of rapid and highly sensitive immunoassays—in some cases for antigens that could not easily be measured by other methods. The development of novel assay formats is extending the utility of antibody reagents both as research tools and as a robust means of measurement for use outside specialist laboratories. While the major use of monoclonal based immunoassays to date has been for clinical testing and research applications, we can expect to see their increasing use in areas such as environmental monitoring and food analysis.

2.2.6. REFERENCES

1. Kohler, G., and Milstein, C. Nature 256, 495 (1975).
2. Yalow, R.S., Berson, S.A. J. Clin Invest. 39, 1157 (1960).
3. Addison, G.M., Hales, C.N. In Kirkham, K.E., Hinks, W.M. (eds.): Radioimmunoassay Methods (p. 481). Edinburgh: Churchill, Livingstone, 1971.
4. Ekins, R.P., Newman, B., O'Riorden, J.L.H. In Hayes, R.C., Goswitz, F.A., Murphy, B.E.P. (eds.): Radioisotopes in Medicine In Vitro Studies. Oak Ridge Symposia (p. 59), 1968.
5. Jackson, T.M., Marshall, N.J., Elkins, R.P. In Immunoassays for Clinical Chemistry (p. 557). Edinburgh: Churchill, Livingstone, 1983.
6. Thompson, R.J., Jackson, A.P. Trends Biochem. Sci. 9, 1 (1984).
7. Schott, M.E., Milenic, D.E., Yokota, T., Whitlow, M., Wood, J.F., Fordyce, W.A., Cheng, R.C., Schlom, J. Cancer Res. 52, 6413 (1992).
8. Abdul-Ahad, W.F., Gosling, J.P. J. Reprod. Fertil. 80, 653 (1987).
9. Chiang, C.S., Grove, T., Cooper, M. Clin. Chem. 35, 946 (1989).
10. Ishikawa, E. Clin. Biochem. 20, 375 (1987).
11. Weng, L., Choo, S., Troiano, J., and Bernardino, R. Clin. Chem. 38, 947 (1992).
12. Thorpe, G.H.G., Kricka, L.J. Methods Enzymol. 133, 331 (1986).
13. Johannsson, A., Ellis, D.H., Bates, D.L. J. Immunol. Methods 87, 7–11 (1986).
14. Obzansky, D.M., Tseng, S.Y., Sererino, D.M. Clin. Chem. 38, 1099 (1992).
15. Barnard, G., Kohen, F., Mikda, H., Lougren, T. Clin. Chem. 35, 555 (1989).
16. Papanastasiou-Diamandi, A., Christopulos, T., Diamandis, E.P. Clin. Chem. 38, 545 (1992).
17. Weeks, I., Sturgess, M., Siddle, K., Jones, M.K., Woodhead, J.S. Clin Endocrinol. 20, 489 (1984).
18. Bender, H., Maier, A., Wiedermann, B. Clin. Chem. 38, 2267 (1992).
19. Hart, R.C., Taafle, L.R., J. Immunonol. Methods 101, 91 (1987).
20. Peters, J., Schmidt-Gayle, H., Peters, B. Clin. Chem. 35, 573 (1989).
21. Sano, T., Smith, C.L., Cantor, C.R. Science 258, 120 (1992).
22. Gibbs, R.A. Anal. Chem. 62, 1202 (1990).
23. Colbert, D.L., Gallacher, G., Mainwaring Burton, R.W. Clin. Chem. 31, 1193 (1985).
24. Calvin, J., Burling, K., Blow, C. J. Immunol. Methods 86, 249 (1986).
25. Barnard, G., Kohen, F., Mikola, H., Lougren, T. Clin. Chem. 35, 555 (1989).
26. Bosworth, N., Towers, P. Nature 341, 167 (1989).
27. Shindelman, J., Singh, H., Hertle, V., Davoudzadeh, F., Lingenfelter, D., Lour, R., Khanna, P. Clin. Chem. 38, 1078 (1992).
28. Perry, M.J. Patent W088/08536 (1988).
29. Kaneko, H., Koiwai, K., Hasegawa, A., Ono, M., Ashihara, Y. Clin. Chem. 38, 1093 (1992).
30. Davidson, I.W. Anal. Proc. 29, 459 (1992).
31. Mayo, C.S., Hallock, R.B. J. Immunol. Methods 120, 105 (1989).
32. Vadgama, P., Crump, P.W. Analyst 117, 1657–1670 (1992).
33. Saller, B., Clara, R., Spottl, G., Siddle, K. Clin. Chem. 36, 234 (1990).

34. Sobey, W.J., Beer, S.F., Carrington, C.A., Clark, P.M., Siddle, K. Biochem. J. 260, 535 (1989).
35. Brown, D. J. Immunol. Methods 154, 87 (1990).
36. Smith, B.J., Perry, M.J. In preparation.
37. Faller, D.C., De La Cruz, V.F. Nature 349, 720 (1991).
38. Groome, N.P. J. Endocrinol. 129, 1 (1991).
39. Groome, N.P., Lawrence, M. Hybridoma 10, 309 (1991).
40. De Kretser, D.M. Endocrinology 69, 17 (1990).
41. McLachland, R.L., Robertson, D.M., De Kretser, D.M., Burger, H.G. Clin. Endocrinol. 29, 77 (1988).

CHAPTER 2.3

IMMUNOAFFINITY PURIFICATION WITH MONOCLONAL ANTIBODIES

C.R. HILL
Celltech Biologics, Slough, Berkshire, SL1 4EN, UK

2.3.1. INTRODUCTION

Affinity chromatography is a form of adsorption chromatography in which the material to be purified undergoes a selective interaction with a specific molecule (ligand) that has been immobilized on a suitable insoluble support. When this interaction has taken place, unreactive impurities can be washed away, leaving the purified material, which can then be recovered by treatment with a reagent that will disrupt the specific interaction with the ligand. Affinity chromatography is a powerful technique that has the potential to produce high degrees of purification with high yield. Over the last 25 years the technique has found increasing application as both a research tool and manufacturing process unit operation. In a survey of purification literature carried out by Dunnill and Lilly [1], published in 1972, it was observed that few affinity unit operations were used and that when they were, they tended to be applied late in the purification scheme. However, in a similar survey carried out in 1986 [2] there was a much wider use of affinity techniques and they were being applied at stages throughout the purification scheme.

Immunoaffinity chromatography is a form of affinity chromatography in which the ligand is an antibody and the separation depends on the selective reaction between an antibody and its specific antigen, the product to be purified. In its general application the antibody is immobilized on a suitable inert support material and packed into a column. A solution containing the specific antigen is passed through the column. The antigen will bind to the immobilized antibody and impurities can be washed away. The captured antigen is then recovered by changing the composition of the mobile phase such that dissociation of the antibody–antigen complex takes place.

Immunopurification has the potential to provide a number of advantages over conventional biochemical methods. The antibody–antigen interaction has a high

Monoclonal Antibodies: Principles and Applications, pages 121–136
© 1995 Wiley-Liss, Inc.

degree of specificity, which enables the development of highly selective separation techniques that can give rise to very high purification factors [3]. Furthermore, as a result of the high purification factors, immunopurification can often replace several steps in a conventional separation method. This can lead to increased speed of separation, which can be important if the product is inherently unstable or is susceptible to degradation during the purification procedure. In addition, fewer purification steps should result in increased recovery [2]. There are few published examples in which a direct comparison has been made between a conventional purification method and a scheme that includes an immunopurification step. However, in a study with carcinoembryonic antigen, Ford et al. [4] demonstrated that, in obtaining the same purity and activity, the immunopurification method resulted in higher yield of product for less expenditure of time and effort.

For these reasons immunopurification is used widely in research laboratories for the rapid separation and purification of a very wide range of antigens. Its application for purification of proteins for human therapeutic use has been limited to products that are of high commercial value, are required in small quantity, and are difficult to purify by conventional means (for example, plasma derived proteins such as Factor VIII [5] and Factor IX [6]). The reasons for this have been based on concerns that leakage of ligand from immunoaffinity columns might lead to the potential for contamination of product, on the regulatory requirements for antibodies derived from animal cells that are used for immunopurification [7,8], and on the cost of manufacturing the antibody ligand. The examples given above demonstrate that the first two of these concerns have been adequately met. The wider application of the technique will be dependent on the future economics of antibody ligand manufacture.

There are a number of key stages in the development of an efficient immunopurification unit operation.

1. Selection of antibody ligand
2. Selection of support matrix
3. Selection of immobilization chemistry
4. Development of an antigen elution procedure
5. Optimization of antibody utilization

Each of these will be dealt with below. Examples will be given of a number of applications of immunopurification, including its use in the manufacture of proteins for human therapeutic use.

2.3.2. SELECTION OF ANTIBODY LIGAND

2.3.2.1. Advantages of Using Monoclonal Rather Than Polyclonal Antibodies

Since the early reports describing immunopurification [9] a large number of applications of the technique have appeared in the literature. However, it was not until the

TABLE 2.3.1. Comparison of the Properties of Monoclonal and Polyclonal Antibodies[1]

	Polyclonal	Monoclonal
Antigen	Need to obtain large quantities of purified antigen	Need to obtain small quantities of less pure antigen
Antibody	Variable affinity, specificity, class	Defined affinity, specificity, class
Quality	Batch to batch variation	Consistent quality
Production and availability	Limited by availability of appropriate animals	Large quantities readily available from large scale fermentation (Chapter 5)

[1]Until recently only polyclonal antibodies were available for immunopurification. However, the properties of monoclonal antibodies make them more suitable as ligands for immunopurification.

discovery of hybridoma technology [10] and the production of monoclonal antibodies that the full potential of the technique could be realized.

A number of advantages can be gained by using monoclonal antibodies. These are summarized in Table 2.3.1. Monoclonal antibodies can be made relatively easily with small quantities of antigens that do not usually need to be highly purified. In contrast, to prepare polyclonal antibodies much larger quantities of antigen are required that need to be highly purified in order to reduce undesirable cross-reactivities. Polyclonal antibody preparations may contain numerous different antibodies that can be directed to different epitopes on the antigen with different affinities, whereas a monoclonal antibody preparation recognizes a single epitope with a defined affinity. In practical terms this should result in the development of a more effective immunopurification reagent. A major advantage that can be gained from using monoclonal antibodies is that a panel of antibodies can be generated that contains antibodies with a range of different properties (e.g., epitope binding and affinity). Application of appropriate screening should then allow selection of an antibody with properties best suited to any particular application. A further major advantage of using monoclonal antibodies, particularly for manufacturing applications, is that they can be prepared in large quantities [11] to a predefined and constant specification.

2.3.2.2. Selection of Monoclonal Antibodies for Immunopurification

The use of hybridoma technology to generate monoclonal antibodies can result in a large panel of antibodies of which only a small minority will have the properties required for any particular application. The key to developing an efficient immunopurification reagent is the selection at this stage of the antibody with the most appropriate properties. This can be achieved by the application of a range of screening assays.

There are a number of stages during the development of an immunopurification reagent at which decisions regarding the selection of an antibody can be made.

TABLE 2.3.2. Criteria for Selection of Monoclonal Antibodies for Immunopurification

Screening Stage	Selection Criteria	Typical Number of Lines at this Stage
Hybridoma cell lines	Antibody specificity[1] Antibody affinity	50–200
Cell culture	Growth kinetics[2] Nutritional requirements Antibody production	10–20
Purification of antibody	Antibody yield[3] Physicochemical properties Ig class Antibody stability	8–10
Immunopurification reagent development	Purity of eluted antigen[4] Elution buffer Reusability	6–8

[1]Select those antibodies that recognize antigen and that have appropriate affinity to allow efficient binding of antigen to the immobilized antibody and recovery of the purified product.
[2]Select only those cell lines that grow well and secrete high levels of antibody.
[3]Select only those antibodies that are straightforward to purify and that are stable.
[4]This is the first opportunity to evaluate the properties of the antibody after it has been immobilized to the selected support. The final selection of antibody will be made following evaluation of the purity and yield of antigen, the elution conditions required to recover the antigen, and the stability of both the immobilized antibody and the antigen under the selected operating conditions.

These are summarized in Table 2.3.2 along with selection criteria that can be used at each stage.

The first step in the selection process is to identify from a large number of cell lines (typically 50–200) the cell lines that secrete antibodies that have appropriate specificity and affinity for the antigen. If the association constant for antibody and antigen is too high, it may be difficult to recover the antigen from the antibody. However, if the association constant is too low, antigen may not bind efficiently to the immobilized antibody. An association constant of 10^4 M to 10^8 M is considered to be ideal by Phillips [12]. Specific assays for screening hybridoma cell lines can be developed to identify those lines that secrete antibody with a suitable affinity to allow efficient binding and elution of antigen. This approach has been taken by a number of workers [13–15]. In some cases it may be possible to identify antibodies that switch from high affinity to low affinity with only a small change in pH [16], which would then allow elution of antigen under relatively mild conditions. At stage 10–20 cell lines may be identified for further development.

The screening of cell lines described above will have been carried out with small quantities of antibody, typically obtained from cultures grown in 96-well microtiter plates. The next stage will require larger quantities of antibody (10–100 mg). During the expansion of the selected cell lines, information can be obtained on growth characteristics, nutritional requirements, and antibody yield. Those lines that exhibit poor growth characteristics or low antibody yield can be eliminated at this stage. During the purification of antibody from the remaining lines, information can be obtained on physicochemical properties such as isoelectric point, solubility, and stability. Antibodies that are difficult to purify or are unstable can be eliminated at this stage.

Typically six to eight cell lines can be identified for the final stage of screening. The final stage of the selection process involves an assessment of the properties of the antibody after it has been immobilized to a suitable matrix. The important criteria are purity of eluted antigen, the elution conditions required [7,16–20], and the stability of both the antigen and the immobilized antibody under the selected operating conditions.

2.3.3. SELECTION OF SUPPORT MATRIX

A wide range of matrix materials is available that can be used for affinity chromatography. Several recent reviews of this subject contain lists of suitable matrices and their suppliers [21–23]. The selection of matrix type will be dictated to a large extent by the use for which the matrix is required. Traditionally, affinity separations have been carried out at low pressure with agarose- or polyacrylamide-based supports. More recently, with the advent of high-performance separations [24] and larger-scale separations, more rigid supports based on synthetic polymers and silica have been developed, although the ideal matrix has yet to be synthesized [21]. The properties that influence the choice of a matrix are listed in Table 2.3.3. The majority of these properties are required of the support material used in any type of chromatographic separation. A specific requirement of matrices to be used for affinity chromatography is that the material be easily chemically derivatized to allow covalent attachment of ligand.

TABLE 2.3.3. Important Factors in the Choice of a Support Matrix for Immunopurification[1]

1. Insolubility
2. Hydrophilicity
3. Porosity/permeability
4. Rigidity
5. Resistance to chemical and microbial degradation
6. Ease of chemical derivatisation
7. Resistance to abrasion damage
8. Cost and lifetime

[1]The majority of these properties are important for a matrix to be used for any form of chromatography. The one property that is specific to a matrix to be used for affinity chromatography is that it can be easily chemically derivatized to allow covalent attachment of the ligand.

2.3.4. SELECTION OF COUPLING CHEMISTRY

A great variety of coupling chemistries are available that can be used to activate the support material to allow covalent attachment of the antibody to be used for immunoaffinity chromatography. These have been extensively reviewed [25,26]. Many of these chemistries are available as preactivated supports from a wide range of commercial suppliers. Lists of the suppliers and preactivated supports available can be found in references Angal and Dean [22] and Kenney et al. [23]. Covalent coupling of proteins to support materials will generally be through available primary amine, sulphydryl, or hydroxyl groups. The pH of the coupling reaction, speed of coupling, the type and stability of the linkage formed, and the potential for nonspecific interactions will be dependent on the coupling chemistry used [21]. Comparative studies of a number of matrices and coupling chemistries can be found in Fowell and Chase [27], Desai and Lyddiatt [28], Highsmith et al. [29], and Thanakan et al. [30].

2.3.5. DEVELOPMENT OF AN ANTIGEN ELUTION PROCEDURE

The forces that hold the antibody–antigen complex together are a combination of weak physical forces. In order of strength these are ion attraction, hydrogen bonds, hydrophobic interaction, and van de Waals forces. These forces must be disrupted to bring about elution of the antigen. This is usually accomplished by a change in pH, ionic strength, dielectric constant, surface tension, or water structure, or by a combination of these [12]. The strength of the antibody–antigen interaction means that generally harsh conditions are required to disrupt the interaction. Typically these are buffers at pH of less than 3 or strong solutions of chaotropic ions such as 3 M sodium thiocyanate. The stability of both the antigen and the immobilized antibody in the presence of the selected elution buffer must be borne in mind at this stage. There are some examples where less severe eluants have been used. Hill [31] and Andersson et al. [32] have described the use of dioxane and ethylene glycol, which act by disrupting hydrophobic attractive forces. Olson [33] has demonstrated that the application of high pressure can lead to elution of antigen. Electrophoretic desorption has been used [34] and there have been several reports describing the use of very low ionic strength buffers [35] or even water [36]. An alternative strategy has been described by Ohlson [37], who identified low-affinity antibodies ($K_a < 10^4$ M) and were able to elute the antigen under mild isocratic conditions. These examples demonstrate that, with selection of appropriate antibodies, relatively mild elution agents can be used. With the development of specific assays it should be possible to identify such antibodies at the stage of screening hybridoma cell lines as described above. Bartholomew [16] used this approach to identify antibodies that switched from high to low affinity with only a small change in pH.

Other examples of mild elution conditions have been described for specific antigens. Conformation-specific antibodies to the calcium-ion-stabilized complexes of Protein C [38] and Factor IX [39] have been used to purify these proteins. Removal of calcium by treatment with EDTA results in a conformation change that leads to elution of the antigen. Elution by competition with excess free antigen (trimethylguanosine) has been used successfully by Krainer [40] to elute ribonucleoprotein particles from an immobilized antitrimethylguanosine monoclonal antibody.

2.3.6. EFFICIENCY OF ANTIBODY UTILIZATION

The efficiency of antibody utilization can be defined as the number of antigen molecules bound per unit volume of matrix (working capacity) divided by the theoretical maximum capacity, expressed as a percentage. The theoretical maximum capacity is the number of molecules of immobilized antibody multiplied by the number of binding sites per molecule of antibody (two for an IgG). The working capacity of the immunopurification reagent is dependent on the accessibility of antibody binding sites to passing antigen molecules. This in turn is dependent on the structure of the matrix material [41], the coupling chemistry used [42], and the amount of ligand immobilized per unit volume of matrix [43–46]. These studies have demonstrated that if a large antigen is to be immunopurified, a matrix with a large pore size is required, whereas pore size is not so important if a small molecule is to be purified. Furthermore they have shown that highest efficiencies are achieved at low immobilized ligand concentrations, as at high ligand concentrations overcrowding takes place on the matrix.

The efficiency of immunopurification reagents is generally about 10%; in other words, 90% of antigen-binding sites are not able to bind antigen. The commonly used coupling methods result in the formation of covalent bonds between activated groups on the matrix and primary amino and hydroxyl groups on the protein. The widespread distribution of these groups over the surface of the protein will lead to a random orientation of immobilized antibody with only a small proportion of antibodies positioned to allow binding of antigen. A number of approaches have been developed to improve efficiency of antibody utilization. The majority of these exploit specific properties of the antibody molecule to form covalent bonds at sites remote from the antigen-binding site. Protein A is a molecule that binds to the Fc region of antibodies, leaving the antigen combining site exposed. Gyka et al. [47] utilized this property by absorbing the antibody to immobilized Protein A and chemically cross-linking the complex formed. This resulted in a 20-fold improvement in the efficiency of antibody utilization. Cress and Ngo [48] and Fleminger et al. [49] found two- to four-fold improvement in efficiency when they used a hydrazide coupling chemistry that leads to formation of covalent bonds with carbohydrate groups located in the Fc region of the antibody molecule. Prisyazhnoy et al. [50] described a method for immobilization of Fab′ fragments by covalent linkage through the sulphydryl groups in the hinge region of the molecule. An alternative approach has been described by Velander et al. [51], who protected the antigen-binding site by interaction with a synthetic peptide antigen during the immobilization reaction.

The weight of antigen that can be purified per unit volume of immunoaffinity

TABLE 2.3.4. Cost Per Cycle of Immunopurification Reagent to Purify 1 g Antigen Per Cycle[1]

	Current	Future prospects
Cost for 20,000-MW antigen	$375–3,075[1]	$8–62
Cost for 80,000-MW antigen	$94–769	$2–15
Assumptions		
Cost of antibody	$400–4,000/g	$40–400/g
Cost of matrix and coupling chemistry	$1,000/L	$100/L
IgG Immobilized per liter	10 gm	10 g
IgG available to bind antigen	10%	50%
Number of cycles the column is used	50	50

[1]Computed as

$$\frac{\text{Cost of immunopurification reagent per liter per cycle}}{\text{Grams of antigen purified per liter per cycle}}.$$

Example:

$$\frac{1/50 \text{ cycles } [(\$400\text{–}4{,}000/\text{g}) \times 10 \text{ g/L} \times 1 \text{ L} + \$1{,}000/\text{L}]}{10 \text{ g/L} \times 10\% \times 2 \times (20{,}000/150{,}000) \times 1 \text{ L}}.$$

matrix will be dependent on the molecular weight of the antigen. An appreciation of the cost of purifying an antigen by immunoaffinity chromatography can be gained if some simple assumptions of the costs and performance of the reagent are made. These are given in Table 2.3.4. The cost of an immunopurification reagent to purify 1 g of antigen of a given molecular weight can then be calculated. Examples for antigens of molecular weight 20,000 and 80,000 are given in Table 2.3.4. Furthermore assumptions can be made of future prospects for both reagent costs and performance, from which potential future reagent costs can be calculated (Table 2.3.4).

2.3.7. STABILITY OF LIGAND–MATRIX BOND

Leakage of antibody from an immunoaffinity reagent may occur by cleavage of the covalent bonds between matrix and antibody, by degradation of the antibody protein and the release of fragments, or by dissolution of matrix material. Ligand leakage must be minimized, as it will result in contamination of the purified antigen and, in the long run, will reduce column life, although immunopurification columns can be used for several hundred consecutive cycles [52,53].

An example showing leakage of radiolabeled antibody for six consecutive cycles of use of an immunopurification reagent is shown in Figure 2.3.1. This reagent was prepared by coupling radiolabeled antibody to Sepharose 4B with the widely used cyanogen bromide coupling chemistry. This pattern of leakage, with higher levels

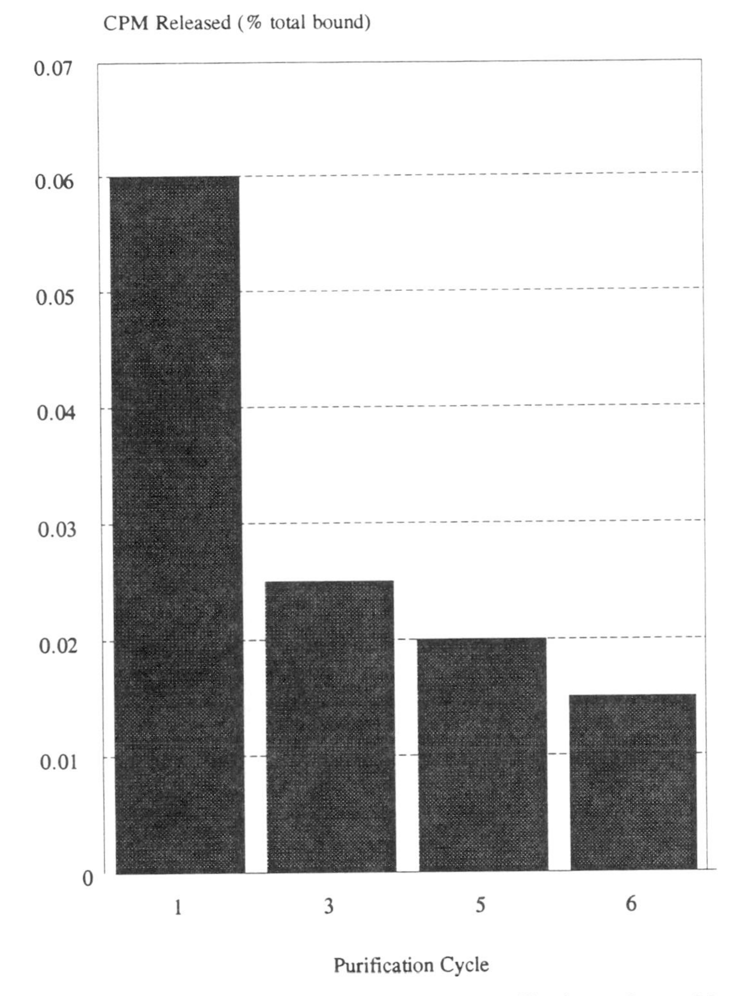

Fig. 2.3.1. Leakage of radiolabeled IgG from immunopurification column. Monoclonal antibody was labeled with iodine 125 and then coupled to cyanogen bromide-activated Sepharose 4B. A column was prepared and six sequential purification cycles were carried out. Radioactivity recovered with the eluted antibody fraction from each cycle was measured.

measured for the early cycles of use falling to low but still detectable levels as the cycle number increases, is typical for protein ligand leakage and has been observed for other, similar systems [54].

The potential for ligand leakage is greatest during elution of the antigen [55]. Leakage can be minimized by selection of an appropriate matrix material and coupling chemistry [55]. In addition the stability of the antibody can be increased by interchain covalent cross-linking of the molecule. Methods describing the use of glutaraldehyde and bismaleimides can be found in Kowal and Parsons [56] and

TABLE 2.3.5. Effect on Ligand Leakage Following Cross-Linking of Immobilized Antibody With Glutaraldehyde

Glutaraldehyde (%)	Leakage (ng IgG/ml matrix)[1]	Leakage (% total IgG × 1000)[2]	Matrix capacity (ng antigen/ml)[3]
0.00	640.0	8.00	0.76
2.00	42.0	0.53	0.66
1.00	39.4	0.49	0.69
0.05	61.3	0.77	0.68
0.01	28.2	0.35	0.73

[1]Leakage expressed as nanograms of IgG ligand recovered in the eluted fraction per 1 ml of matrix packed in the column.
[2]Leakage of total IgG ligand recovered in the eluted fraction expressed as a percentage of the total IgG immobilized on the matrix.
[3]The ligand leakage and antigen capacity were measured for the first cycle of use for each column.

Goldberg et al. [57]. The effect of cross-linking the immobilized antibody with glutaraldehyde on both the antibody activity and ligand leakage is shown in Table 2.3.5 for an antibody developed for the immunopurification of γ-interferon. An alternative approach designed to protect the antibody from proteolytic degradation has been described by Kondo et al. [58], who modified the antibody by covalent attachment of polyethylene glycol, which has been shown to protect proteins from proteolytic digestion.

2.3.8. APPLICATIONS OF IMMUNOAFFINITY PURIFICATION

Immunoaffinity purification has become a widely used technique and has found many applications in the research laboratory where its high selectivity can be exploited to give rapid purification of antigens [53,59,60]. Its high selectivity has also been exploited in the development of manufacturing processes for human therapeutic proteins [61], for example, interferons [62], Factor VIII [5,63], and Factor IX [6]. The high selectivity and affinity of the antibody–antigen interaction can also be exploited to remove low levels of contaminants from otherwise pure proteins in what is termed negative or reverse immunopurification [64]. Examples of the application of this technique include the removal of bovine serum albumin [65], the removal of the highly toxic contaminant ricin B from ricin A chain for use in immunotoxins [66], and a potential therapeutic application for the removal of IgE from plasma of patients with an allergy or other hyper-IgE syndrome [67].

Immunopurification is being increasingly used for the extraction and purification of metabolites, drug substances, and their derivatives from biological samples prior to their analysis by other techniques. The technique provides a rapid, high-yielding purification that removes substances that can interfere with analytical techniques such as gas chromatography and mass spectrometry [68–70].

Traditionally immunopurification is carried out with an antibody immobilized on a suitable support retained in a column, although other solid phases such as membranes can be used [71]. Applications that do not require a solid phase have also been described. These include aqueous two-phase systems in which the antibody is coupled to a soluble polymer that selectively partitions into one of the phases [72] and affinity ultrafiltration systems [73] in which the antibody is retained within the device by immobilization to a water-soluble high-molecular-weight polymer. Recently, Welling et al. [74,75] have demonstrated that very small fragments of an antibody (miniantibody) can be used for purification of an antigen. A 10 amino acid fragment that mimicked the antigen-binding site of the antibody was used to immunopurify lysozyme.

2.3.9. CONCLUSIONS AND FUTURE PROSPECTS

The application of immunoaffinity techniques will continue to grow in research laboratories, where their high selectivity can be exploited to give rapid extraction and purification of antigens. Examples ranging from purification of high-molecular-weight proteins to low-molecular-weight drugs and metabolites have been given in this chapter. In the research laboratory the use of the technique is limited only by the availability of suitable antibodies.

Large-scale application of the technique for the purification of human therapeutic proteins has been limited due to concerns over ligand stability, regulatory requirements, and cost of antibody manufacture. The recent licences granted by the regulatory authorities for manufacturing processes that include an immunoaffinity chromatography step have demonstrated that these issues can be adequately addressed. However, all the licenses to date have been granted for processes used to manufacture relatively low-volume, high-value products. The wider application of immunoaffinity chromatography as a manufacturing unit operation is still limited by the cost of manufacturing and gaining regulatory approval for antibody immunoaffinity ligands derived from mammalian cell lines. However, the recent demonstration that miniantibodies can be used as immunoaffinity ligands introduces the possibility of a major step forward in the development of the technique. Small, genetically engineered antibody fragments (see Chapter 3) can potentially be produced in large quantities by microbial fermentation processes. This could have a significant impact on the economics of antibody production from the point of view of both the cost of manufacture and simplification of the regulatory requirements. Furthermore, the possibility exists for modifying, or tailoring, the antibody at the molecular level to generate (for example) antibodies with appropriate affinity for individual applications.

2.3.10. REFERENCES

1. Dunnill, P., Lilly, M.D. Continuous enzyme isolation. Biotechnol. Bioeng. Symp. 3, 97–101 (1972).

2. Bonnerjea, J., Oh, S., Hoare, M., Dunnill, P. Protein purification: The right step at the right time. Biotechnology 4, 954–958 (1986).
3. Hornsey, V.S., Griffin, B.D., Pepper, D.S., Micklem, L.R., Prouse, C.V. Immunoaffinity purification of Factor VIII complex. Thromb. Haemost. 57, 102–105 (1987).
4. Ford, C.H.J., Macdonald, F., Griffin, J.A., Life, P., Bartlett, S.E. Immunoadsorbent purification of carcinoembryonic antigen using a monoclonal antibody: A direct comparison with a conventional method. Tumour Biol. 8, 241–250 (1987).
5. Griffith, M. Ultrapure plasma factor VIII produced by anti-F VIIIc immunoaffinity chromatography and solvent/detergent viral inactivation. Characterisation of the Method M process and Hemofil M antihemophilic factor (human). Ann. Hematol. 63, 131–137 (1991).
6. Kim, H.C., McMillan, C.W., White, G.C., Bergman, G.E., Horton, M.W., Saidi, P. Purified factor IX using monoclonal immunoaffinity technique: Clinical trials in hemophilia B and comparison to prothrombin complex concentrates. Blood 79, 568–575 (1992).
7. Committee for Proprietary Medicinal Products. Guidelines on the preclinical biological safety testing of medicinal products derived from biotechnology. Trends Biotechnol. 7, G13–G16 (1988).
8. Office of Biologics Research and Review. Points to Consider in the Manufacture and Testing of Monoclonal Antibody Products for Human Use. Bethesda, MD: Food and Drug Administration, 1987.
9. Campbell, D.H., Luescher, E., Lerman, L.S. Proc. Natl. Acad. Sci. USA 37, 575–578 (1951).
10. Kohler, G., Milstein, C. Derivation of specific antibody-producing tissue culture and tumour cell lines by cell fusion. Eur. J. Immunol. 6, 511–519 (1976).
11. Birch, J.R., Lambert, K., Thompson, P.W., Kenney, A.C., Wood, L.A. Antibody production with airlift fermenters. In Lydersen, B.K. (ed.): Large Scale Cell Culture Technology. Munich, Vienna, New York: Hanser Publishers, 1987, pp. 1–20.
12. Phillips, T.M. High-performance immunoaffinity chromatography. Adv. Chromatogr. 29, 133–173 (1989).
13. Novick, D., Eshhar, Z., Rubinstein, M. Monoclonal antibodies to human α-interferon and their use for affinity chromatography. J. Immunol. 129, 2244–2247 (1982).
14. Novick, D., Eshhar, Z., Gigi, O., Marks, Z., Revel, M., Rubinstein, M. Affinity chromatography of human fibroblast interferon (IFNβ_1) by monoclonal antibody columns. J. Gen. Virol. 64, 905–910 (1983).
15. Hall, R.A., Coelen, R.J., Mackenzie, J.S. Immunoaffinity purification of the NS1 protein of Murray Valley encephalitis virus: Selection of the appropriate ligand and optimal conditions for elution. J. Virol. Methods 32, 11–15 (1991).
16. Bartholomew, R.M. Method of affinity purification employing monoclonal antibodies. World patent WO83/03678 (1983).
17. Noll, F., Handschack, W., Eichmann, E., Gaestel, M., Schneider, F., Kuhl, R., Loster, K. Production and characterisation of four monoclonal antibodies specific for human interferon-alpha-1 and -alpha-2. Biomed. Biochem. Acta 48, 165–176 (1989).
18. Iman, S.A., Mills, L.A., Taylor, C.R. Optimum dissociating condition for immunoaffinity and preferential isolation of antibodies with high specific activity. J. Immunol. Methods 138, 291–299 (1991).

19. Bessos, H., Prouse, C.V. Immunopurification of human coagulation factor IX using monoclonal antibodies. Thromb. Haemost. 56, 86–89 (1986).
20. Kobiler, D., Grosfeld, H., Leitmer, M., Monzain, R., First, C., Seri, T., Cohen, C., Velan, B., Shafferman, A., Gozes, Y. Production of monoclonal antibodies toward bovine interferons-α suitable for immunopurification. J. Interferon. Res. 9, 189–193 (1989).
21. Narayanan, S.R., Crane, L.J. Affinity chromatography supports: A look at performance requirements. Tibtech 8, 12–16 (1990).
22. Angal, S., Dean, P.D.G. Purification by exploitation of activity. In Harris, E.L.V., Angal, S. (eds.): Protein Purification Methods—A Practical Approach (pp. 245–262). Oxford, New York, Tokyo: IRL Press.
23. Kenney, A.C., Goulding, L., Hill, C.R. The design, preparation and use of immunopurification reagents. In Walker, J.M. (ed.): Methods in Molecular Biology (pp. 99–110). Clifton, NJ: Humana, 1988.
24. Ohlson, C. Hansson, L., Larsson, P.O. High performance liquid affinity chromatography (HPLAC) and its application to the separation of enzymes and antigens. FEBS Lett. 93, 5–9 (1978).
25. Lowe, C.R. The chemical technology of affinity chromatography. In Work, T.S., Work, E. (eds.): An Introduction to Affinity Chromatography (pp. 344–400). Amsterdam: Elsevier, 1979.
26. Cabrera, K.E., Wilchek, M. Trends Analyt. Chem. 7, 58–63 (1988).
27. Fowell, S.L., Chase, H.A. A comparison of some activated matrices for preparation of immunoadsorbents. J. Biotechnol. 4, 355–368 (1986).
28. Desai, M.A., Lyddiatt, A. Comparative studies of agarose and kieselguhr-agarose composites for the preparation and operation of immunoadsorbents. Bioseparation 1, 43–58 (1990).
29. Highsmith, F., Regan, T., Clark, D., Drohan, W., Tharakan, J. Evaluation of CNBr, FMP and hydrazide resins for immunoaffinity purification of Factor IX. Biotechniques 12, 418–423.
30. Tharakan, J., Highsmith, F., Clark, D., Drohan, W. Physical and biochemical characterisation of five commercial resins for immunoaffinity purification of Factor IX. J. Chromatog. 595, 103–111 (1992).
31. Hill, R.J. Elution of antibodies from immunoadsorbents: Effect of dioxane in promoting release of antibody. J. Immunol. Methods 1, 231–245 (1972).
32. Andersson, K., Benyamin, Y., Douzou, P., Balny, C. Organic solvents and temperature effects on desorption from immunoadsorbents. DNP-BSA anti-DNP as a model. J. Immunol. Methods 23, 17–21 (1978).
33. Olson, W.C., Leung, S.K., Yarmush, M.L. Recovery of antigens from immunoadsorbents using high pressure. Bio/Technology 7, 369–373 (1989).
34. Morgan, M.R.A., Kerr, E.J., Dean, P.D.G. Electrophoretic desorption: Preparative elution of steroid specific antibodies from immunoadsorbents. J. Steroid Biochem. 9, 767–770 (1978).
35. Hardie, G., van Regenmortel, M.H.V. Isolation of specific antibody under conditions of low ionic strength. J. Immunol. Methods 15, 305 (1977).
36. Bureau, D., Daussant, J. Efficiency of a smooth desorption procedure for the purifica-

tion of barley β-amylase using immunoaffinity chromatography. Biochimie 65, 361–365 (1983).

37. Ohlson, S., Lundblad, A., Zope, D. Novel approach to affinity chromatography using "weak" monoclonal antibodies. Anal. Biochem. 169, 204–208 (1988).
38. Nakamura, S., Sakata, Y. Immunoaffinity purification of protein C by using conformation-specific monoclonal antibodies to protein C-calcium ion complex. Biochim. Biophys. Acta 925, 85–93 (1987).
39. Limentani, S.A., Furie, B.C., Poiesz, B.J., Montagna, R., Wells, K., Furie, B. Separation of human plasma Factor IX from HTLV-1 or HIV by immunoaffinity chromatography using conformation-specific antibodies. Blood 70, 1312–1315 (1987).
40. Krainer, A.R. Pre-mRNA splicing by complementation with purified human U1, U2, U4/U6 and U5 snRNPs. Nucleic Acids Res. 16, 9415–9429 (1988).
41. Hearn, M.T., Davies, J.R. Evaluation of factors which affect column performance with immobilised monoclonal antibodies. Model studies with a lysozyme-antilysozyme system. J. Chromatogr. 512, 23–39 (1990).
42. Pfeiffer, N.E., Wylie, D.E., Schuster, S.M. Immunoaffinity chromatography utilizing monoclonal antibodies. Factors which influence antigen-binding capacity. J. Immunol. Methods 97, 1–9 (1987).
43. Fowell, S.L., Chase, H.A. Variation of immunosorbent performance with the amount of immobilised antibody. J. Biotechnol. 4, 1–13 (1986).
44. Liapis, A.I., Anspach, B., Findley, M.E., Davics, J., Hearn, M.T.W., Unger, K.K. Biospecific adsorption of lysozyme onto monoclonal antibody ligand immobilised on nonporous silica particles. Biotechnol. Bioeng. 34, 467–477 (1989).
45. Comoglio, S., Massaglia, A., Rollcri, E., Rosa, A. Factors affecting the properties of insolubilised antibodies. Biochim. Biophys. Acta 420, 246–257 (1976).
46. Orthner, C.L., Tharakan, J., Highsmith, F.A., Madurawe, R.D., Morcol, T., Velander, W.H. Comparison of the performance of immunoadsorbents prepared by site-directed or random coupling of monoclonal antibody. J. Chromatogr. 558, 55–70 (1991).
47. Gyka, G., Ghetie, V., Sjoquist, J. Crosslinkage of antibodies to staphylococcal protein A matrices. J. Immunol. Methods 57, 227–233 (1983).
48. Cress, M.C., Ngo, T.T. Site specific immobilisation of immunoglobulins. Am. Biotechnol. Lab. Feb. 21, 16–19 (1989).
49. Fleminger, G., Hadas, E., Wolf, T., Solomon, B. Oriented immobilisation of periodate-oxidised monoclonal antibodies on amino and hydrazide derivatives of Eupergit C. Appl. Biochem. Biotechnol. 2, 123–127 (1990).
50. Prisyazhnoy, V.S., Fusek, M., Alakhov, Y.B. Synthesis of high capacity immunoaffinity sorbents with oriented immobilised immunoglobulins or their Fab′ fragments for isolation of proteins. J. Chromatogr. 424, 243–253 (1988).
51. Velander, W.H., Subramanian, A., Madurawe, R.D., Orthner, C.L. The use of Fab-masking antigens to enhance the activity of immobilised antibodies. Biotechnol. Bioeng. 39, 1013–1023 (1992).
52. Janson, J-C. Large-scale affinity purification–State of the art and future prospects. Trends Biotechnol. 2, 31–38 (1984).
53. Calton, G.J. In Chaiken, E.I. (ed.): Affinity Chromatography and Biological Recognition. London, New York: Academic, 1983.
54. Francis, R., Bonnerjea, J., Hill, C.R. Validation of the re-use of Protein A Sepharose for

the purification of monoclonal antibody. In Pyle, D.L. (ed.): Separations for Biotechnology (pp. 491–498). London, New York: Elsevier, 1990.

55. Peng, L., Calton, G.J., Burnett, J.W. Stability of antibody attachment in immunosorbent chromatography. Enzyme Microb. Technol. 8, 681–685 (1986).
56. Kowal, R., Parsons, R.G. Stabilization of proteins immobilised on Sepharose from leakage by glutaraldehyde crosslinking. Anal. Biochem. 102, 72–76 (1980).
57. Goldberg, M., Knudsen, K.L., Platt, D., Kohen, F., Bayer, E.A., Wilchek, M. Specific interchain cross-linking of antibodies using bismaleimides. Bioconjug. Chem. 2, 275–280 (1991).
58. Kondo, A., Kishimura, M., Katoh, S., Sada, E. Improvement of proteolytic resistance of immunoadsorbents by chemical modification with polyethylene glycol. Biotechnol. Bioeng. 34, 532–540 (1989).
59. Hill, C.R., Birch, J.R., Benton, C. Affinity chromatography using monoclonal antibodies. In Bioactive Microbial Products. III. Downstream Processing (pp. 175–190). (Special publication of the Society for General Microbiology.) London: Academic, 1986.
60. Hill, C.R., Thompson, L.G., Kenney, A.C. Immunopurification. In Harris, E.L.V., Angal, S. (eds.): Protein Purification Methods–A Practical Approach (pp. 282–290). Oxford, New York, Tokyo: IRL Press, 1989.
61. Jack, G.W., Wade, H.E. Immunoaffinity chromatography of clinical products. Tibtech 5, 91–95 (1987).
62. Tarnowski, S. J., Liptak, R. A. In Mizrake, A., van Wezel, A.L. (eds.): Advances in Biotechnological Processes–2 (pp. 271–287). New York: Alan R. Liss, 1980.
63. Schreiber, A.B., Hrinda, M.E., Newman, J., Tarr, G.C., D'Alisa, R., Curry, M. Removal of viral contaminants by monoclonal antibody purification of plasma proteins. In Morgenthaler, J-J. (ed.): Virus inactivation in Blood Products (pp. 146–153). Basle: Karger, 1989.
64. Lee-Huang, S. Reverse immunoaffinity chromatography purification method. U.S. patent 4 568 488 (1986).
65. Hill, C.R., Kenney, A.C., Goulding, L. The design, development and use of immunopurification reagents. Int. J. Biotechnol. 4, 167–170 (1987).
66. Fulton, R.J., Blakeley, D.C., Knowles, P.P., Uhr, J.W., Thorpe, P.E., Vitetta, E.S. Isolation of pure ricin A_1, A_2, and B chains and characterisation of their toxicity. J. Biol. Chem. 261, 5314–5319 (1986).
67. Sato, H., Watanabe, K., Azuma, J., Kidaka, T., Hori. Specific removal of IgE by therapeutic immunoadsorption system. J. Immunol. Methods 118, 161–168 (1989).
68. Bagnati, R., Castelli, M.G., Airoldi, L., Paleologo Oriundi, M., Ubaldi, A., Fanelli, R. Analysis of diethylstilbestrol, dienestrol and hexeatrolin biological samples by immunoaffinity extraction and gas chromatography-negative-ion chemical ionization mass spectrometry. J. Chromatogr. 527, 267–278 (1990).
69. Batnati, R., Paleologo Oriundi, M., Russo, V., Danese, M., Berti, F., Fanelli, R. Determination of zeranol and beta zeranol in calf urine by immunoaffinity extraction and gas chromatography-mass spectrometry after repeated administration of zeranol. J. Chromatogr. 564, 493–502 (1991).
70. Ishibashi, M., Watanabe, K., Ohyama, Y., Mizugaki, M., Hayashi, Y., Takasaki, W. Novel derivatisation and immunoextraction to improve microanalysis of 11-dehydrothromboxane B2 in human urine. J. Chromatogr. 562, 613 624 (1991).

71. Yan Pak, K., Randerson, D.H., Blaszczyk, M., Sears, H.F., Steplewski, Z., Koprowski, H. Extraction of circulating gastrointestinal cancer antigen using solid-phase immunoadsorption system of monoclonal antibody-coupled membrane. J. Immunol. Methods 66, 51–58 (1984).

72. Stocks, S.J., Brooks, D.E. Development of a general ligand for immunoaffinity partitioning in two phase aqueous polymer systems. Anal. Biochem. 173, 86–92 (1988).

73. Male, K.B., Nguyen, A.L., Luong, J.H.T. Isolation of urokinase by affinity ultrafiltration. Biotechnol. Bioeng. 35, 87–93 (1990).

74. Welling, G.W., Geurts, T., van Gorkum, J., Damhof, R.A., Drijfhout, J.W., Bloemhoff, W., Welling-Wester, S. Synthetic antibody fragment as ligand in immunoaffinity chromatography. J. Chromatogr. 512, 337–343 (1990).

75. Welling, G.W., van Gorkum, J., Damhof, R.A., Drijfhout, J.W., Bloemhoff, W., Welling-Wester, S. A ten residue fragment of an antibody (mini-antibody) directed against lysozyme as ligand in immunoaffinity chromatography. J. Chromatogr. 548, 235–242 (1991).

CHAPTER 3

GENETIC MANIPULATION AND EXPRESSION OF ANTIBODIES

E. SALLY WARD
Department of Microbiology, Cancer Immunobiology Center, University of Texas Southwestern Medical Center, Dallas, TX 75235-8576

C.R. BEBBINGTON
Celltech Ltd., Slough, Berkshire, SL1 4EN, UK

3.1. PROTEIN ENGINEERING

The techniques for genetic manipulation have expanded enormously over the past decade. As a result of the development of these new techniques it is now possible, for example, to change single amino acids of a protein selectively and to generate protein fragments with precision by inserting translational stop codons into the appropriate genes. The use of recombinant DNA methods frequently offers attractive alternatives to time-consuming and often inaccurate protein chemistry. A major step forward in technology has been made by the development of the now widely used polymerase chain reaction (PCR) [1,2]. For example, the use of this technique allows thc isolation of a chosen gene from the genome or cDNA of a species in a matter of hours, and the only requirement for this to be feasible is that there be some preexisting knowledge of the nucleotide sequences that flank the gene to be isolated at either, or preferably both, of the 5′ and 3′ ends. Previously, the generation and screening of cDNA libraries for the isolation of a gene for which the sequence was known would have taken weeks or months.

Concomitantly with the development of techniques of genetic manipulation, the methods for expression of the proteins encoded by the engineered genes have been expanded and improved. During the early 1980s, mammalian cell expression systems based on lymphoid cells were developed for the efficient production of immunoglobulin molecules [3–5], and these systems have been widely used for the production of recombinant antibodies as both intact molecules and chimeras [6–13].

Monoclonal Antibodies: Principles and Applications, pages 137–185
© 1995 Wiley-Liss, Inc.

It is now also possible to produce immunoglobulin fragments in both prokaryotic [14,15] and eukaryotic hosts in high yields. The expression hosts that can currently be used to produce antibodies or antibody fragments range from tobacco plants [16] to *Escherichia coli,* and include several different mammalian cell hosts (see section 3.6) and baculovirus-infected insect cells [17].

Thus, genes can be altered with precision and efficiency, and the encoded proteins can be expressed and purified for characterization. In the field of antibody engineering, immunoglobulins can now be tailor-made, and recently developed technology allows the isolation of antibody fragments that bind to almost any antigen. In addition, the affinity of a recombinant antibody for cognate antigen can be increased by using current techniques of molecular biology. The wide range of uses of antibodies in medicine and biology makes antibody engineering a particularly attractive prospect. In this review, recent developments in gene technology will be described. Subsequently, the application of these techniques to the engineering of immunoglobulins will be discussed (for reviews of these topics, see Morrison [18] and Winter and Milstein [19]).

3.1.1. The Polymerase Chain Reaction (PCR)

The PCR [1,2] is a method that is designed to selectively amplify discrete segments of DNA. The method usually consists of three steps and is shown schematically in Figure 3.1: (1) denaturation, (2) annealing of oligonucleotide primers which are designed to flank the DNA sequence to be amplified, and (3) extension of the primers by a processive polymerase (a processive polymerase synthesizes relatively long, continuous strands of single-stranded DNA, and its use is therefore important for the production of full-length products) in the presence of deoxynucleotide triphosphates (dNTPs). It is a widely used technique and in the past few years has been greatly simplified by the development of automated methods of temperature cycling, which are manifested in the large number of cycling blocks that are now commercially available. Temperature cycling is essential for the PCR, since the process comprises (1) a denaturation step at high temperature (usually about 95°C, to ensure that the template DNA, which is frequently double-stranded, is completely denatured into separate strands), followed by (2) annealing of the template DNA to the primer DNA at lower temperature (usually 30–60°C, depending on the length and nature of the oligonucleotide primers; generally the shorter the oligonucleotides, the lower the annealing temperature), and then (3) extension of the annealed products at the temperature for optimal activity of thermostable polymerases (usually about 72°C). Each cycling temperature is maintained for about 0.5 to 3 min. The use of a thermostable polymerase derived from the thermophilic bacterium *Thermophilis aquaticus,* and called Taq polymerase [2], has also greatly facilitated the PCR. Prior to this, the Klenow fragment of *E. coli* polymerase [1] was used, and the thermolability of Klenow polymerase meant that fresh enzyme had to be added at the end of each denaturation step. Furthermore, Klenow polymerase is optimally active at 37°C, and this low temperature (compare with 72°C for Taq polymerase) results, in some cases at least, in poor specificity of the primers

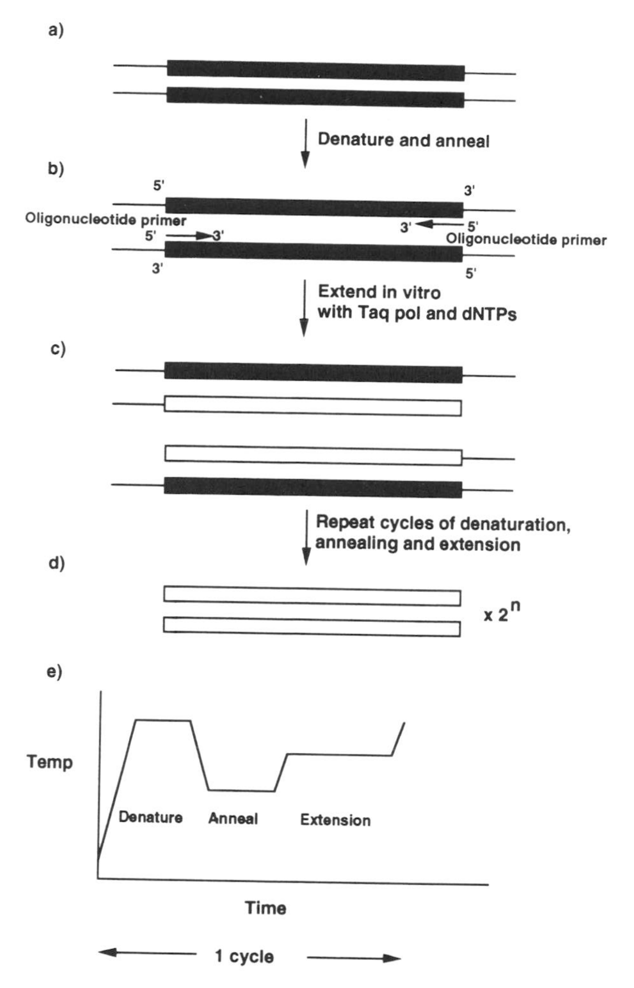

Fig. 3.1. The polymerase chain reaction (PCR) [1,2]. **a.** The region of DNA to be amplified is shown by filled-in boxes, and flanking sequences by single lines. **b, c.** The first round of PCR results in extension of the primers beyond the region to be amplified. **d.** Subsequent rounds result in extension within the region delimited by the primers. During each cycle the amount of DNA delimited by the primers is increased twofold, so that the total number of DNA molecules at the end of the cycling is $2^n \times$ starting number, where n = number of cycles. **e.** Schematic representation of the temperature changes that occur during a single PCR cycle. Denaturation is usually carried out at 94°C, annealing at 25–70°C (depending on the required specificity) and extension at 72°C (the optimal temperature for Taq polymerase activity).

due to annealing and extension at low temperatures. With the development of temperature cycling blocks and thermostable polymerases, the PCR can now be set up by mixing the components and then placing the reaction tube in a cycling block for a programmed period of time. This is therefore much less labor-intensive than the earlier method of carrying out the PCR, which involved transferring the reaction tube between water baths at different temperatures at 0.5- to 3-min intervals and adding fresh enzyme at each cycle; the total number of cycles (usually about 30) meant that this was tedious and also error-prone.

More recently, polymerases with improved properties (for example, with proof-reading activity and with higher processivity) have been developed [20] and the use of these polymerases improves the fidelity of the PCR and therefore increases its utility further. The higher processivity allows longer stretches of DNA to be synthesized during the PCR. In addition to being used for the isolation of genes for which knowledge of the 5′ and 3′ sequences is already available, the PCR has the applications discussed below.

Tailoring Genes for Ligation Into Vectors With "Add-On" Oligonucleotides. Genes can be amplified with oligonucleotide primers that have 5′ sequences that are not complementary to the gene to be amplified, and contain sequences encoding restriction sites [21]. These oligonucleotides are often called "add-on" oligonucleotides because they add extra sequences to the ends of the target DNA during the PCR. The sites are usually chosen so that the recognition enzyme makes staggered cuts, usually with 5′ or 3′ overhangs of two to four nucleotides, rather than blunt ended, or "flush" cuts, which have no overhang. This facilitates the ligation (i.e., "sticking" the PCR product into a vector by using an enzyme called T4 DNA ligase to catalyze the reaction) of the genes into plasmid vectors following the PCR, since the PCR products can be cut with restriction enzymes to produce staggered ends that are complementary to the ends of a plasmid vector cut with the same restriction enzyme. The DNA therefore has cohesive ends and can be ligated into expression vectors that have been cut with the same restriction enzymes. This ligation event occurs with much higher efficiency than the ligation of the untreated PCR products into a vector that has been restricted with an enzyme that makes blunt-ended cuts. The relative inefficiency of the latter process is usually due, in part at least, to the fact that PCR products have ends that are not completely flush, that is, have a one- or two-base overhang. These ends can be converted into flush ends with no overhang by using T4 DNA polymerase to fill in, or S1 nuclease to remove, the single-stranded overhangs. It is generally more efficient, however, to incorporate restriction sites that can be cleaved to produce overhangs that are compatible with the vector cloning sites.

By the same approach, other useful sequences can be added to the PCR product by means of "add-on" oligonucleotides, which extend beyond the ends of the template DNA and encode additional sequences. With suitably designed oligonucleotides, therefore, DNA sequences other than those encoding restriction sites can be added at the 5′ or 3′ ends of the amplified DNA. For example, sequences encoding peptide tags that are in translational frame could be added to facilitate detection or purification of the encoded protein, such as the c-myc peptide epitope [22]. Alternatively, promoter sequences for transcriptional initiation [23] or G + C clamps [24] can be inserted by judicious design of "add-on" oligonucleotides. G + C clamps are used in denaturing gel electrophoresis of lengths of DNA that differ by one or more bases; the addition of a GC-rich DNA segment (called a G + C clamp) results in enhancement of mobility differences.

Isolation of Genes for Which Sequences at Only the 3′ End Are Known. A number of PCR techniques now are available that allow the isolation of genes for which sequence information is available only for the 3′ end of the gene to be isolated [25,26]. Examples of such genes are the variable domain genes of immunoglobulins and T-cell receptors (TCRs). In both these cases, the sequences of the constant domains are well known for a number of species. For immunoglobulin genes it is possible to design degenerate and family-specific primers for their isolation [27–36], due to a reasonable amount of conservation of sequence in framework 1 and/or the secretion leader sequence (see Chapter 1). In contrast, TCR genes are considerably more variable in sequence at their 5′ ends [37], and although a number of reports of the use of family-specific primers have now been documented [38–42], the isolation of TCR variable domain genes can be carried out by inverse PCR [25,43] or anchor PCR [26,44].

Inverse PCR is a technique that has been developed to isolate genes for which sequence information is available only for either the 5′ or 3′ end of the coding region, or the sequences that flank the 5′ or 3′ sides of the coding region. Inverse PCR (Fig. 3.2) involves the use of three (or at least two) different primers, which in the case of TCR gene isolation anneal to the constant region of the gene to be isolated. cDNA synthesis is primed with oligonucleotide I, and second-strand cDNA synthesis is then carried out with RNase H and *E. coli* DNA polymerase I [25]. The resulting double-stranded cDNA is circularized and used as a template in the PCR with oligonucleotides II and III (Fig. 3.2). This will result in the isolation of a PCR product that contains the 5′ region of the variable domain gene, plus leader sequence and 5′ untranslated region. Oligonucleotides II and III can be designed to encode restriction sites within their sequence, to allow the restriction and ligation of the PCR products into appropriate vectors following amplification. The ligation of the PCR products into vectors facilitates DNA sequencing to determine the nucleotide sequence of the gene(s) that have been isolated. Alternatively, the PCR products can be sequenced by direct PCR sequencing [45,46], which avoids the need to ligate the amplified DNA into vectors prior to sequence analysis. To use direct PCR sequencing, however, the PCR products must be homogeneous so as to obtain unambiguous sequence information, and for this reason it cannot be used in situations where members of gene families are being isolated. An example of this is the isolation of repertoires of immunoglobulin or TCR variable domain genes, where the sequence differences from one variable domain gene to another mean that the genes have to be ligated into vectors and individual recombinant clones obtained prior to sequence analysis.

Anchor PCR is another technique that results in the isolation of genes for which

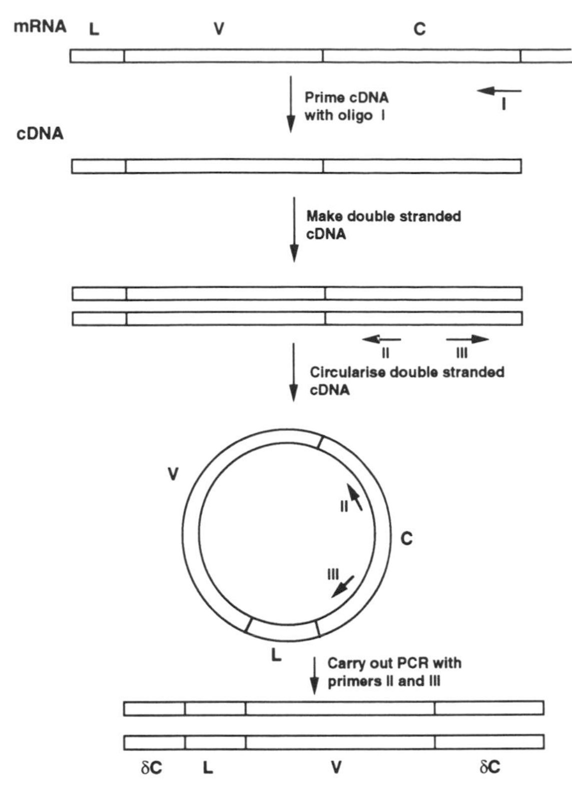

Fig. 3.2. Inverse PCR [25,43]. For the isolation of immunoglobulin or T-cell receptor genes, cDNA synthesis is usually primed with an oligonucleotide (I) that is specific for the constant (C) region. Double-stranded cDNA is synthesized in a reaction containing RNase H and *E. coli* DNA polymerase I and is then circularized with ligase. The resulting circular DNA is used as a template in the PCR with primers II and III to product the product shown. Note that restriction sites can be incorporated into the primers to facilitate cloning of the amplified genes. L, leader sequence gene; V, variable domain sequence gene; δC, part of constant region gene between priming site of oligonucleotide II and variable domain gene, and priming sites of oligonucleotides I and III.

sequence information is only available for either the 5′ or 3′ ends of the coding sequences or the sequences that flank the coding region. For anchor PCR (Fig. 3.3), cDNA synthesis is primed with a constant region primer (analogous to that used for inverse PCR). The single-stranded cDNA is subsequently tailed with deoxyguanosine triphosphate and terminal deoxynucleotidyl transferase (TdT). This tailing results in the addition of a run (of variable length) of G residues to the 3′ end of the first-strand cDNA. The tailed cDNA is then used as a template in the PCR with the cDNA synthesis primer and the anchor primer (see fig. 3.3). The anchor primer comprises a run of approximately 14 C's with an "add-on" sequence encoding a restriction site [26]. The cDNA synthesis primer can also have an internal restriction

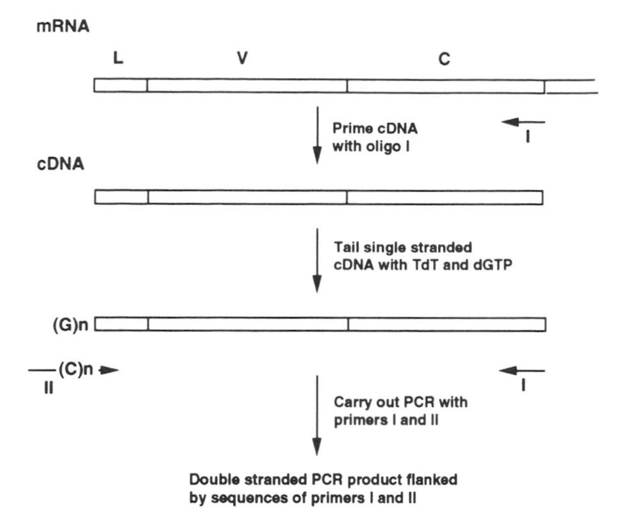

Fig. 3.3. Anchor PCR [26,44]. For the isolation of immunoglobulin or T-cell receptor genes, cDNA synthesis is primed with oligonucleotide I, and the single-stranded cDNA is tailed with TdT (terminal deoxynucleotidyl transferase) and dGTP. The resulting tailed cDNA is then used in the PCR with primers I and II, which usually have internal restriction sites to facilitate cloning of the PCR products. L, leader sequence gene; V, variable domain sequence gene; C, constant region gene.

site, so that the resulting PCR products can be restricted and efficiently ligated into vectors for sequencing.

Splicing by Overlap Extension. The PCR allows genes to be spliced together [47], using designed primers as shown in Figure 3.4. A situation in which it is desirable to splice genes together is in the generation of chimeric antibodies, in which one constant region gene is spliced to a particular variable domain gene. Splicing by overlap extension has the advantage that the splice site can be generated accurately at the nucleotide level, by design of suitable splicing oligonucleotides. The primers are designed to overlap the 3′ end of one gene and the 5′ end of the gene that is to be spliced downstream, as indicated in Figure 4. This approach was initially used to splice together genes encoding different domains of an HLA molecule [48]. More recently it has been extended to the generation of repertoires of randomly combined heavy and light chain immunoglobulin variable domain genes [34, 49] (Fig. 4). This is described in more detail in section 3.5, but it results in large numbers of different heavy chain variable domain genes being randomly combined with a similarly large number of light chain variable domain genes and is a critical step in the production of *in vitro* repertoires of genes encoding Fv fragments, from which fragments with the desired binding specificity can be selected.

Random and Site-Directed Mutagenesis of Genes. It is sometimes desirable to be able to change a particular amino acid in a protein by making a change at the corresponding codon or, alternatively, to insert mutations in a random way into a

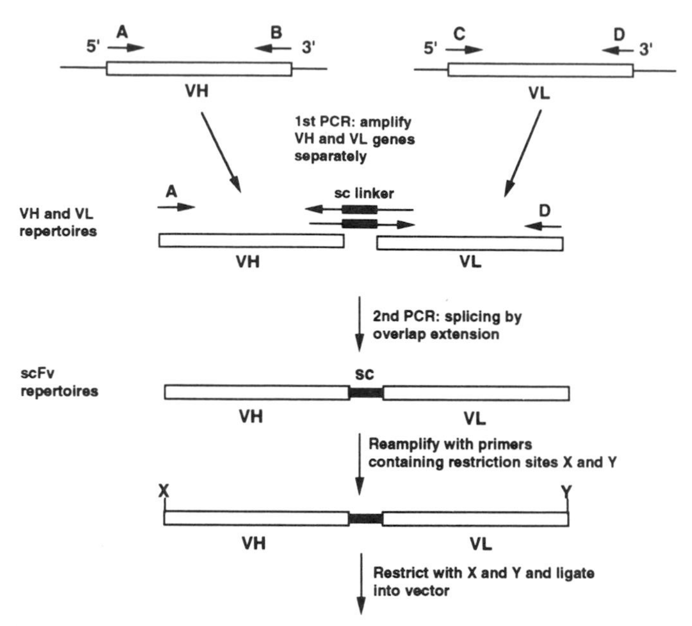

Fig. 3.4. Schematic representation of splicing by overlap extension (SOE) [47] for the generation of scFv genes from random combinations of repertoires of VH and VL domain genes [47,49]. The VH genes are isolated from antibody-producing cells by the PCR and primers A (5′) and B (3′). The VL genes are isolated from antibody-producing cells in a separate PCR with primers C (5′) and D (3′). The VH and VL gene repertoires are randomly combined by splicing together in a PCR with primers A and D, together with single chain (sc) linker primers. These sc linker primers encode a synthetic peptide linker [49,104] and also have the same or complementary regions as primers B (3′ of VH genes) and C (5′ of VL genes). Open boxes represent the VH and VL domain genes, and stippled boxes represent the sc linker sequence. Following the SOE reaction the assembled genes can be restricted with enzymes X and Y and ligated into an appropriately restricted vector.

gene. Site-directed mutagenesis, that is, alteration of a chosen amino acid, may be useful if this amino acid is believed to be critical for the function of a protein. It is clearly important that the function of the protein and the corresponding mutant can be assayed in a binding or enzymatic assay. Random mutagenesis is useful in cases where it is convenient to insert mutations at random sites within genes and then to screen the expressed proteins for altered properties; in the case of an antibody molecule, random mutagenesis could be used to generate mutants that have increased affinity for binding to antigen, for example. The PCR can be used to insert both random and specific mutations into genes. A requirement for this to be possible is that a convenient and preferably unique restriction site flank the region to be mutated; if no such site exists, it can generally be inserted by oligonucleotide-directed mutagenesis (Fig. 3.5) [50] without loss of translational sense. Oligonucleotide primers can then be designed to overlap the restriction site and to change one or more codons at the desired position in the gene. Mutations can be to a

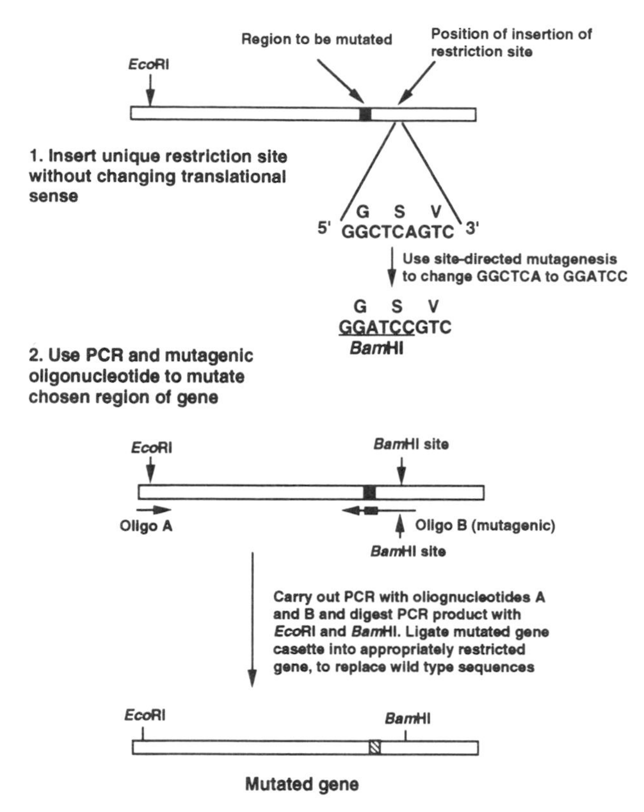

Fig. 3.5. Strategy for targeted mutagenesis of genes, using the PCR. First, a unique restriction site is inserted by oligonucleotide-directed mutagenesis in the vicinity of the region of the gene that is to be mutated. (This step is not necessary if such a site already exists.) The site is inserted without loss of translational sense (the corresponding amino acid sequences in one-letter code are shown above the nucleotide sequences). The wild-type gene containing the restriction site is then used as a template for the PCR with oligonucleotides A and B. Oligonucleotide A is complementary to the 5′ end of the gene and contains an *Eco*RI restriction site (which is present in the wild-type gene). Oligonucleotide B contains a *Bam*HI restriction site and overlaps the region of the gene to be mutated. For this primer (shown by an arrow) the solid box represents regions that are mutagenic—that is, not complementary to the gene that is to be mutated, but encoding the desired mutations. The gene is represented by an open box, the region to be mutated by a filled-in box, and the mutated region in the context of the wild-type gene by a hatched box.

predetermined codon, or mixed oligonucleotides can be made that will randomize the sequence of the gene at the desired codon(s). One of the problems of this approach is that since some amino acids are encoded by as many as six different codons (e.g., serine), whereas others are encoded by only one (methionine), it is difficult to generate a random oligonucleotide that encodes the 20 different amino acids at the same frequency. Thus, the codons of the random primer will be biased towards encoding amino acids that have the largest number of possible codons in the genetic code, for example, arginine and serine. Theoretically, it is possible to design the primers in such a way that there is an equal likelihood that each amino acid is encoded at each codon position. Whether it is necessary to design such oligonucleotides depends on the number or random mutants that can be screened, since the higher the number of mutants screened, the more likely it is that every possible amino acid at each mutated position will be inserted. A second problem is that if the random section of the primer is designed so that A, T, C, and G are incorporated in equimolar amounts, a significant number of the codons in the random primer will be stop codons. Thus, to minimize this, it is advisable to design the primers to insert only A and C at the third position of each codon in the sense strand, since two or three of the stop codons end in A (C is also not inserted at this position, to minimize the occurrence of biases towards particular amino acids in the random primer).

Error Prone PCR for Targeted Random Mutagenesis. Although for most purposes the PCR is carried out under conditions designed to minimize the error frequency, conditions have been developed to reduce the fidelity of Taq polymerase for the purpose of random mutagenesis [51]. These conditions involve the addition of low concentrations of manganese chloride and lowering the dATP concentration relative to the other three nucleotide triphosphates. Such conditions result in error frequencies of up to 2% at each base position [51,52]; that is, for a gene of length 400 bases, eight mutations on average can be inserted. The mutations are more frequently transversions/transitions than deletions/insertions [51], and this is of particular significance if the goal is to insert random point mutations in the region between the PCR primers. Deletions and insertions are undesirable in such random mutagenesis experiments, since they produce frameshifts that result in gross structural alterations of the encoded protein.

3.2. GENETICALLY ENGINEERED ANTIBODIES

3.2.1. Structure of the Immunoglobulin Molecule

The IgG molecule comprises two heavy and two light chains, which are linked to each other by an –S–S– bridge located at the C termini of the CH1 domain and the Cκ/Cλ domain [53] (Fig. 3.6; see also Chapter 1). The two heavy chains are also covalently linked to each other by one or more –S–S– bridges that are located in the hinge region. Crystallographic analyses of a number of antibodies [54–60] indicate that the immunoglobulin molecule is made up of strings of discrete domains, and each domain comprises two β sheets that pack against each other and are pinned together by an intramolecular –S–S– bridge. The individual β strands are connected to each other by relatively exposed loops. The peptide loops at the tips of the heavy and light chain variable domains (designated VH and VL domains respectively; shown schematically in Fig. 3.7) are hypervariable in sequence and are

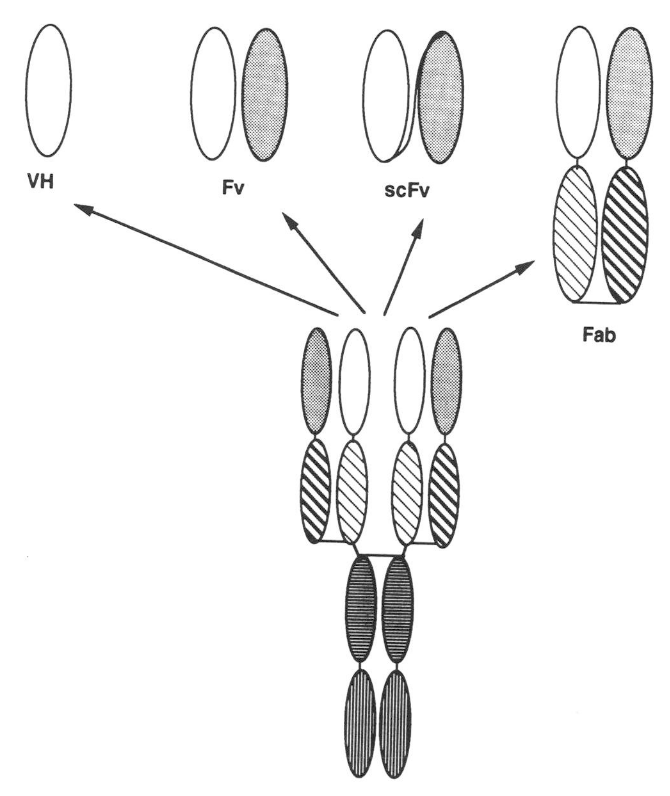

Fig. 3.6. Schematic representation of the immunoglobulin molecule. The immunoglobulin IgG molecule consists of strings of discrete domains, and comprises two heavy chains and two light chains [53]. The heavy and light chains are linked by an intermolecular disulfide bridge (indicated by a horizontal line between the hatched CH1 and CL domains). The heavy chains are also linked to each other by one or more intermolecular disulfide bridges. These –S–S– bridges are in the hinge region, which links the CH1 domain to the CH2 domain (indicated by horizontal hatching). The CH3 domain is indicated by vertical hatching. Immunoglobulin-derived fragments that can be expressed and secreted from recombinant *E. coli* cells are shown at the top of the figure. VH, heavy chain variable domain; Fv, VH and light chain variable (VL) domains; scFv, VH and VL domains linked by a synthetic peptide linker [104,105]; Fab, Fd (VH linked to CH1 domain) and paired light chain.

involved in contacting antigen during antibody–antigen interaction. In other words, it is residues within and flanking these loops that determine the specificity and affinity of a particular antibody for interaction with cognate antigen. Despite the variation in the sequences of the variable domains, and in particular their hypervariable loops, the overall main chain conformation is relatively well conserved [61,62]. This conservation is of relevance in CDR grafting, in which hypervariable loops are grafted at the genetic level from one antibody to another. The purpose of such CDR grafting experiments is to confer the binding specificity of an antibody of

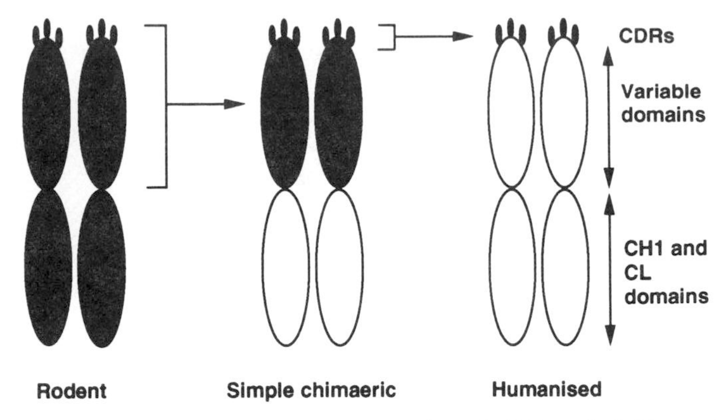

Fig. 3.7. Schematic representation of simple chimeric and CDR-grafted ("humanized") antibodies [9,11–13,70,71]. Rodent-derived regions are shown by filled ellipsoids, and human-derived regions by open ellipsoids. Only the Fab fragment of the immunoglobulin molecule is shown. For the simple chimeric molecule, the variable domains and CDR loops are derived from rodents, and the remainder of the antibody molecule is of human origin. In contrast, for humanized antibodies, only the CDR loops are derived from rodents.

rodent origin onto a human antibody framework, to produce "humanized" antibodies that are less immunogenic in human therapy. This is discussed in more detail in section 3.2.3. below.

The immunoglobulin can be fragmented into Fab and Fc portions, by virtue of the proteolytic sensitivity of the hinge region. For example, papain can be used to cleave the molecule into Fc and monovalent Fab fragments. It is more difficult, however, to produce Fv fragments (comprising VH and VL domains) by proteolysis, and although a method has recently been reported for this for a particular immunoglobulin molecule [63], it is generally more reliable and easier to generate these fragments by genetic engineering [15,64].

3.2.2. Gene Structure of Immunoglobulins

The domain structure of immunoglobulins is mirrored in the genetic organization, since each domain is encoded within a separate exon [65]. This allows domain-swapping experiments to be carried out with relative ease [6,9,11–13]. Domain-swapping experiments can be used to link antibody variable domains to constant regions of the isotype of choice, so that the effector functions of the antibody can be selected at will. In addition, it may be useful to link rodent variable domains to human constant regions to produce simple chimeric antibodies (discussed in section 3.2.3). For example, prior to the development of methods for the PCR and site-directed mutagenesis, restriction sites located within the introns were used to splice immunoglobulin genes together to produce simple chimeric antibodies. The relative advantage of the immunoglobulin gene organization, compared with many other genes that encode multidomain proteins, has now been reduced by the development of methods that allow genes to be spliced together with precision and rapidity in the absence of conveniently located restriction sites.

3.2.3. Production of Chimeric Immunoglobulins and CDR Grafting

One of the problems associated with the use of rodent antibodies in human therapy is their immunogenicity, and until recently, it was difficult to produce human-derived monoclonal antibodies for both ethical and technical reasons [66–68]. The use of rodent antibodies in therapy results in an anti-immunoglobulin immune response (human anti-mouse antibody response, or HAMA response) [69,70]. This response is directed mainly towards the rodent constant regions and is highly undesirable, since it causes toxic reactions and rapid elimination/neutralization of the antibody. In the 1980s therefore, considerable effort was put into genetically manipulating rodent antibodies to make them less immunogenic for use in therapy [9,11–13,71,72]. The production of the genetically manipulated antibodies was made possible by the development of mammalian-based expression systems for efficient secretion of recombinant immunoglobulins [3–5]. The PCR-based approaches described above can all be applied to the manipulation of immunoglobulins. However, these techniques have been available only for the past few years. Prior to this, other methods of genetic manipulation were used to generate chimeric and humanized antibodies.

Simple Chimeric Immunoglobulins. In the early mid-1980s, the availability of the genes encoding antibody molecules allowed the formation of human–mouse chimeric molecules by the replacement of the rodent constant domains with human constant domains by relatively simple genetic manipulations. These spliced genes were then expressed in mammalian cells as secreted, chimeric proteins [9,11–13]. This approach links the specificity (determined by the variable domains) of the rodent antibody to human constant domains, and has been carried out for a variety of antibodies of potential therapeutic value. Such antibodies include those that recognize carcinoma antigen 17-1A [11] and L6.6 [9]. The binding specificity, as would be predicted from the discrete domain structure of the immunoglobulin molecule, is not affected by the chimerization process. Moreover, the replacement of the rodent constant domains with those of human origin allow the isotype and therefore the effector functions of the therapeutic antibody to be optimized [73] (see Chapter 1).

The use of human constant regions may not only generate less immunogenic antibodies for therapy, but the chimeras may have increased *in vivo* stability [74]. This is probably for two reasons: first, during prolonged therapy the HAMA response will increase the clearance rate of the therapeutic antibody. Second, the Fc region appears to be involved in determining the *in vivo* stability of the immunoglobulin molecule [75,76], and it is conceivable that human Fc regions confer higher stability in humans than murine Fc regions, owing to the presence of as yet unidentified species-specific residues in this region of the molecule that are involved in the control of immunoglobulin catabolism.

CDR-Grafted or Humanized Antibodies. For the chimeric antibodies described above, the rodent variable domains may also be seen as foreign, although they are less immunogenic than the rodent constant regions [74]. A refinement of the production of chimeric antibodies is the generation of humanized, or CDR-grafted, antibodies [71,72,77–83] (Fig. 3.7). The hypervariable, or CDR loops, of the immunoglobulin variable domains not only are the most variable regions of sequence from one antibody to another but, together with the flanking framework regions, they also determine the specificity and affinity of the antibody for binding to antigen [56,58–60].

To test the feasibility of transplanting the CDR loops, and concomitantly the binding specificity of a rodent antibody to a human framework, Jones and colleagues [71] initially "CDR-grafted" an antihapten antibody and demonstrated that the binding specificity could be transferred onto a human antibody with retention of affinity. This work was subsequently extended to the humanization of Campath-1 [72] and the antilysozyme D1.3 antibody [77]. Campath-1 has applications in therapy, and the humanized version has been used in the treatment of non-Hodgkin's lymphoma [84], systemic vasculitis [85], and rheumatoid arthritis [86]. More recently a number of other antibodies of possible therapeutic importance have been humanized—for example, anti-Tac [78], anti-HER2 [79], anti-respiratory syncytial virus [80], anti-herpes simplex [81], anti-HIV [82], and anti-EGF receptor [83] antibodies. All these humanized antibodies await the results of their use in controlled clinical trials to test their efficacy and immunogenicity.

In carrying out CDR grafting, it appears to be important for the retention of binding activity that the contacts of the CDRs with the donor framework residues be maintained in the acceptor framework. Thus, judicious choice of the human acceptor framework is essential if the binding affinity of the humanized antibody is to match that of the parent rodent antibody. The importance of maintaining the correct framework residues is exemplified in the work of Riechmann and colleagues [72], who showed that changing a single amino acid in the framework region of a humanized version of the Campath-1 changed the antibody from being almost inactive in binding antigen to having activity similar to that of the parent rodent Campath-1. Consistent with this, the importance of framework residues in orienting the CDR loops and/or contacting the antigen is demonstrated in the structures of antibody–antigen complexes which have now been solved by x-ray crystallography [56–60]. In addition, Tramontano and colleagues [87] have recently shown that framework residue 71 of the VH domain plays a major role in determining the conformation and orientation of the second CDR loop of this domain. The importance of this residue in maintaining the "correct" conformation of the CDRs in humanized antibodies, together with several other framework residues, has been analyzed extensively using the antilysozyme D1.3 system [88]. This study describes the effect of several framework changes on the binding affinity of the humanized antibody. In this respect it is interesting that in the antibody that was humanized by Riechmann and colleagues [72], the VH domain residue 71 is arginine in the parent antibody and valine in the human acceptor framework. Despite this, the humanized version maintains the binding affinity of the rat antibody; as suggested by Tramontano and colleagues [87], the smaller size of valine relative to arginine may allow the transplanted CDR2 loop to adopt the "correct" conformation in the recombinant antibody.

At the technical level, CDR grafting can now be carried out in several different ways: (1) Synthetic DNA that encodes the CDR loops can be grafted from the donor to the acceptor framework by using designed synthetic oligonucleotides and site-directed mutagenesis [50]. This involves the design of oligonucleotides that are complementary to the 5′ and 3′ framework regions that flank each CDR of the human acceptor antibody. The oligonucleotides have sequences that encode the rodent CDRs, and these sequences are flanked by the human framework sequences. (2) The PCR can be used, provided that restriction sites (that are preferably unique) flank the CDRs in the genes encoding the donor framework regions. Such sites can be inserted by site-directed mutagenesis (Fig. 3.5).

3.3. GENERATING MONOCLONAL ANTIBODIES *IN VITRO*

Recent developments in recombinant DNA technology, including the PCR, can be applied to the manipulation of antibody genes. The PCR can be used to isolate diverse repertoires of immunoglobulin variable domain genes from antibody-producing cells such as peripheral blood lymphocytes (PBLs) and splenocytes. This has led to the development of a new area of recombinant antibody work that involves the generation of vast numbers of different antibody variable domain genes in a relatively small number of PCRs. The genes can then be used as templates for the expression of recombinant antibody fragments in *E. coli*. The development of ways of efficiently expressing antibody fragments in this bacterial host allows the isolation of recombinant antibody fragments that have binding activities towards a chosen antigen. Thus, libraries of the variable domains can be expressed as either single VH domains (comprising a single heavy chain variable domain; Fig. 3.6), Fv fragments and Fab fragments using *E. coli* as a host, and the desired binding activity screened for [31,64]. More recently ways have been developed for the expression of recombinant antibodies on the surface of bacteriophage, which allows the selection of bacteriophage that bear immunoglobulin fragments that have antigen-binding activities using antigen coated surfaces [49,89,90] (Fig. 3.8). In this context, the term *screening* is used to describe the analysis of large numbers of bacterial colonies or plaques for the expression of the desired antibody fragments, using either ELISAs [64] or plaque screening [31]. Selection is used to describe the surface expression of antibodies on bacteriophage, and this has the advantage that the bacteriophage that bear fragments with the desired binding specificity can be isolated by selection ("panning") on antigen-coated surfaces (Fig. 3.8). The selected bacteriophage can then be used to reinfect *E. coli* to produce clones of cells that contain immunoglobulin genes with the desired antigen-binding activity (usually several rounds of selection are required, but this depends on the nature of the starting material; for example, if libraries of antibody genes generated from spleen cells and containing many different genes are expressed on the surface of bacteriophage, the

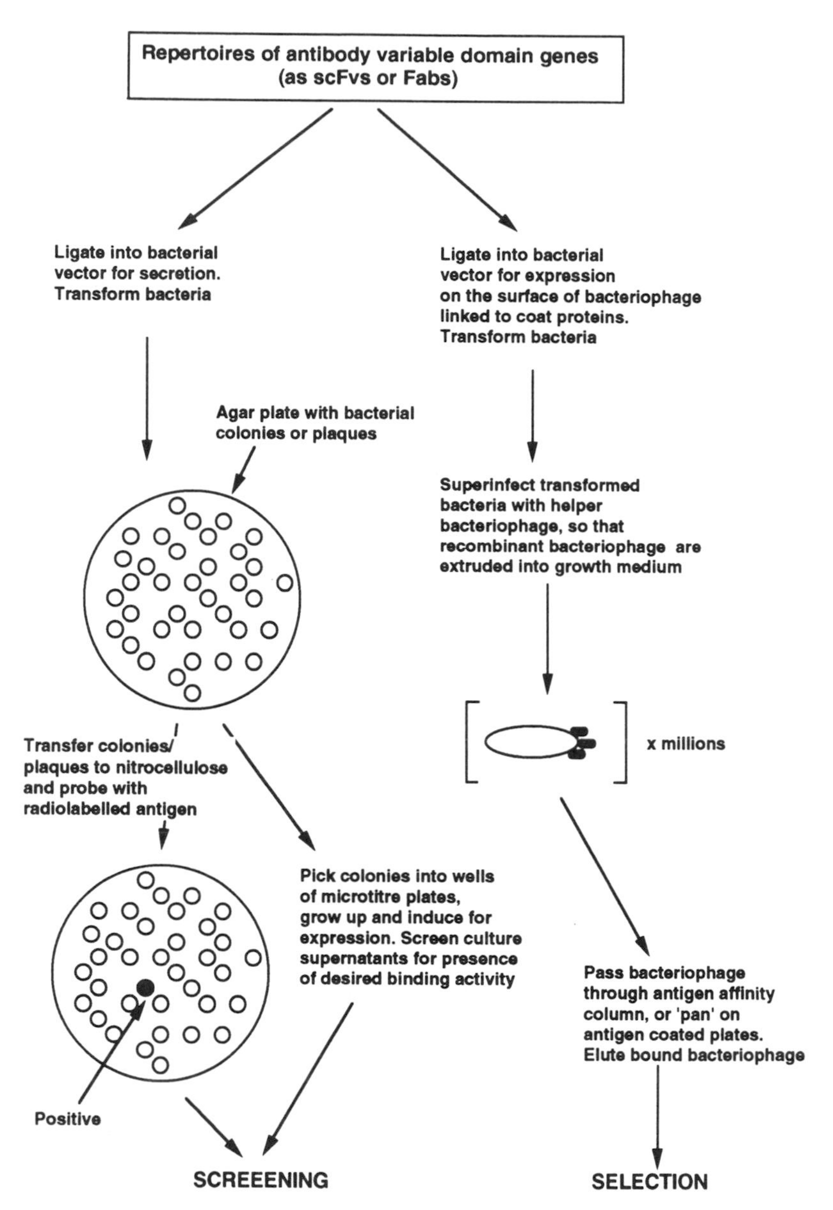

Fig. 3.8. Schematic representation of the screening and selection of immunoglobulin gene repertoires. The VH and VL gene repertoires are generated by the PCR and assembled as either scFv's or Fab's as described in the text. The genes can then be ligated into vectors for secretion or for surface expression on bacteriophage. The use of the former allows the antibody fragments that have the desired binding activity to be identified by screening (by either transferring the colonies to nitrocellulose and probing with radiolabeled antigen or by screening bacterial culture supernatants by ELISA). The use of bacteriophage surface expression allows the antibodies that have binding activity to be selected. The open ellipsoid represents a bacteriophage and the small filled ellipsoids represent antibody fragments expressed at the tip of the bacteriophage as fusion proteins with the gene III coat protein.

number of rounds of selection is usually three to four). Thus, selection clearly has advantages over screening, in that it is relatively easy to pan millions of bacteriophage on antigen-coated surfaces and, by reinfection of *E. coli* with the panned bacteriophage, to enrich for the desired clones. In contrast, screening of millions of clones is extremely tedious and would take weeks. It is now possible to isolate and express immunoglobulin gene repertoires in a matter of days, and this technology is an attractive alternative to the more time-consuming hybridoma technology [91]. This new approach and hybridoma technology each have their relative merits. Although the *in vitro* route is much more rapid, for the study of the development of an immune response *in vivo* the hybridoma route is still preferable, since the pairing between immunoglobulin heavy and light chains, as they exist within a cell *in vivo,* is maintained.

First, the developments that have led to this new technology will be described; they are (1) the use of the PCR to isolate repertoires of immunoglobulin genes, and (2) the generation of suitable prokaryotic expression systems for the production of the immunoglobulin fragments, and the screening or selection of clones that express fragments with the desired binding activity. Once fragments with the desired binding activity have been isolated, it is relatively straightforward to use recombinant DNA methods to link them to the constant regions of immunoglobulin molecules (for which the genes of many species and isotypes are now available) or to other proteins such as toxins (discussed in more detail in section 3.5). This combines the binding functions of the fragments with the effector functions of the desired properties. Second, the applications of these recombinant antibodies will be briefly discussed.

3.3.1. Use of the PCR

For the isolation of immunoglobulin variable domain genes (designated VH and VL for heavy and light chain variable respectively), several different types of PCR primers can be used [27–36]. As indicated in section 3.1.1, for the PCR it is necessary to have prior knowledge of the nucleotide sequences at the 5′ or 3′ ends of the genes (or flanking regions) that are to be isolated. The sequence knowledge does not have to be complete, as the PCR can be used with partially degenerate primers. The use of partially degenerate primers allows members of gene families to be isolated in a single PCR. These members may differ in one or more bases in the region to which the primer anneals. Clearly, the higher the primer degeneracy, however, the greater the degree of nonspecific binding and priming. Since immunoglobulin variable domain genes share considerable homology at their 5′ and 3′ ends, yet are not perfectly matched in sequence, they can be isolated en masse by the PCR and partially degenerate primers. This overcomes the problem of choosing a unique primer, and it is straightforward in programming oligonucleotide synthesizers to insert two or more bases at appropriate positions during syntheses.

There are now a number of different data bases that document the known sequences for the immunoglobulin genes of a variety of species. For example, the Kabat data base [37] contains sequence information for nine species including

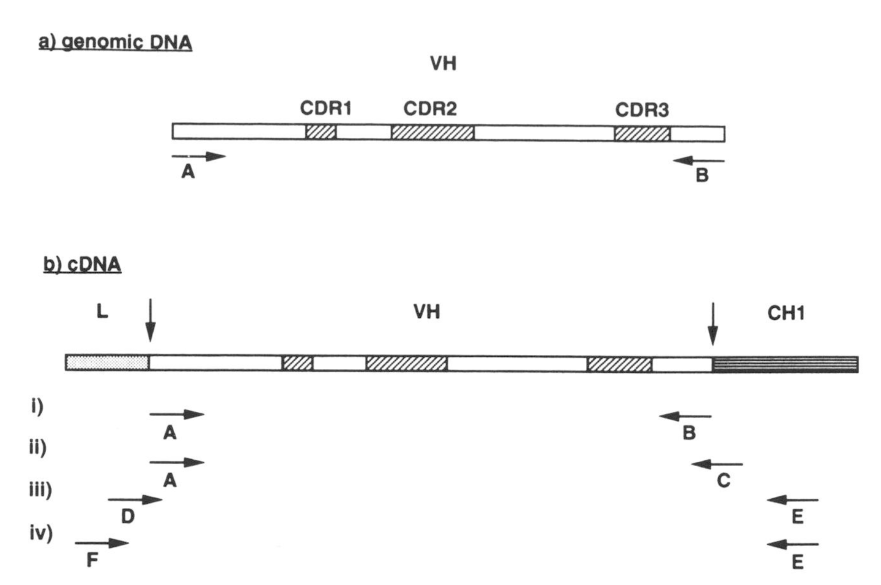

Fig. 3.9. Strategies for PCR amplification of immunoglobulin VH domain genes: **a.** From genomic DNA isolated from antibody-producing cells, primers A (5′) and B (3′) can be used to isolate the rearranged VH genes [27,34,64]. These primers anneal to the 5′ and 3′ ends of the coding sequences of the VH domain genes, and both productively and nonproductively rearranged genes (productively rearranged genes are those that give rise to complete open reading frames for translation) will be isolated. **b.** From cDNA isolated from antibody-producing cells, (i) primers A (5′) and B (3′) can be used, as in **a;** (ii) primers A (5′) and C (3′) can be used on cDNA only, since primer C overlaps both the J region and CH1 domain [33]; (iii) primers D (5′) and E (3′) can be used on cDNA only, since D overlaps the leader sequence and 5′ end of the VH domain genes, and E anneals to the CH1 domain [31,35,36]; and (iv) primers F (5′) and E (3′) can be used on cDNA only, since F anneals to the leader sequence (L) and E anneals to the CH1 domain [29,30,32]. All primers can be partially degenerate or family-specific. Hatched boxes represent CDR1, 2, and 3. The leader exon is represented by a stippled box, and the 5′ end of the CH1 domain by horizontal hatching. Vertical arrows indicate the location of intron–exon boundaries in the corresponding genomic DNA, which encodes rearranged immunoglobulin genes. Similar strategies can be used for the isolation of VL domain genes.

human, rat, mouse, and rabbit. This data base has been used to design oligonucleotide primers that anneal to the 5′ and 3′ ends of the variable domain genes of the mouse [27]. In addition, the primers were designed to have internal restriction sites, so that the resulting PCR products could be cloned directly into expression plasmids that have compatible restriction sites following amplification [27,64]. As an alternative approach, PCR primers have been designed for mouse and man that are specific for different VH and VL domain gene families at the 5′ ends and/or anneal to the constant domain genes [33] (Fig. 3.9). In addition, primers that are complementary to the leader sequences, or part of the leader sequence and the 5′ end of the VH domain gene, can be used [29–32,35,36].

Thus, there are a variety of primers that can be used in the PCR to isolate VH and VL domain genes as diverse repertoires from antibody-producing cells such as splenocytes or PBLs, or more simply, from homogenous populations of cells such as hybridomas. These genes can be cloned for expression as either Fv or Fab fragments, and the desired specificity isolated by screening or selection (described below).

3.3.2. Use of the PCR for the Isolation of the Immunoglobulin Genes

A variety of sources of immunoglobulin-producing cells can be used for the isolation of VH and VL domain genes. In the simplest case, isolation of the genes from the genomic DNA of preexisting hybridomas involves first washing the cells in phosphate-buffered saline [92]. Around 10^6 cells are the optimal number, but substantially smaller numbers can be used; in theory, one cell could be sufficient, but use of exceptionally small numbers of cells frequently results in both inefficient PCR and contamination problems. Following washing, the cells are resuspended in sterile water and boiled. Debris is pelleted by centrifugation, and aliquots of the supernatant containing the DNA are used in a standard PCR with the appropriate primers [92]. This method can clearly be used only with primers that anneal to positions within the VH and VL domain genes, that is, within one exon. In addition, the use of these primers on genomic DNA can result in the isolation of nonproductively rearranged genes [93]. In other words, during immunoglobulin gene rearrangement, due to the random nature of the process, some V, (D), and J elements are combined in a way that results in an interrupted (by a stop codon) reading frame. These genes will be present in the genomic DNA, but they are not efficiently transcribed into mRNA and would therefore be poorly represented in a cDNA library. Clearly the presence of a high proportion of nonproductively rearranged genes is a disadvantage for the generation of expression libraries of immunoglobulin genes derived from heterogenous populations of antibody-producing cells, and therefore for this purpose the use of cDNA may be preferable. As pointed out by Ghcrardi and Milstcin [94], however, the use of cDNA (mRNA) biases the libraries toward isolation of the VH and VL genes from plasma cells at the expense of memory cells, and this may also be a disadvantage in some cases. The choice between using cDNA or genomic DNA depends on the goals of the individual experiment. For example, if the animal or human from which the genes are being isolated has been immunized (in the case of humans, by immunizations such as tetanus or HIV infection), it is preferable to use mRNA if the goal is to isolate antibodies that recognize the immunogen. This is because during active immune response, expanding B cells will be expressing high levels of antigen-specific mRNA.

To generate cDNA for use in the PCR, several approaches can be used. Lymphocytes are first isolated on Ficoll-Hypaque from either PBLs or splenocytes and then total RNA purified. The cDNA synthesis can be primed with either poly-dT, immunoglobulin constant region specific primers, or random hexanucleotide primers [32]. Random hexanucleotides are available commercially and are hexamers that anneal randomly throughout the genome. Priming with constant region specific primers may be preferable in many cases, as it will result in the production of only immunoglobulin-derived cDNA for use in the PCR. If cDNA derived from all the different isotypes is desired, however, it may be preferable to prime cDNA with poly-dT or random hexanucleotides.

3.3.3. Expression of the Antibody Genes Using *E. coli* as a Host

In the early 1980s considerable effort was put into expressing immunoglobulin fragments in *E. coli,* and the protein was expressed as intracellular inclusion bodies [95–99]. Antibody fragments could be obtained from these inclusion bodies by solubilization and refolding, frequently with low yields of functional protein. Improved methods for refolding immunoglobulin fragments that are isolated from inclusion bodies are not available [100,101], but these methods are generally more tedious than the purification of fragments from either the culture supernatant or periplasmic space [14,15]. In addition, the secretion of immunoglobulin fragments into the periplasmic space, whence it leaks into the culture supernatant during prolonged induction periods, greatly facilitates the screening of clones for the production of binding activities [64].

Secretion systems for the production of antibody Fab and Fv fragments were first reported in 1988 [14,15] and involved the linkage of the antibody genes in translational frame to prokaryotic leader sequences such as OmpA, phoA, and pelB (the former two leaders are derived from *E. coli* genes, and the latter leader from the pectate lyase gene of *Erwinia carotovora*). These leaders direct the expressed protein into the periplasmic space, where the leader sequences are cleaved from the recombinant protein by signal peptidase. By means of such systems, yields of 2–5 mg/L and 5–10 mg/L of culture for Fab and Fv fragments, respectively, can be obtained [14,15,64]. Plasmids for the expression and secretion of antibody fragments are shown in Figure 3.10. Although Fv fragments may have advantages due to their relatively small size (approximately one sixth of the size of a complete immunoglobulin molecule), the stability of the association of the VH and VL domain appears to vary considerably from one Fv to another. In contrast, for Fab fragments the association of the CH1 and CL (Cκ or Cλ) domains appears to be stable, which in turn acts to stabilize the VH–VL domain interaction. The variable stability of the VH–VL interaction in Fv fragments is presumably due to the differences in the sequences of the CDR3s from one antibody to another, as these loops form the core of the antibody–antigen combining site and bridge the interface of the VH/VL domain interaction [102]. The sequences of these loops may be directly involved in the interaction of the two domains or, alternatively, may affect the conformation of the flanking framework residues, which in turn affects the stability of the interaction of the VH and VL domains. In a high proportion of cases the genes encoding the antibody fragments will be used to rebuild complete antibody molecules, in which the VH–VL domain interaction is stable. Thus, the instability is usually only a problem during screening or selection for the Fv fragment of the desired specificity. In some instances, however, it may be that smaller fragments such as Fv's are attractive for use. For example, it is conceivable that the rapid

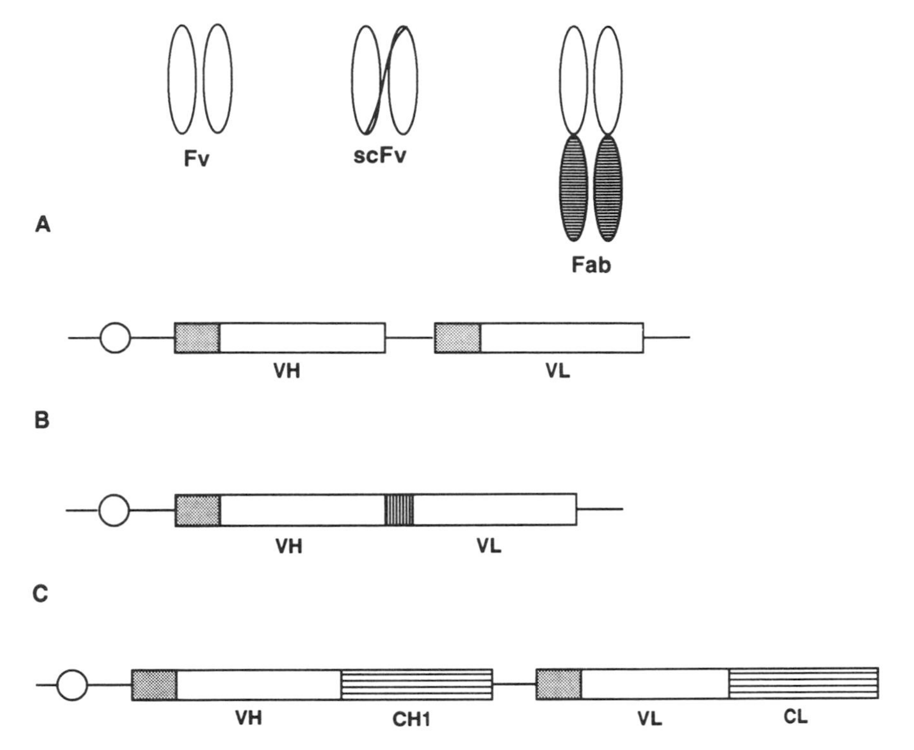

Fig. 3.10. Plasmic vectors for the secretion of immunoglobulin fragments [14,64]. **A.** Fv fragments. **B.** scFv fragments. **C.** Fab fragments. The restriction sites in these plasmids can be modified to accommodate genes tailored with different restriction sites. These sites are incorporated during the PCR [27,31,64]. The pelB leader sequences [14] are represented by stippled boxes, VH and VL genes by open boxes, the single chain linker sequence [104] by vertical hatching, and the antibody CH1 and CL domains by horizontal hatching. Open circles represent the *lacz* promoter.

clearance of the Fv fragment may make it a useful reagent for binding to and rapidly clearing toxic substances from the body—for example, in the case of drug overdoses. There are several possible ways of stabilizing the VH–VL domain interaction, as follows.

1. An intramolecular –S–S– bridge can be inserted between the two domains by genetic engineering [103]. Thus, light chain residue 55 and heavy chain residue 108, or light chain residue 56 and heavy chain residue 106, can be changed to cysteine residues that are close together in the three-dimensional structure of the associated VH–VL domains. This has been carried out for an antiphosphorylcholine Fv and shown to result in the expression of functionally active Fv fragments that are stably associated [103]. Due to the conservation of the structure of the immunoglobulin fold, this approach is probably general for a wide range of Fv fragments; it is difficult, however, to see how such an –S–S– bridge can be incorporated into repertoires of antibody genes during the generation of expression libraries.

2. Glutaraldehyde treatment has been used to chemically cross-link the VH and VL domains of an antiphosphorylcholine Fv fragment, to improve the stability without loss of functional activity [103].

3. An approach that can readily be incorporated into synthetic libraries of antibody genes, unlike (2) and (2) above, is to express the VH and VL domains as single chain Fv fragments (scFv's) (Fig. 3.6). In scFv's, the VH domain is linked by a peptide linker to the VL domain [104,105]. The use of a number of different peptide linkers has now been reported, and the scFv's can either be expressed as intracellular inclusion bodies or be secreted into the periplasmic space [101,104–110]. The secretion yields are usually lower than those of inclusion bodies, but the ease of purification when secretion systems are used may offset the advantage of high yields. For most purposes, the secretion yields are sufficient for further analyses of binding specificity and affinity of the scFv fragments prior to their being used to rebuild complete antibodies (section 3.5). The presence of the linker peptide may result in a lowering of affinity of the antibody fragment for interaction with antigen, possibly due to steric interference of the linker with the binding site and/or distortion of the CDR loops by the linker peptide. The isolation of the scFv's from inclusion bodies by denaturation followed by renaturation may also result in a decrease in affinity due to incomplete refolding. The potential disadvantage of a reduction in affinity is usually outweighed by the advantage of producing the Fv as a stably associated VH–VL heterodimer, and if binding activities are being selected from recombinant libraries, this becomes irrelevant, since it is possible to select the highest-affinity scFv from a background of lower-affinity fragments (section 3.5 below).

For optimization of the expression yield and affinity, testing of a variety of peptide linkers and/or switching the order of the VH and VL domains in the expression plasmid is advisable. For an anticarbohydrate Fv fragment, much higher levels of secretion were obtained when the VL domain was located 5′ to the VH domain gene [110], indicating that the order of VH and VL domain genes with respect to each other may have a drastic effect on the expression levels.

In summary, for the production and isolation of Fv's with the desired binding characteristics from libraries of antibody genes, the insertion of a single chain linker peptide is useful for the screening/selection steps as this stabilizes the VH–VL domain interaction. If the ultimate aim is to rebuild a complete immunoglobulin with the desired effector functions (section 3.5), it is easy to remove the synthetic linker by using designed synthetic primers and the PCR.

3.3.4. Small Units of Antigen Binding

Single VH Domains ("dAbs"). The observation that the VH domain of the antilysozyme D1.3 antibody has high affinity for lysozyme binding in the absence of the paired VκD1.3 domain prompted the generation of diverse repertoires of immunoglobulin VH domain genes by using the PCR [64]. These genes were isolated from the splenocytes of mice that had been immunized with hen egg lysozyme or keyhold limpet hemocyanin (KLH). The VH domain genes were cloned into expression plasmids containing the pelB leader for secretion, and the supernatants of the clones screened for the presence of VH domains with antigen-binding activities. The diversity of the repertoires were checked by nucleotide sequencing; and of more than 50 VH gene sequences, all were unique, and members of each of the murine DH (diversity segment) and JH (joining segment) families were represented [64,93]. The recombinant *E. coli* clones were grown up and induced for expression, and the culture supernatant was screened with ELISA for the presence of binding activities towards lysozyme and KLH. VH domains with antigen-binding activities against these two antigens were isolated, and two of the antilysozyme VH domains characterized further in terms of affinity, nucleotide sequence, and specificity. These VH domains had affinities of the order of 10^{-8} M for binding lysozyme in solution, and also did not show binding affinities for a variety of antigens other than hen egg lysozyme. This affinity is reasonable, and is similar to that of a significant proportion of complete immunoglobulin molecules. More recently, VH domains that have specificity for mucin and for influenza neuraminidase have also been isolated (D. Allen and P. Hudson, personal communication).

The expression levels of VH domains as secreted proteins from recombinant *E. coli* cells are significantly lower than those of Fv and Fab fragments, and this is probably a reflection of their rather hydrophobic and insoluble character [64]. The exposure of the conserved hydrophobic residues of VH domains that are buried in an Fv or Fab fragment by association with the paired VL domain presumably contributes to the rather insoluble characteristics of these single domains. It may also be that the VH domains, in the absence of associated VL domains, are rather unstable and denature easily. Thus, these rather unattractive features may offset the advantages of the small size of VH domains. It may be possible, however, to use protein engineering to improve the properties of these domains. Such engineering would involve the insertion of residues to stabilize the immunoglobulin fold and residues of more hydrophilic character at the exposed positions. An alternative approach may be to design a stable framework structure onto which one or more CDR loops, which are known to play an important role in binding to cognate antigen, could be mounted in a suitable conformation for binding. This approach requires a significant amount of protein modeling and design, and the technology for the generation of such "de novo" antibodies may be available in the near future.

Minimal Recognition Units (MRUs). CDR-derived peptides have recently been isolated from antibodies and shown to retain binding affinity for cognate antigen [111,112]. These peptides were designed on the basis of modeling and/or sequence analysis and have potential as therapeutic reagents as blocking peptides. This approach will probably work only for antibodies for which one of the six CDRs makes the major contribution to the binding of the antigen to the antibody, and for this reason it may not be general. It does, however, offer an new immunotherapeutic approach in the cases in which individual CDRs have significant binding affinities.

3.3.5. Combinatorial Libraries of Immunoglobulin Genes

One of the potential problems that may occur during the isolation of antibody VH and VL domain genes from antibody-producing cells is that the matching that exists between a VH and a VL domain within a particular B cell is lost once the cells have been lysed (as a mixed population) and their nucleic acids extracted. To reconstitute an Fv or Fab fragment with binding activity, the genes need to be recombined in such a way that functional binding units can be produced. This may not necessitate the matching of VH and VL domains as they existed *in vivo,* as VH domains appear to be able to combine with a number of different VL domains and still retain antigen-binding activity [113]. There is therefore a need to be able to recombine the libraries (or repertoires) of VH and VL domain genes in as random a way as is possible. Recombinant DNA technology now allows the *in vitro* generation of randomly combined repertoires of VH and VL domain genes. The VH and VL domain genes can be combined either by ligation at a unique restriction site [31], or by splicing by overlap extension (SOE) (Fig. 3.4) [49,114]. The expression of randomly combined VH and VL domain genes as Fab fragments using the lambda Zap expression plasmid was reported by Huse and colleagues [31]. In this study, Fab fragments with reactivities towards the immunogen p-nitrophenyl phosphonamidate (NPN; a transition state analogue for carboxamide hydrolysis) were isolated. More recently, the approach has been used to isolate Fabs from both murine and human-derived repertoires [31,35,36,116]. In all cases, the genes were isolated from "immunized repertoires"; that is, even for the human-derived fragments the human donor had recently been boosted with tetanus toxoid, the antigen used in these studies for subsequent selection for binding fragments. For the isolation of Fv or Fab fragments from unimmunized or naive repertoires, it is generally predicted that extremely large numbers of recombinant clones will have to be screened. This is due to the fact that exposure of the immune system to antigen results in the clonal expansion and affinity maturation of the antigen-specific B-cell clones with extremely high efficiency [116,117]. Thus, for the isolation of very-low-frequency clones from naive libraries, the use of selection systems is desirable to avoid tedious screening.

A major step in the development of selection systems was reported by McCafferty and colleagues in 1990 [89] (Fig. 3.8). This system involves the expression of antibody scFv fragments on the surface of the filamentous bacteriophage fd. Genes encoding the D1.3 scFv were inserted in translational frame into the gene III coat protein of the bacteriophage. The recombinant bacteriophage were shown to bind specifically to lysozyme-coated surfaces, and this opened the door for the surface expression of libraries of immunoglobulin variable domain genes, followed by selection of phage bearing antigen-binding scFv's.

The phage expression systems have now been extended to the expression of libraries of scFv's derived from immunized mice, and scFv's that recognize the immunogen pHOx (2-phenyloxazol-5-one) have been isolated by selection by binding to antigen-coated surfaces (panning) and characterized [49]. Interestingly, from a library size of 10^6 clones derived from a naive mouse, no pHOx-specific clones could be identified, indicating that for the isolation of antigen-specific clones from naive libraries the numbers of clones generated should be extremely large. In this respect, Marks and colleagues [34] have recently extended the use of the phage expression system to the isolation of anti-turkey-egg lysozyme and anti-pHOx clones from naive human-derived libraries. The library sizes in this study were 2.9×10^7 to 1×10^8, and four rounds of panning (i.e., bacteriophage isolated from the first round of panning were used to reinfect *E. coli,* the resulting clones grown up, and the extruded bacteriophage panned again on antigen-coated surfaces) were required to isolate antigen-specific, and presumably low-frequency, clones.

Phage expression systems similar to those used for the display of scFv fragments have been reported for the surface expression of Fab fragments using the gene III and gene VIII coat protein genes of filamentous bacteriophage such as fd or M13 [90,118,119]. Using these phage display systems, Fab's with binding specificities toward tetanus toxoid (human-derived) [119], HIV (human-derived) [120], hepatitis antigens (human-derived [121], and progesterone (murine) [52] have been isolated. The gene VIII coat protein is expressed in up to 24 molecules per phage particle, in contrast to the gene III coat protein that can be expressed in up to 4 copies per particle. Thus, for the isolation of low-affinity, high-avidity clones it may be preferable to use the gene VIII system, in which Fabs are expressed in multivalent form on the surface of the bacteriophage. In contrast, for the selection of higher-affinity, low-avidity clones, surface expression using antibody–gene III fusions is suitable.

The bacteriophage display systems greatly facilitate the isolation of low-frequency, low-affinity clones from immunoglobulin fragment expression libraries. Prior to the development of these selection systems, it was impractical to screen the extremely large numbers of clones necessary to isolate the desired specificity from naive repertoires. The use of the phage display systems for the isolation of human-derived antibodies is particularly attractive, since to date the production of human monoclonal antibodies by hybridoma technology has proved difficult due to both technical and ethical barriers [66–68].

For the purification and further characterization of the antibody fragments, following identification of phage that bear scFv's or Fab's with the desired binding activity, it is facile to express the fragments as secreted proteins in one of the two following ways.

1. The pHEN1 vector [122] has a suppressible amber codon that is located between the 3′ end of the antibody gene and the 5′ end of the gene III coat protein gene. Thus, by switching between a suppressor and a nonsuppressor strain of *E. coli,* the fragment can be surface-expressed on bacteriophage or secreted as soluble fragments into the culture supernatant.

2. The Fab fragment-gene III/VIII genes in the pCOMB vectors can be modified by restriction and religation, to remove the gene III/VIII gene, causing the Fab fragment to be secreted as a soluble protein, rather than attached to the surface of bacteriophage particles in linkage to coat protein genes [119].

3.3.6. Improvement of the Affinities of the Fragments Isolated From Naive Repertoires

The binding affinity of the antibody fragments derived from naive libraries will generally be lower than those derived from immunized repertoires, due to the lack of affinity maturation [116,117] for naive repertoires. There are several ways in which the affinities could be improved. The genes encoding the low-affinity fragments could be used as templates for rounds of *in vitro* mutagenesis followed by surface expression and selection [52] (Figs. 3.8 and 3.11), in an attempt to mimic somatic mutation *in vivo*. Alternatively, severe combined immunodeficient (SCID) mice that have been populated with human PBLs [123] or transgenic mice that have the human immunoglobulin loci [124] can be used. These mice can be immunized and the antibody-producing cells used to generate VH and VL gene libraries for surface expression and selection. In addition, *in vitro* immunization of human PBLs

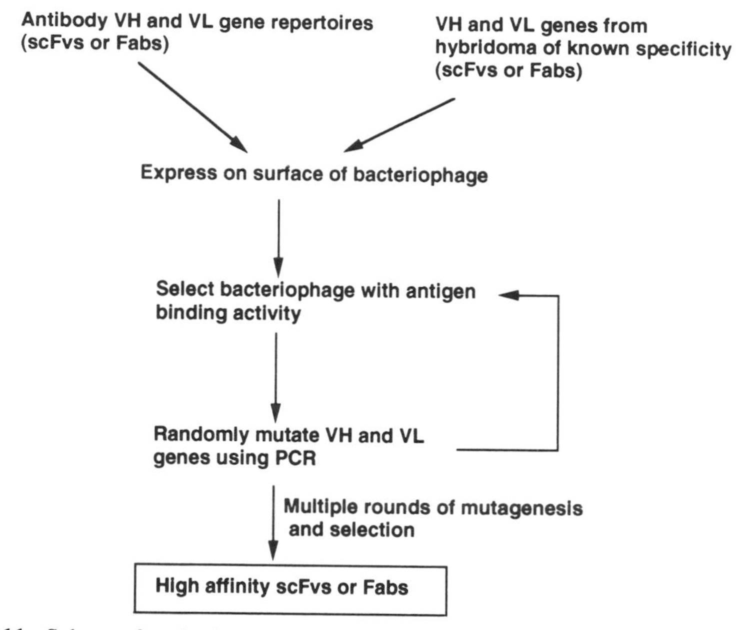

Fig. 3.11. Scheme for the improvement of antigen-binding activities of immunoglobulin scFvs or Fabs expressed on the surface of bacteriophage. The scFv or Fab can be derived from either VH and VL gene libraries or from a hybridoma of defined specificity. For the former, antibodies that have the desired binding activity have to be first selected from the libraries. For the latter, selection of binding fragments from a background of nonbinders is unnecessary due to the clonal nature of hybridomas. The VH and VL genes can then be randomly mutated by a number of different methods (see text) and the antibody variants that have improved affinity can be selected from a background of lower-affinity variants, by bacteriophage surface expression. After each round of mutagenesis and selection, the fragments can be characterized as secreted scFvs or Fabs (see text) and further rounds of mutagenesis/selection can be carried out until the desired affinity is reached.

[123,125] can be used to increase the proportion of antigen-specific B cells that are used to repopulate the SCID mice. These approaches are described in more detail below.

Error-Prone PCR. A possible route for the production of high-affinity antibodies from naive repertoires has recently been reported by Gram and colleagues [52]. This involved the isolation of antiprogesterone (IgM) antibodies in the form of Fab fragments from the bone marrow cells of unimmunized mice using the high-avidity pCOMB8 (gene VIII coat protein) vector. The genes encoding several anti-progesterone specificities were then expressed as scFv's using the low-avidity pCOMB3 (gene III coat protein), and the scFv genes were used as templates for error-prone PCR [51,52]. By this approach, it was demonstrated that the affinity could be improved 13- to 30-fold after one round of mutagenesis and selection. Further rounds of mutagenesis and selection could presumably be used to increase the affinity further. What the limits to improvement of affinity by this approach might be is not clear at present.

Immunization of Transgenic Mice and Severe Combined Immunodeficient (SCID) Mice

Transgenic Mice. Transgenic mice have recently been described that have a human immunoglobulin heavy chain minilocus (an incomplete human heavy chain locus in germline configuration [124]). These genes can be rearranged in functionally active form, presumably due to the conservation of the recombinase activity between mouse and man. Immunoglobulin molecules that make up human heavy chains and murine light chains can be isolated from the serum of the mice. This is a step towards creating the entire human immunoglobulin repertoire in mice; it would provide an invaluable tool for generation of "pure" human antibodies. Moreover, the transgenic mice could be immunized with antigens of choice, allowing affinity maturation of the immune response to occur *in vivo,* rather than using the *in vitro* approaches described above. The main advantage of using the *in vivo* route rather than *in vitro* mutagenesis is that it is clearly less labor-intensive. There may, however, be disadvantages, insofar as the extent of somatic mutation *in vivo* depends on the immunogenicity of the antigen. In contrast, *in vitro* mutagenesis is not limited by the immunogenicity of the antigen for which high-affinity antibodies are being generated.

SCID mice. It has recently been demonstrated that it is possible to repopulate SCID mice with human PBLS [123,126,127], and to isolate human-derived antibodies from these mice by the *in vitro* approaches described above. In this approach, human PBLs were immunized *in vitro* prior to being injected into mice, and then the mice were immunized with antigen *in vivo* after transfer of the PBLs [123]. Fabs with specificities for hepatitis B core antigen and tetanus toxoid were isolated and characterized further. This route promises to have potential in the isolation of antibodies from PBLs derived, for example, from HIV-seropositive donors and from autoimmune patients, and the resulting antibodies have obvious applications in therapy and diagnosis.

Semisynthetic Immunoglobulins. For the structures of antibody–antigen complexes solved to date, the CDR3 region of the immunoglobulin VH domain usually plays a significant, if not major, role in the interaction [56,59,60,128]. Thus hypermutation of this region of an immunoglobulin may not only result in clones that have higher affinity for antigen binding, but may also result in the generation of novel specificities. In this respect, hypermutation of the CDR3 of the antilysozyme D1.3 VH domain, using the PCR and a degenerate oligonucleotide (with equimolar mixes of A, T, C, and G inserted at each position corresponding to CDR3 residues during synthesis), resulted in the generation of higher-affinity variants [129]. These were identified by screening. It is now much more convenient to combine random PCR-directed mutagenesis with the phage display systems, and to use this combination of PCR-based technology with selection to identify phage that bear novel binding specificities.

Recently, the PCR, in combination with a highly degenerate oligonucleotide primer ("randomized" in the region corresponding to CDR3 and complementary at the 5′ and 3′ ends to the flanking framework regions; Fig. 3.12), has been used to generate a library of human-derived antibodies that differ only in their VH CDR3 region [130]. The maximum number of different CDR3 sequences in this library was estimated to be 10^{20}. For this study, the gene encoding an anti-tetanus-toxoid Fab was used as the template, and the mutated genes were cloned for phage display using pCOMB3. The resulting phage were panned on fluorescein-coated plates, and phage bearing antifluorescein specificities were selected and isolated under a variety of elution conditions. Thus, this approach can be used to generate new specificities from an antibody fragment that originally had a specificity towards tetanus toxoid.

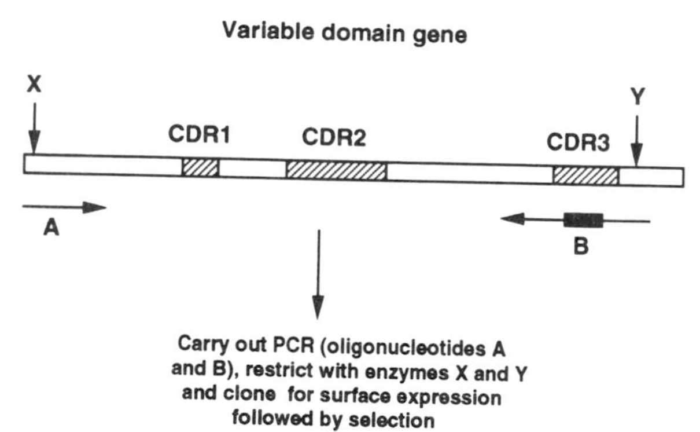

Fig. 3.12. Strategy for the random mutagenesis of a VH domain gene. The PCR is carried out with two primers: A anneals to the 5′ end of the VH domain gene and has an internal restriction site X. B is complementary to the framework regions that flank CDR3 and has a region of random nucleotides (shown schematically by the filled box) that overlaps, but is not complementary, to CDR3. B also has an internal restriction site Y.

To date this approach has been used to generate antihapten antibodies, and haptens are known to interact with substantially smaller numbers of CDR residues than protein antigens. Because of the crystallographic structures of antibody–antigen complexes [56,58–60], protein antigens usually contact residues that are located in all six CDRs, and the surface area of the interaction is extensive. It is probable that to generate fragments that have specificities for protein antigens (as opposed to haptens) rather more extensive mutagenesis of the CDR loops will have to be carried out—that is, up to all six CDRs may have to be hypermutated. This is technically possible using the PCR, for example, but the numbers of mutants that will have to be generated is enormous; thus, if this is to be feasible, the bacteriophage surface expression system needs to be able to select higher-affinity variants out of a large background of lower-affinity variants, and as yet the limits of this system in this respect have not been rigorously tested.

3.3.7. Immunogenicity of Antibodies Generated *In Vitro*

One of the goals of engineering antibodies for use in therapy is to reduce their immunogenicity. Initially, simple chimeric antibodies were made [9,11–13], followed by CDR-grafted or humanized immunoglobulins, in which only the CDR loops are of rodent origin [71,72,77–83]. Using current developments in technology, it is possible to isolate antibody fragments of human origin, and these have considerable potential as therapeutic reagents. However, hypermutation of human antibody variable domains with the aim of improving the binding affinity may generate sequence motifs that are immunogenic *in vivo*. The likelihood that this will occur probably depends on the extent of hypermutation and on the regions of the variable domains that are hypermutated. *A priori* it would be predicted that mutations in the CDR loops (particularly the highly diverse CDR3) are less likely to be immunogenic than mutations in the more conserved framework regions. The latter mutations are less likely to be selected, since mutations in these regions have (1) a lower probability of improving the binding affinity and (2) a higher probability than mutations in the CDRs of disrupting the immunoglobulin fold structure. In addition, the production of antibodies with completely novel CDR loops may generate fragments that have autoreactivities *in vivo,* and are representative of specificities that are normally "forbidden." Whether these concerns are valid awaits the testing of these new recombinant antibodies *in vivo*.

3.4. FRAGMENTS FOR THERAPY AND DIAGNOSIS

The rapid *in vivo* clearance rates [131–135] of immunoglobulin Fv and Fab fragments makes them attractive reagents for use in clinical situations in which short half-lives are either advantageous or required. A prerequisite for an Fv or Fab to be useful *in vivo* is for it to be retained in the location where it is needed, and this will clearly depend on factors such as the affinity of the fragment for antigen, and the tissue in which the target antigen is located. The fragments, like the complete immunoglobulin molecule, have biphasic clearance curves. For an Fv fragment the half-life for the α phase is of the order of 0.15 h and that for the β phase is 1.4 h; and after 4 h, less than 1% of the injected dose is present in the serum [135]. In contrast, the complete IgG molecule has a serum half-life that is of the order of 6–9 days [136,137]. The reasons for the rapid clearance of the fragments are that (1) they do not contain the Fc portion of the molecule, and a considerable amount of experimental data indicate that this is important in stabilizing the immunoglobulin molecule *in vivo* [75,76]; and (2) they are small relative to a complete immunoglobulin. The role of size in determining *in vivo* clearance rates is indicated by the finding that an $(Fab)_2$ fragment is cleared twice as slowly as an Fab fragment [131]. The *in vivo* half-life of the $(Fab)_2$ fragment, however, is still substantially less than that of the complete immunoglobulin molecule and even that of the smaller Fc fragment derived by proteolysis [138], indicating that sequences in the Fc region are involved in stabilizing immunoglobulins *in vivo*. Fab fragments may be preferable to Fv fragments for *in vivo* use, as Fv fragments have a tendency to dissociate (see section 3.3 above). For example, dissociation of the D1.3 Fv fragment was observed in an *in vivo* clearance study carried out in rats [135]. Using a combination of genetic manipulation and protein chemistry [135], a bivalent Fv fragment has been constructed that has improved stability *in vivo*. In addition, the bivalent nature of this protein may improve the avidity of the antibody–antigen interaction. An alternative way to prevent dissociation of the VH and VL domains of Fv fragments for *in vivo* use might be to covalently link the domains using recently described peptide linkers to generate scFv's (see section 3.3). However, these linkers may be immunogenic, which is clearly a disadvantage for use in therapy in which repeated doses are necessary.

Rapid clearance of immunoglobulin fragments is particularly attractive if they are to be used as reagents for the imaging of tumors [131,132]. Clearly for an Fab to be a useful imaging reagent, it needs to localize efficiently to the tumor site. The rapid clearance reduces the immunogenicity, which is of significance if the imaging is to be followed by therapy using the same antibody as a complete immunoglobulin molecule. A further advantage of the use of Fab or Fv fragments in imaging is that they appear to give high tumor : normal tissue localization ratios [131,132], and they may be more penetrative into tumor masses [139]. The relatively high tumor : normal tissue localization is presumably due to the absence of Fc receptor-mediated binding [140,141] and rapid clearance. Interestingly, a recent study [132] has indicated that scFv fragments may be preferable to the use of Fab or $(Fab)_2$ fragments for imaging, since the smaller fragments do not accumulate in the kidneys.

Fragments of immunoglobulins may also have uses in situations in which binding followed by rapid clearance is desirable, and such a situation occurs in the treatment of drug overdoses. The fragments would be predicted to bind to the drug and clear it rapidly from the circulation. For such an antibody fragment, it would be attractive if it were also to block the binding of the drug to its receptor, and the new *in vitro* approaches to generate antibody fragments of defined specificities may facilitate the development of such reagents.

3.5. REBUILDING THE FRAGMENTS AND EXPRESSION AS COMPLETE IMMUNOGLOBULIN MOLECULES OR AS ANTIBODY-CHIMERAS

Clearly antibody Fv and Fab fragments have no effector functions, and therefore for effective treatment of tumors and infections it is necessary to link the fragments to moieties such as the immunoglobulin Fc fragment or bacterial/plant toxins. In addition, the fragments may be linked to radioisotopes, but as recombinant DNA techniques are not used to generate radiolabeled antibody (fragments), this will not be discussed further. The Fc portion of the antibody is known to be involved in the effector functions such as antibody-dependent cell-mediated cytotoxicity (ADCC) and complement fixation [140–142]. The Fv and Fab fragments that are generated by the methods described above can be used as building blocks for complete antibody molecules. Fabs may, however, have uses in the treatment of infections in situations in which neutralization of the infectious agent is all that is required to block progression to disease. To use Fv or Fab fragments as building blocks, the genes that encode high-affinity Fv/Fab fragments of the desired specificity can be linked to the genes encoding the desired isotype. The choice of isotype depends on the effector functions that are required. For example, using a set of matched chimeric antibodies that differ only in their constant regions, it has been shown that the human IgG1 isotype is the most effective in complement-mediated lysis and ADCC [73,143,144], and this may therefore be the one of choice for human therapy. In this respect, the humanized version of the antibody Campath-1 was linked to the human IgG1 isotype prior to effective use in the therapy of non-Hodgkin's lymphoma [72,84]. The complete antibodies can be expressed in a variety of mammalian expression hosts, and these are described in more detail in section 3.6 below.

The antibody fragments can also be linked to toxins such as *Pseudomonas* exotoxin [145,146], ricin A chain [147], or phospholipase C from *Clostridium perfringens* [148] to generate immunotoxins (ITs). For example, in the case of *Pseudomonas* exotoxin, a truncated form that lacks the cell-binding domain and therefore has no toxicity if used unlinked to an antibody has been linked to a scFv fragment. This scFv recognizes a carbohydrate antigen expressed on the surface of many human carcinomas [107]. The exotoxin also has a lysine–aspartic acid–glutamic acid–leucine (KDEL) tetrapeptide at the carboxy terminus, and this peptide has been demonstrated to improve the efficacy of the IT, presumably due to an effect on the intracellular routing of the toxin [149]. This Fv–toxin chimera can be expressed in high yields in recombinant *E. coli* cells, and it has recently been demonstrated that it is effective in the treatment of tumors in immunodeficient mice [107]. A potential caveat about the use of ITs in prolonged therapy is that problems may be associated with the immunogenicity of the toxin moiety. It may be possible to overcome this by using different toxin moieties in cycles, and this depends on the availability of toxins with the desired therapeutic efficacies. Clearly, if the IT has immunosuppressive activity, the immunogenicity is not of as much concern as for ITs without this activity.

Generally the toxins that are attached to antibodies or antibody fragments have extremely high potencies, and for this reason it is particularly important to ensure that the antibody fragment to which they are attached is specific, to avoid the potentially disastrous consequences of nonspecific cytotoxicity. For example, neurological toxicity occurred when a therapeutic IT was constructed using a monoclonal antibody that unexpectedly cross-reacted with a component of the central nervous system [150]. This problem could be circumvented by the generation of antibodies of high affinity and specificity to epitopes that are not shared by other tissues in the body. This is frequently difficult to do, however, since in practice it is obviously very difficult to screen all the tissues of a human or animal for cross-reactivities.

A further route for the attachment of effector functions to an antibody fragment is an indirect one. Shalaby and colleagues [79] recently reported the construction of a bispecific Fab, for which one arm is specific for the HER2 protooncogene and the other arm is specific for the CD3 complex expressed on T cells. The bispecific antibody has the capacity both to bind to tumor cells and to capture T cells in the vicinity of the tumor cells. This bispecific reagent was produced by expressing the two Fabs individually in *E. coli* cells, followed by purification and chemical linkage [79]. This approach may be generally applicable for the production of bispecific antibodies, and is particularly attractive since it is much more rapid than the more conventional route of fusion and screening of hybridomas [151]. In addition, it allows an Fab of a particular specificity to be coupled to a range of Fabs of different specificities to generate a panel of bispecifics with relative ease and rapidity. Thus, mixing and matching experiments can be carried out with relative ease using purified Fabs of different specificities.

3.6. EXPRESSION OF ANTIBODY GENES IN MAMMALIAN CELLS

Rebuilt antibody genes, in which the V-regions are linked to a C-region with the desired effector function, cannot be efficiently expressed in *E. coli*. Several attempts have been made to produce intact immunoglobulins in bacteria (reviewed in Better and Horwitz [152], and in all cases inefficient reassembly of the heavy and light chains *in vitro* was required to produce active antibody, since the immunoglobulins were not correctly assembled or secreted from the cell. Recombinant antibodies have, on the other hand, been successfully produced from the cells of diverse eukaryotes, including yeast [153], insects [17], and plants [16,154], generally by secretion into the medium. However, mammalian cell expression currently provides the most efficient route for production of intact antibodies. Fragments such as Fabs, and perhaps some toxin conjugates, can be secreted from either *E. coli* or mammalian cells, and in some cases the productivities of these fragments of intermediate size may be sufficiently similar from the two host cell types that the choice of expression system is not entirely straightforward. In such cases, other factors such as the end use of the antibody and the downstream processing required may have to be taken into consideration. In this section, the technology for expression in mammalian cells is first described and then a comparison of expression systems is made.

Cloned DNA can be introduced into mammalian cells by a number of transfection techniques and can enter the nucleus, where it may persist for a period of a few days in a high proportion of the cell population and result in transient expression of genes present on the vector. Alternatively, low-frequency integration events can be selected, by using an appropriate marker gene, in which the introduced DNA becomes inserted into apparently random sites in the host cell genome. In this case, the DNA replicates along with the rest of the genome and a permanently transfected cell line is established.

3.6.1. Transient Expression

The ability to produce small amounts of antibody (up to about 1 mg) within a few days from transient expression experiments is very valuable in the analysis of multiple genetically engineered forms of an antibody. Recombinant antibodies have been shown to assemble efficiently in a number of nonlymphoid as well as lymphoid types of cells and this has provided the possibility of exploiting a number of transient expression systems for Ig production (reviewed in Bebbington [155]). COS cells (monkey kidney cells transformed with SV40 T-antigen) support the replication of virtually any bacterial plasmid containing a 340-bp sequence of SV40 including the origin of replication (provided that particular sequences from pBR322 that inhibit replication are removed). A strong promoter such as the MIE promoter from cytomegalovirus (CMV) is used to direct transcription of a heavy chain gene and a light chain gene on separate vectors, which are cotransfected into the same COS cell population. This system can be used to produce recombinant antibodies secreted into the culture medium at levels of about 1 μg/ml within 2–3 days from transfection from cells grown to confluence (i.e., about (2–5) $\times$ 10^6 cells/ml) [156].

The CMV–MIE promoter can also be used effectively for transient expression in an adenovirus-transformed human embryo cell line, 293. One of the early proteins of adenovirus (an E1A protein) enhances transcription from the CMV–MIE promoter when coexpressed in the same cell, and yields of 7–15 μg/ml of recombinant antibody have been obtained a week after transfection by this system [157].

3.6.2. Establishing Permanent Transfected Cell Lines

To select for retention of the vector within the host genome, a number of marker genes exist. The most widely used are two bacterial genes provided with mammalian transcription signals: neo, conferring resistance to an antibiotic, G418 [158]; and gpt, conferring resistance to mycophenolic acid, which is an inhibitor of IMP dehydrogenase, an enzyme involved in nucleoside biosynthesis [159]. Vectors containing such selectable marker genes can be introduced into mammalian cells by various techniques. The method that is most appropriate depends largely on the cell type but electrophoretic methods (in which a millisecond pulse of high-voltage electricity is applied across the suspension of cells and DNA) tend to be highly efficient for many types of cells and are the preferred choice for myeloma cells [160]. However, fragments of cloned genomic DNA containing complete Ig genes that are introduced in this way into the genome of a myeloma cell are inefficiently expressed relative to the same genes in their natural chromosomal locations in a hybridoma cell. Thus a 9.5-Kb genomic fragment containing a mouse γ2b gene has been expressed at only up to 20% of the level in the parental hybridoma [161], and genomic light chain genes tend to be even more poorly expressed [162,163]. Expression levels are also highly variable among different transfectant lines. Such low and variable expression levels are generally ascribed to two factors: the likelihood that not all the DNA sequences necessary for complete expression are isolated when the genes are cloned, and the profound influence on gene expression of the chromosomal structure surrounding the integration site [164].

Use of Immunoglobulin Promoters and Enhancers. In the early 1980s, several transcription control elements were characterized in immunoglobulin genes. Promoter regions were identified and tissue-specific downstream enhancers were isolated from the heavy chain J–C intron [161,165] and the κ-gene intron [166] (see Chapter 1). This enabled versatile expression vectors to be constructed with immunoglobulin promoters and enhancers to direct transcription of cDNA or genomic clones encoding antibodies or their fragments in myeloma-type cells. Chimeric antibody expression vectors assembled from murine genomic V-region clones, fused upstream of human constant region sequences, make use of the natural promoter and enhancer present in the V-region DNA (Fig. 3.13A). The use of a genomic V-region fragment also serves to provide an intron, which is an absolute requirement for the formation of stable mRNA from a transcript initiated at an immunoglobulin promoter [167]. Typically the genes for the heavy and light chains have been placed on separate vectors and cotransfected into a myeloma cell line [3–13,168,169], most commonly the SP2/0 hybridoma cell line. Expression levels from such cotransfected vectors can be up to 35 μg/ml of antibody accumulated in the medium once the cells have overgrown and died [168], although much lower figures are more common. Specific production rates of 1–7 $\mu g/10^6$ cells per day have been obtained [168] or, in one case, up to 10 $\mu g/10^6$ cells per day [169]. This is considerably less than the levels at which hybridomas typically express antibody from genes in their natural chromosomal loci, as hybridomas have production rates between 10 and 100 $\mu g/10^6$ cells per day and accumulate between 50 and 500 μg/ml, depending on the culture medium used and hence the cell biomass attainable [170] (and see Chapter 5).

A better approach may be to use sequential transfection, in which a light-chain-producing cell line is established first and the heavy chain vector is subsequently introduced into this cell line. This approach has yielded 32 μg/ml at small scale and up to 100 μg/ml in serum-free fermentation [171]. Although it is not possible to compare expression levels for different antibodies and different fermentation conditions, yields resulting from retransfection can be expected to be higher due to the importance of balancing the levels of synthesis of the two chains. Many heavy chains are not secreted in the absence of light chain but are retained in the endoplasmic reticulum (ER) by association with the ER protein grp78 [172]. Accumulation of certain heavy chains in the ER can be toxic to the cell [173], so synthesis of

A) Chimeric antibody vectors for myelomas

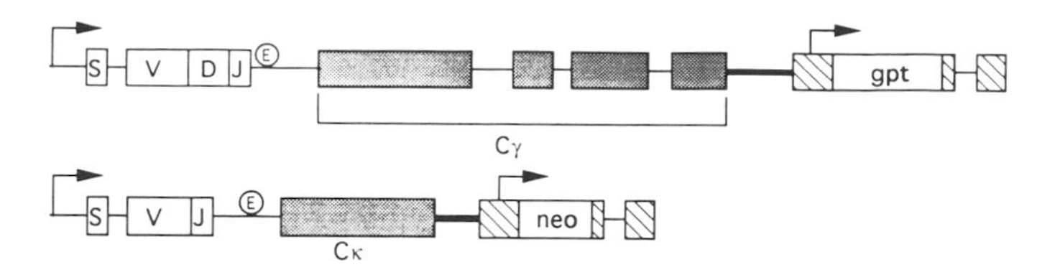

B) Gene targeting vector

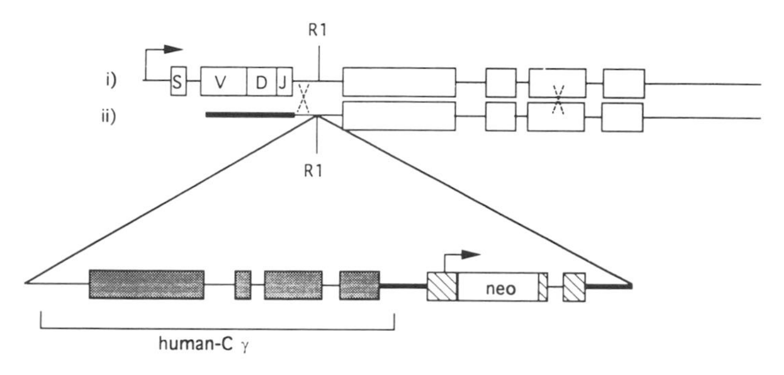

C) GS-gene amplification vector

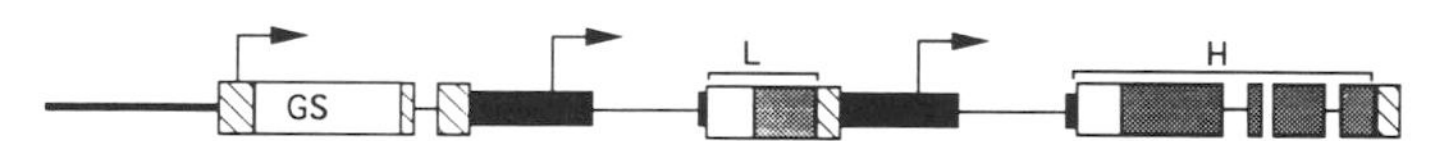

Fig. 3.13. Mammalian expression vectors for recombinant antibody genes. Vectors are shown as linear molecules, as they appear in the host genome. Promoters are marked with arrows. Immunoglobulin enhancers are marked E. R1 represents a convenient restriction enzyme site. Thin lines denote introns and plasmid backbones. Hatched segments are from SV40. Speckled segments are human immunoglobulin DNA regions. Black segments are from the CMV-MIE upstream region. L and H represent light and heavy chain coding sequences. In **B** the chromosomal allele is marked (i) and the vector DNA (ii). Representative recombination sites are marked with crosses.

excess heavy chain may limit expression. In a cotransfection experiment, the probability that both vectors will integrate such that synthesis of the two chains is balanced will inevitably be small, whereas by selecting light-chain-producing lines first, it is relatively straightforward to screen heavy chain transfectants for high levels of intact antibody. Nevertheless, any such vectors are still limited by the inefficient transcription of the transfected genes. There are, however, several newly identified elements from regions distant from the genes in the heavy chain [174], κ [175], and λ [176] loci, which show enhancer activity. These have yet to be evaluated fully for the expression of recombinant antibodies in cultured cells, and it therefore remains to be seen whether it will be possible to assemble all the control elements necessary for full expression of immunoglobulin genes on a practically useful vector.

Targeting Genes to Transcriptionally Active Sites in the Genome. An attractive alternative to the use of expression vectors containing immunoglobulin transcription control elements might be to target the required coding regions to an active, rearranged immunoglobulin locus so that the control elements in their natural chromosomal locations are exploited to achieve full expression. If there is an extended region of homology between sequences on a vector and a region of the host cell genome, homologous recombination can occur between them to introduce the DNA at a precisely defined site in a chromosome. An example of a targeting vector is shown in Figure 3.13B. Although random integration by nonhomologous recombination still predominates, a number of reports of detection of the rare gene-targeting events exist, leading to high-level expression of heavy and light chain genes [177–179]. In principle, separate targeting events could be carried out to introduce light and heavy chain genes into their respective loci, but such experiments have not been reported, perhaps because the procedures are so lengthy and laborious. Nor has the technology been extended to introduce DNA sequences other than constant regions into immunoglobulin loci.

Vector Amplification. A more versatile approach for the efficient expression of immunoglobulin-type molecules has been to use selection for gene amplification to obtain increased copy numbers of vectors randomly integrated into the genome. Gene amplification is a common mutation in mammalian tissue culture cells in which a large region of a chromosome comes to be represented in multiple tandem repeats as a result of a poorly understood sequence of events (perhaps associated with "illegitimate" recombination between regions of two sister chromatids that are not precisely homologous [180]). Amplification events can be readily detected in regions of the genome containing a gene that encodes an essential enzyme whose activity can be inhibited by a tight-binding selective inhibitor. Mutants arising in a cell population with increased resistance to the inhibitor commonly result from overproduction of the enzyme and this in turn most commonly results from amplification of the corresponding structural gene. Thus a gene coding for such an enzyme can act as a selectable marker for amplification events. An example of a suitable enzyme is dihydrofolate reductase (DHFR), a key enzyme in nucleoside biosynthesis, which can be inhibited by methotrexate (MTX). DHFR has been widely exploited as an amplifiable marker in a DHFR-deficient mutant chinese hamster ovary (CHO) cell line (reviewed in Kaufman [181]). Transfectants are selected in nucleoside-free medium and amplification of the region of the chromosome containing the vector can be selected with MTX. The procedure has also been adapted for use in other cell types, such as myelomas, that have endogenous DHFR activity, either by using a mutant DHFR marker gene encoding an enzyme partially resistant to MTX or by using a DHFR gene expressed from a strong promoter [181]. In either case, transfectants are selected on the basis that the vector confers resistance to elevated levels of MTX and amplification is selected by using progressively higher levels of MTX.

A MTX-resistant DHFR gene has been used as the selectable marker to produce a chimeric antibody in SP2/0 cells by vector amplification [182]. In this case, genes for the heavy and light chains expressed from the promoters and enhancers present in the murine V-region genomic DNA were inserted into separate vectors, each containing a DHFR gene. After cotransfection of both plasmids and multiple rounds of selection in gradually increasing concentrations of MTX, cell lines were obtained that secreted up to 35 μg/10^6 cells per day of antibody. This is similar to the productivity of typical hybridomas, but the selection procedures were comparatively lengthy (taking up to 6 months) and the levels of MTX required are exceptionally high. No data on stability are given but many amplification events are known to be unstable if MTX is removed. If high levels of MTX are indeed required continuously to maintain productivity, the high cost of MTX may cause problems in deriving an economically feasible process for producing antibody on a large scale. In other cases, MTX selection has been reported to lead to increased productivity of recombinant antibody without gene amplification [183,184]. The mechanism is unclear but will presumably also require the continued presence of the drug.

An alternative amplifiable marker that is particularly attractive for use in myeloma cells is glutamine synthetase (GS) [155,185]. GS provides the only pathway for synthesis of glutamine in mammalian cells, so for cells in glutamine-free culture media it is an essential enzyme. The levels of GS vary between different cell types and are particularly low in myelomas. Consequently, such cells are unable to grow without glutamine in the culture medium, and a GS gene can act as a selectable marker in these cells by conferring glutamine-independence [185]. Vector amplification can subsequently be selected using a the specific GS inhibitor, methionine sulfoximine (MSX). Using a GS vector as shown in Figure 3.13C, in with the immunoglobulin genes are under the control of CMV-MIE promoters, a chimeric B72.3 antibody has been produced in the NSO myeloma cell line at secretion rates of 10–15 μg/10^6 cells per day in exponentially growing cultures and up to 35 μg/10^6 cells per day towards the end of a batch fermentation [185]. By using an enriched serum-free medium to grow the cells in a fed-batch air lift fermenter (see Chapter 5), it is possible to obtain accumulated yields of 560 mg/L of chimeric antibody from this system. This is higher than typically found for murine hybridomas grown in similar fermentation systems, suggesting that it will be readily feasible to obtain an economically viable process from such cell lines. Only a single round of selection for amplification was required to obtain these high-producing lines, in contrast to the multiple rounds of selection usually required with DHFR selection. The productivity of the highest-producing cell line was also maintained for at least 30 cell generations in the absence of MSX, which may be a significant advantage in the development of a manufacturing process. More recently, cell lines have been obtained with this system, which can accumulate up to 1 g/L antibody in enriched serum-free medium after a single round of selection for GS gene amplification (H. Finney, unpublished) or 500 mg/L without selection for gene amplification (C.R.B., unpublished).

Recombinant antibodies have also been produced from CHO cells. There is extensive experience in the use of this cell type for the production of other recombinant proteins for therapeutic use. Efficient expression systems based mainly on

DHFR gene amplification are well established in these cells and have the attraction that, at least in some cases, gene amplification is stable for extended periods in the absence of continued MTX selection. Like myeloma cells, CHO cells can be grown on a large scale in suspension culture using serum-free media. Immunoglobulin transcription control elements do not function efficiently in CHO cells but there are a number of strong promoters available. For instance, a humanized antibody has been produced from CHO cells by placing the heavy and light chain genes under the control of separate β-actin promoters [186]. In this case, the DHFR gene and both immunoglobulin genes were all on the same plasmid. Although MTX selection did lead to increased productivity, the levels of antibody achieved after a single round of amplification were low (4 μg/ml). Better results were obtained in this case by cotransfection of a plasmid containing just the β-actin/heavy chain transcription unit with a separate plasmid containing the DHFR gene and the β-actin/light chain gene. Selection for amplification of the light chain plasmid led to cell lines secreting up to 100 μg/10^6 cells per day and accumulating 200 μg/ml of antibody. It is not reported whether the heavy chain gene was coamplified.

GS gene amplification has also been used to express immunoglobulin genes efficiently in CHO cells, by using the CMV-MIE promoter to express the immunoglobulin genes [155]. The vector used is essentially similar to that shown in Figure 3.13C, but a GS minigene (containing part cDNA and part genomic sequences) expressed from an SV40 late promoter is used in CHO cells, since this seems to amplify more readily. Cell lines secreting up to 15 μg/10^6 cells per day and accumulating 200 μg/ml of a chimeric B72.3 antibody have been obtained (see Table 3.1).

An alternative approach is to use two amplifiable markers, one to select for efficient expression of the light chain gene and another for the heavy chain gene. An IgM antibody has been made (although without the J chain) by selecting for amplification of a light chain vector containing a DHFR gene and amplification of the heavy chain with an adenosine deaminase-selectable marker in a separate transfection [192]. A cell line making the complete antibody was then formed by fusion of

TABLE 3.1. Expression of Recombinant B72.3 Antibodies

Molecule	Promoter	Host	Selection	Yield (mg/L)	Reference
IgG1	Ig	SP2/0	gpt + neo	60	[184]
	Ig	SP2/0	gpt + neo	20	[187]
	Ig	SP2/0	DHFR[1]	150	[184]
IgG4	CMV-MIE	CHO	gpt + neo	100	[188, 189]
	CMV-MIE	CHO	GS[1]	200	[155]
	CMV-MIE	NSO	GS[1]	560	[185]
Fab	CMV-MIE	CHO	gpt + neo	120	[190]
Fv	CMV-MIE	CHO	gpt	2	[191]

[1]Selection for gene amplification.

the cell lines expressing heavy and light chains. The fusion product secreted approximately 30 μg/10^6 cells per day of assembled antibody. In a variation of this procedure, Fouser et al. [193] used sequential amplification of these two marker genes in the same cell line, thus avoiding the necessity of carrying out cell fusion. In a further modification, they also included untranslated regions of the κ gene (from the J/C intron and the 3′-untranslated region of the mRNA), which enhanced expression even in the CHO cells. By this means, cell lines were generated after amplification of each selectable marker that secreted 80–110 μg antibody per 10^6 cells per day. Any approach that uses two sequential amplification regimes is clearly rather time-consuming, and the stability of these cell lines in the absence of continued selection is not reported. However, it is clear from these very high expression levels that CHO cells can secrete antibody at least as efficiently as myeloma-type cells. Figures of 100–110 μg/10^6 cells per day [186,193] are in fact higher than have been reported for either natural hybridomas or transfected myelomas. Whether these high production rates measured at laboratory scale will allow CHO cell expression to compete with the yields obtained from transfected myelomas at manufacturing scale remains to be seen.

Other Factors Influencing Expression in Mammalian Cells. Mammalian expression systems continue to be developed. One factor that influences expression levels but has frequently been overlooked is the possibility of interference between closely adjacent genes (reviewed in Bebbington [155]). Although genes in the expression vectors are equipped with polyadenylation signals, they do not generally act as transcription terminators, so transcription from one promoter may extend beyond the end of the desired gene. Interference between neighboring promoters may therefore contribute to imbalanced synthesis of the two chains. If, however, all genes are present on a single plasmid, they may be arranged so as to minimize this effect.

Although much of the emphasis has been placed on vectors that are directed to optimizing transcription, the effect on protein synthesis of postranscriptional processes should also be considered. Indeed gene amplification can be used to increase vector copy numbers to such an extent that transcription rates are no longer limiting. The stability of the immunoglobulin mRNA will of course affect its steady-state concentration in the cytoplasm and hence the rate of protein synthesis. RNA stability can be influenced by sequences in the 3′-untranslated region, often in a cell-type-specific manner [194], as well as by the presence of introns. The influence of particular introns and untranslated RNA sequences on antibody expression has been suggested as a contributing factor in the high yields of recombinant antibody obtained from vectors containing immunoglobulin κ genomic sequences in CHO cells [193]. There is also an absolute requirement for introns in immunoglobulin mRNA if the Ig H promoter is used but not if the CMV-MIE promoter directs transcription in myeloma cells [168]. In high-yielding cell lines it is possible that the secretory pathway may ultimately limit the rate of antibody production from the cell. At present, however, little is known about precisely which parts of the secretion apparatus might limit immunoglobulin secretion, apart from the involvement of the ER protein grp78 described above in section 3.6.2.1. Lastly it should be noted that fermentation conditions can have a profound effect on the amount of product obtained and there is often considerable scope for increasing the concentration of cells in the fermentation system and hence the overall yield (see Chapter 5).

3.6.3. Choosing an Expression System

Comparisons between different reports of antibody expression in mammalian cells must be treated with caution because of the effects of random integration on expression levels ("position effects"), variability in secretion of different antibodies, and the effects of culture conditions on antibody yield. Nevertheless, from the results reported for expression of various genetically engineered antibodies containing V-region sequences of the murine B72.3 antibody (Table 3.1), it is clear that gene amplification systems currently provide the most efficient expression of whole antibodies in both CHO and myeloma cells. On a per cell basis it is possible to produce antibodies equally efficiently from CHO or myeloma hosts. Indeed the highest quoted production rates are from CHO cells. However, the process advantages associated with myeloma cells (higher biomass attainable in currently available media) mean that these have so far produced the highest reported yields in batch fermentations. It may also be worth noting when designing an expression system that vectors using CMV promoters and containing a selectable marker gene expressed from a SV40 early promoter can generally also be used in COS cells or 293 cells for transient expression. This facilitates the analysis of novel engineered molecules before the establishment of transfected cell lines is undertaken.

Because of the higher yields, mammalian-cell expression systems are generally preferable for producing whole antibodies. For the production of fragments or other engineered molecules, it is often less clear which system will provide the highest yields. Fv fragments have been produced with CMV-MIE promoters and gpt selection in NS0 myeloma cells at yields of 8 mg/L [195], but it is likely that *E. coli* expression systems will be more suitable for such molecules, since yields of up to 500 mg/L Fv have been obtained from bacterial expression [196]. For Fab fragments, it is not clear whether bacterial or mammalian expression systems will prove more productive: mammalian expression systems can produce at least 100 mg/L of Fab but the highest yields so far reported are from *E. coli* expression [196,197]. However, it should be borne in mind that comparisons between different antibodies and fragments are not always valid. Secretory pathways in mammalian cells, for instance, may recognize certain recombinant proteins as foreign and fail to secrete them efficiently. Even single amino acid changes can profoundly affect the secretion of immunoglobulin chains [198], so for any chosen molecule it may make sense to try more than one expression system.

It may also be appropriate to consider a number of other factors in addition to productivity if, for instance, the antibody is intended for therapeutic use in humans. Here we consider briefly two significant issues that are discussed further in Chapters 5 and 6: the effect of host cell type and culture conditions on glycosylation, and the implications of host cell type on the downstream processing of the product.

Recombinant antibodies differ from their natural counterparts with respect to the detailed oligosaccharides attached because of host cell differences in glycosylation patterns and possibly the effects of different culture conditions. For instance, mouse cells possess a glycosyl transferase, not present in human or CHO cells, that adds a terminal α (1,3)-gal residue [199,200]. The effects of such differences on the behavior of antibodies *in vivo* have so far not been characterized. Experience with other recombinant proteins suggests that carbohydrate microheterogeneity in general may not significantly affect the properties of the molecules [199]. Furthermore, for most antibodies, the carbohydrate is buried between the two heavy chains and so may not be accessible, for instance, to immune complex formation. Nevertheless, it remains to be seen whether antibodies produced in different host cell types differ in their immunogenicity, rates of clearance from the circulation, or effector functions.

Purification procedures are likely to differ significantly for molecules produced from different host cell types and this may be a contributory factor in the choice of expression system. For instance, antibodies derived from *E. coli* require downstream processing steps to eliminate bacterial toxins such as lipopolysaccharide. Proteins produced intracellularly in bacteria may also be subject to proteolytic degradation. On the other hand, an important issue in the purification of antibodies from mammalian cells is the ensuring of removal of infectious agents. Both CHO and murine myeloma cell lines contain endogenous retroviruses [201,202,203], which can produce viral particles shed into the culture medium. Although there is no evidence that such retroviruses could be harmful to humans, downstream processing steps to remove these and other adventitious agents will inevitably be required.

Although for many purposes, none of the expression systems described would appear to be excluded, the choice of production process will inevitably take into account the costs associated with the entire expression system, including the requirements for downstream processing.

3.7. SUMMARY

It is now possible to use genetic manipulation to generate antibodies with improved properties. For example, rodent antibodies can be humanized to produce antibodies that promise to be less immunogenic than their rodent counterparts in human therapy. Recently developed technology involving the use of *E. coli* as an expression host provides almost unlimited potential for the isolation and generation of antibody fragments of the desired specificity and affinity. Such recombinant antibody fragments can be produced efficiently from *E. coli* or can be linked at the genetic level to an isotype that carries the effector function of choice or to a toxin molecule. The rebuilt antibody genes can then be introduced into mammalian cells. Advances in expression systems for mammalian cells, particularly using vector amplification in rodent myeloma or CHO cells, means that it is now feasible to produce many engineered antibodies for human therapeutic use economically from mammalian cell fermentations.

The "designer" antibodies that result from the application of these recombinant techniques could provide invaluable reagents for therapy and diagnosis. The next few years promise to be an exciting era for the field of antibody engineering.

3.8. REFERENCES

1. Saiki, R.K., Scharf, S., Faloona, F., Mullis, K.B., Horn, G.T., Erlich, H.A., Arnheim, N. Science 230, 1350 (1985).
2. Saiki, R.K., Gelfand, D.H., Stoffel, S., Scharf, S.J., Higuchi, R., Horn, G.T., Mullis, K.B., Erlich, H.A. Science 239, 487 (1988).
3. Neuberger, M.S. EMBO J. 2, 1373 (1983).
4. Oi, V.T., Morrison, S.L., Herzenberg, L.A., Berg, P. Proc. Natl. Acad. Sci. USA 80, 825 (1983).
5. Ochi, A., Hawley, R.G., Hawley, T., Shulman, M.J., Traunecker, A., Köhler, G., Hozumi, N. Proc. Natl. Acad. Sci. USA 80, 6351 (1983).
6. Neuberger, M.S., Williams, G.T., Fox, R.O. Nature 312, 604 (1984).
7. Schnee, J.M., Runge, M.S., Matsueda, G.R., Hudson, N.W., Seidman, J.G., Haber, E., Quertemous, T. Proc. Natl. Acad. Sci. USA 84, 6904 (1987).
8. Sharon, J., Gefter, M.L., Manser, T., Morrison, S.L., Oi, V.T., Ptashne, M. Nature 309, 364 (1984).
9. Liu, A.Y., Robinson, R.R., Hellström, K.E., Murray, E.D., Jr., Chang, C.P., Hellström, I. Proc. Natl. Acad. Sci. USA 84, 3439 (1987).
10. Morrison, S.L. Science 229, 1202 (1985).
11. Sun, L.K., Curtis, P., Rakowicz-Szulczynska, E., Ghrayeb, J., Chang, N., Morrison, S.L., Koprowski, H. Proc. Natl. Acad. Sci. USA 84, 214 (1987).
12. Boulianne, G.L., Hozumi, N., Shulman, M.J. Nature 312, 643 (1984).
13. Morrison, S.L., Johnson, M.J., Herzenberg, L.A., Oi, V.T. Proc. Natl. Acad. Sci. USA 81, 6851, (1984).
14. Better, M., Chang, C.P., Robinson, R.R., Horwitz, A.H. Science 240, 1041 (1988).
15. Skerra, A. Plückthun, A. Science 240, 1038 (1988).
16. Hiatt, A., Cafferkey, R., Bowdish, K. Nature 342, 76 (1989).
17. Hasemann, C.A., Capra, J.D. Proc. Natl. Acad. Sci. USA 87, 3942 (1990).
18. Morrison, S.L., Oi, V.T. Adv. Immunol. 44, 65 (1989).
19. Winter, G., Milstein, C. Nature 349, 293 (1991).
20. Mattila, P., Korpela, J., Tenküanen, T., Pitkänen, K. Nucleic Acids Res. 19, 4967 (1991).
21. Higuchi, R. In Erlich (ed.): "PCR Technology: Principles and Applications for DNA Amplification (p. 61). New York, London, Tokyo, Melbourne, Hong Kong: Stockton Press, (1989).
22. Evan, G., Lewis, G.K., Ramsay, G., Bishop, J.M. Mol. Cell. Biol. 5, 3610 (1985).
23. Stoflet, E.S., Koeberl, D.D., Sarkas, G., Sommer, S.S. Science 239, 491 (1988).
24. Scheffield, V.C., Cox, D.R., Lerman, L.S., Myers, R.M. Proc. Natl. Acad. Sci. USA 86, 232 (1989).

25. Ochman, H., Gerber, A.S., Hartl, D.L. Genetics 120, 621 (1988).
26. Loh, H.Y., Elliott, J.F., Cwirla, S., Lanier, L.L., Davis, M.M. Science 242, 217 (1989).
27. Orlandi, R., Güssow, D.H., Jones, P.T., Winter, G. Proc. Natl. Acad. Sci. USA 86, 3833 (1989).
28. Chiang, Y.L., Sheng-Dong, R., Brow, M.A., Larrick, J.W. BioTechniques 7, 360 (1989).
29. Larrick, J.W., Danielsson, L., Brenner, C.A., Wallace, E.F., Abrahamson, M., Fry, K.E., Borrebaeck, A.K. Biotechnology 7, 934 (1989).
30. Larrick, J.W., Danielsson, L., Brenner, C.A., Abrahamson, M., Fry, K.E., Borrebaeck, C.A.K. Biochem. Biophys. Res. Commun. 160, 1250 (1989).
31. Huse, W.D., Sastry, L., Iverson, S.A., Kang, A.S., Alting-Mees, M., Burton, D.R., Benkovic, S.J., Lerner, R.A. Science 246, 1275 (1989).
32. Sanz, I., Kelly, P., Williams, C., School, S., Tucker, P., Capra, J.D. EMBO J. 8, 3741 (1989).
33. Marks, J.D., Tristem, M., Karpas, A., Winter, G. Eur. J. Immunol. 21, 985 (1991).
34. Marks, J.D., Hoogenboom, H.R., Bonnert, T.P., McCafferty, J., Griffiths, A.D., Winter, G. J. Mol. Biol. 222, 581 (1991).
35. Mullinax, R.L., Gross, E.A., Amberg, J.R., Hay, B.N., Hogrefe, H.H., Kubitz, M.M., Greener, A., Alting-Mees, M., Ardourel, D., Short, J.M., Sorge, J.A., Shopes, B. Proc. Natl. Acad. Sci. USA 87, 8095 (1990).
36. Persson, M.A.A., Caothien, R.H., Burton, D.R. Proc. Natl. Acad. Sci. USA 88, 2432 (1991).
37. Kabat, E.A., Wu, T.T., Perry, H.M., Gottesmann, K.S., Foeller, C. Washington, DC: U.S. Department of Health and Human Services, U.S. Government Printing Office, 1991.
38. Choi, Y., Kotzin, B., Herron, L., Callahan, J., Marrack, P., Kappler, J. Proc. Natl. Acad. Sci. USA 86, 8942 (1989).
39. Oksenberg, J.R., Stuart, S., Begovich, A.B., Bell, R.B., Erlich, H.A., Steinman, L., Bernard, C.C.A. Nature 345, 344 (1990).
40. Bragado, R., Lauzurica, P., López, D., López de Castro, J.A. J. Exp. Med. 171, 1189 (1990).
41. Wucherpfennig, K.W., Ota, K., Endo, N., Seidman, J.G., Rosenzweig, A., Weiner, H.L., Hafler, D.A. Science 248, 1016 (1990).
42. Panzara, M.A., Gussoni, E., Steinman, L., Oksenberg, J.R. BioTechniques 12, 728 (1992).
43. Uematsu, Y., Wege, H., Straus, A., Ott, M., Bannwarth, W., Lanchbury, J., Panayi, G., Steinmetz, M. Proc. Natl. Acad. Sci. USA 88, 8534 (1991).
44. Ohara, O., Dorit, R.L., Gilbert, W. Proc. Natl. Acad. Sci. USA 86, 5673 (1989).
45. Innis, M.A., Myambo, K.B., Gelfand, D.H., Brow, M.A.D. Proc. Natl. Acad. Sci. USA 85, 9436 (1988).
46. Lee, J.-S. DNA Cell Biol. 10, 67 (1991).
47. Horton, R.M., Hunt, H.D., Ho, S.N., Pullen, J.K., Pease, L.R. Gene 77, 61 (1989).
48. Horton, R.M., Cai, Z., Ho, S.N., Pease, L.R. BioTechniques 8, 528 (1990).
49. Clackson, T.C., Hoogenboom, H., Griffiths, A.D., Winter, G.P. Nature (London) 352, 624 (1991).

50. Carter, P., Bedouelle, H., Winter, G. Nucleic Acids Res. 13, 4431 (1985).
51. Leung, D.W., Chen, E., Goeddel, D.V. J. Methods Cell. Mol. Biol. 1, 11 (1989).
52. Gram, H., Marconi, L.-A., Barbas, C.F. III, Collet, T.A., Lerner, R.A., Kang, A.S. Proc. Natl. Acad. Sci. USA 89, 3576 (1992).
53. Porter, R.R. Science 180, 713 (1973).
54. Saul, F.A., Amzel, L.M., Poljak, R.J. J. Biol. Chem. 253, 585 (1978).
55. Marquart, M., Deisenhofer, J., Huber, R. J. Mol. Biol. 141, 369 (1980).
56. Amit, A.G., Mariuzza, R.A., Phillips, S.E.V., Poljak, R.J. Science 233, 747 (1986).
57. Satow, Y., Cohen, G.H., Padlan, E.A., Davies, D.R. J. Mol. Biol. 190, 593 (1986).
58. Colman, P.M., Laver, W.G., Varghese, J.N., Baker, A.T., Tullock, P.A., Air, G.M., Webster, R.G. Nature 326, 358 (1987).
59. Sheriff, S., Silverton, E.W., Padlan, E.A., Cohen, G.H., Smith-Gill, S.J., Finzel, B.C., Davies, D.R., Proc. Natl. Acad. Sci. USA 84, 8075 (1987).
60. Padlan, E.A., Silverton, E.W., Sheriff, S., Cohen, G.H., Smith-Gill, S.J., Davies, D.R. Proc. Natl. Acad. Sci. USA 86, 5938 (1989).
61. Chothia, C., Lesk, A.M. J. Mol. Biol. 196, 901 (1987).
62. Chothia, C., Lesk, A.M., Tramontano, A., Levitt, M., Smith-Gill, S.J., Air, G., Sheriff, S., Padlan, E.A., Davies, D., Tulip, W.R., Colman, P.M., Spinelli, S., Alzari, P.M., Poljak, R.J. Nature 342, 877 (1989).
63. Inbar, D., Hochman, J., Givol, D. Proc. Natl. Acad. Sci. USA 2659 (1972).
64. Ward, E.S., Güssow, D.H., Griffiths, A.D., Jones, P.T., Winter, G. Nature 341, 544 (1989).
65. Blackwell, T.K., Alt, F.W. In Hames, Glover (eds.): Molecular Immunology (p. 1), Oxford, Washington, DC: IRL Press, 1988.
66. Carson, D.A., Freimark, B.D. Adv. Immunol. 38, 275 (1986).
67. Borrebaeck, C.A.K. Immunol. Today 9, 355 (1988).
68. Thompson, K.M. Immunol. Today 6, 113 (1988).
69. Meeker, T., Lowder, J., Maloney, D., Miller, R., Thielemans, K., Wainke, R., Levy, R. Blood 65, 1349 (1985).
70. Khazaeli, M., Saleh, M., Wheeler, R., Huster, W., Holden, H., Carrano, R., LoBuglio, A. J. Natl. Cancer Inst. 80, 937 (1988).
71. Jones, P.T., Dear, P.H., Foote, J., Neuberger, M.S., Winter, G. Nature 321, 522 (1986).
72. Riechmann, L., Clark, M., Waldmann, H., Winter, G. Nature (London) 332, 323 (1988).
73. Brüggeman, M., Williams, G.T., Bindon, C.I., Clark, M.R., Walker, M.R., Jefferis, R., Waldmann, H., Neuberger, M.A. J. Exp. Med. 166, 1351 (1987).
74. LoBuglio, A.F., Wheeler, R.H., Trang, J., Haynes, A., Rogers, K., Harvey, E.B., Sun, L., Ghrayeb, J., Khazaeli, M.B. Proc. Natl. Acad. Sci. USA 86, 4220 (1989).
75. Mueller, B.M., Reisfeld, R.A., Gillies, S.D. Proc. Natl. Acad. Sci. USA 87, 5702 (1990).
76. Ellerson, J.R., Yasmeen, D., Painter, R.H., Dorrington, K.J. J. Immunol. 116, 510 (1976).
77. Verhoeyen, M., Milstein, C., Winter, G. Science 239, 1534 (1988).

78. Queen, C., Schneider, W.P., Selick, H.E., Payne, P.W., Landolfi, N.F., Duncan, J.F., Avdalovic, N.M., Levitt, M., Junghans, R.P., Waldmann, T.A. Proc. Natl. Acad. Sci. USA 86, 10029 (1989).
79. Shalaby, M.R., Shepard, H.M., Presta, L., Rodrigues, M.S., Beverley, P.C.L., Feldmann, M., Carter, P. J. Exp. Med. 175, 217 (1992).
80. Tempest, P.R., Bremner, P., Lambert, M., Taylor, G., Furze, J.M., Carr, F.J., Harris, W.J. BioTechnology 9, 266 (1991).
81. Co, M.S., Deschamps, M., Whitley, R.J., Queen, C. Proc. Natl. Acad. Sci. USA 88, 2869 (1991).
82. Maeda, H., Matsushita, S., Eda, Y., Kimachi, K., Tokiyoshi, S., Bendig, M.M. Hum. Antibodies Hybridomas 2, 124 (1991).
83. Kettleborough, C.A., Saldanha, J., Heath, V.J., Morrison, C.J., Bendig, M.M. Protein Eng. 4, 773 (1991).
84. Hale, G., Dyer, M.J., Clark, M.R., Phillips, J.M., Riechmann, L., Waldmann, H. Lancet 2, 1394 (1988).
85. Mathieson, P.W., Cobbold, S.P., Hale, G., Clark, M.R., Oliveira, D.B.G., Lockwood, C.M., Waldmann, H. N. Engl. J. Med. 323, 250 (1990).
86. Isaacs, J.D., Watts, R.A., Hazleman, B.L., Hale, G., Keogan, M.T., Cobbold, S.P., Waldmann, H. Lancet 340, 748 (1992).
87. Tramontano, A., Chothia, C., Lesk, A.M. J. Mol. Biol. 215, 175 (1990).
88. Foote, J., Winter, G. J. Mol. Biol. 224, 487 (1992).
89. McCafferty, J., Griffiths, A.D., Winter, G., Chiswell, D.J. Nature (London) 348, 552 (1990).
90. Kang, A.S., Barbas, C.F., Janda, K.D., Benkovic, S.J., Lerner, R.A. Proc. Natl. Acad. Sci. USA 88, 4363 (1991).
91. Kohler, G., Milstein, C. Nature 256, 52 (1984).
92. Jackson, R.H., McCafferty, J., Johnson, K.S., Pope, A.R., Roberts, A.J., Chiswell, D.J., Clackson, T.P., Griffiths, A.D., Hoogenboom, H.R., Winter, G. In Rees, Sternberg, Wetzel (eds.): Protein Engineering (p. 277). Oxford, Washington, DC: IRL Press, 1992.
93. Güssow, D., Ward, E.S., Griffiths, A.D., Jones, P.T., Winter, G. Cold Spring Harbor Laboratory Quant. Biol. 54, 265 (1989).
94. Gherardi, E., Milstein, C. Nature 357, 201 (1992).
95. Cabilly, S., Riggs, A.D., Pande, H., Shively, J.E., Holmes, W.E., Rey, M., Perry, L.J., Wetzel, R., Heyneker, H.L. Proc. Natl. Acad. Sci. USA 81, 3273 (1984).
96. Boss, M.A., Kenten, J.H., Wood, C.R., Emtage, J.S. Nucleic Acids Res. 12, 3791 (1984).
97. Kurokawa, T., Seno, M., Sasada, R., Ono, Y., Onda, H., Igarashi, K., Kikuchi, M., Sugino, Y., Honjo, T. Nucleic Acids Res. 11, 3077 (1983).
98. Ishizaka, T., Helm, B., Hakimi, J., Niebyl, J., Ishizaka, K., Gould, H. Proc. Natl. Acad. Sci. USA 83, 8323 (1986).
99. Liu, F., Albrandt, K.A., Bry, C.G., Ishizaka, T. Proc. Natl. Acad. Sci. USA 81, 5369 (1984).
100. Buchner, J., Rudolph, R. BioTechnology 9, 157 (1991).

101. Whitlow, M., Filpula, D. Methods: A Companion to Methods in Enzymology (Vol. 2, 97), 1991.

102. Chothia, C. Novotny, J., Bruccoleri, R., Karplus, M. J. Mol. Biol. 186, 651 (1985).

103. Glockshuber, R., Malia, M., Pfitzinger, I., Plückthun, A. Biochemistry 29, 1362 (1990).

104. Huston, J.S., Levinson, D., Mudgett-Hunter, M., Tai, M., Novotny, J., Margolies, M.N., Ridge, R.J., Bruccoleri, R.E., Haber, E., Crea, R., Oppermann, H. Proc. Natl. Acad. Sci. USA 85, 5879 (1988).

105. Bird, R.E., Hardman, K.D., Jacobson, J.W., Johnson, S., Kaufmann, B.M., Lee, S.L., Pope, S.H., Riordan, G.S., Whitlow, M. Science 242, 423 (1988).

106. McCartney, J.E., Lederman, L., Drier, E.A., Cabral-Denison, N.A., Wu, G.-M., Batorsky, R.S., Huston, J.S., Opperman, H. J. Protein Chem. 10, 669 (1991).

107. Brinkmann, U., Pai, L.H., FitzGerald, D.J., Willingham, M., Pastan, I. Proc. Natl. Acad. Sci. USA 88, 8616 (1991).

108. Gibbs, R.A., Posner, B.A., Filpula, D.R., Dodd, S.W., Finkelman, M.A.J., Lee, T.K., Wroble, M., Whitlow, M., Benkovic, S.J. Proc. Natl. Acad. Sci. USA 88, 4001 (1991).

109. Laroche, Y., Demaeyer, M., Stassen, J.-M., Gansemans, Y., Demarsin, E., Matthyssens, G., Collen, D., Holvoet, P. J. Biol. Chem. 266, 16343 (1991).

110. Anand, N.N., Mandal, S., MacKenzie, C.R., Sadowska, J. Sigurskjold, B., Young, N.M., Bundle, D.R., Narang, S.A. J. Biol. Chem. 266, 21874 (1991).

111. Williams, W.V., Moss, D.A., Kieber-Emmons, T., Cohen, J.A., Myers, J.N., Weiner, D.B., Greene, M.I. Proc. Natl. Acad. Sci. USA 86, 5537 (1989).

112. Taub, R., Gould, R.J., Garsky, V.M., Ciccarone, T.M., Hoxie, J., Friedman, P.A., Shattil, S.J. J. Biol. Chem. 264, 259 (1989).

113. Kang, A.S., Jones, T.M., Burton, D.R. Proc. Natl. Acad. Sci. USA 88, 11120 (1991).

114. Mullinax, R.L., Gross, E.A., Hay, B.N., Amberg, J.R., Kubitz, M.M., Sorge, J.A. BioTechniques 12, 864 (1992).

115. Caton, A.J., Koprowski, H. Proc. Natl. Acad. Sci. USA 87, 6450 (1990).

116. Berek, C., Milstein, C. Immunol. Rev. 105, 5 (1988).

117. French, D.L., Laskov, R., Scharff, M.D. Science 244, 1152 (1989).

118. Chang, C.N., Landolfi, N.F., Queen, C. J. Immunol. 147, 3610 (1991).

119. Barbas, C.F. III, Kang, A.S., Lerner, R.A., Benkovic, S.J. Proc. Natl. Acad. Sci. USA 88, 7978 (1991).

120. Burton, D.R., Barbas, C.F. III, Persson, M.A.A., Koenig, S., Chanock, R.M., Lerner, R.A. Proc. Natl. Acad. Sci. USA 88, 10134 (1991).

121. Zebedee, S.L., Barbas, C.F. III, Hom, Y.-L., Caothien, R.H., Graff, R., DeGraw, J., Pyati, J., LaPolla, R., Burton, D.R., Lerner, R.A., Thornton, G.B. Proc. Natl. Acad. Sci. USA 89, 3175 (1992).

122. Hoogenboom, H.R., Griffiths, A.D., Johnson, K.S., Chiswell, D.J., Hudson, P., Winter, G. Nucleic Acids Res. 19, 4133 (1991).

123. Duchosal, M.A., Eming, S.A., Fischer, P., Leturcq, D., Barbas, C.F. III, McConahey, P.J., Caothien, R.H., Thornton, G.B., Dixon, F.J., Burton D.R. Nature 355, 258 (1992).

124. Brüggemann, M., Caskey, H.M., Teale, C., Waldmann, H., Williams, G.T., Surani, M.A., Neuberger, M.S. Proc. Natl. Acad. Sci. USA 86, 6709 (1989).
125. Borrebaeck, C.A.K., Danielsson, L. Möller, S.A. Proc. Natl. Acad. Sci. USA 85, 3995 (1988).
126. McCune, J.M., Namikawa, R., Kaneshima, H., Shultz, L.D., Lieberman, M., Weissman, I.L. Science 241, 1632 (1988).
127. Carlsson, R., Martensson, C., Kalliomaki, S., Ohlin, M., Borrebaeck, C.A.K. J. Immunol. 148, 1065 (1992).
128. Stanfield, R.L., Fieser, T.M., Lerner, R.A., Wilson, I.A. Science 248, 712 (1990).
129. Ward, E.S., Güssow, D.H., Griffiths, A., Jones, P.T., Winter, G.P. In Melchers, et al. (eds.): Progress in Immunology (p. 1144). Berlin: Springer-Verlag, 1989.
130. Barbas, C.F. III, Bain, J.D., Hoekstra, D.M., Lerner, R.A. Proc. Natl. Acad. Sci. USA 89, 4457 (1992).
131. Covell, D.G., Barbet, J., Holton, O.D., Black, C.D.V., Parker, R.J., Weinstein, J.N. Cancer Res. 46, 3969 (1986).
132. Colcher, D., Bird, R., Roselli, M., Hardman, K.D., Johnson, S., Pope, S., Dodd, S.W., Pantoliano, M.W., Milenic, D.E., Schlom, J. J. Natl. Cancer Inst. 82, 1191 (1990).
133. Sealy, D., Nedelman, M., Tai, M.-S., Huston, J.S., Berger, H., Lister-James, J., Dean, R.T. J. Nuclear Med. 31, 776 (1990).
134. King, D., Mountain, A., Harvey, A., Weir, N., Owens, R., Proudfoot, K., Phipps, A., Adair, J., Lisle, H., Bergin, S., Lawson, A., Rhind, S., Pedley, B., Boden, J., Begent, R., Yarranton, G. Antibody Immunoconj. Radiopharm. 4, 210 (1991).
135. Cumber, A.J., Ward, E.S., Winter, G., Parnell, G.D., Wawrzynczak, E.W. J. Immunol. 149, 120 (1992).
136. Pollock, R.R., French, D.L., Metlay, J.P., Birshtein, B.K., and Scharff, M.D. Eur. J. Immunol. 20, 2021, (1990).
137. Tao, M-H., and Morrison, S.L. J. Immunol. 143, 2595 (1989).
138. Spiegelberg, H.L., Fishkin, B.G. Clin. Exp. Immunol. 10, 599 (1972).
139. Sutherland, R., Buchegger, F., Schreyer, M., Vacca, A., Mach, J. Cancer Res. 47, 1627 (1987).
140. Burton, D.R. In Calabi, Neuberger (eds.): Molecular Genetics of Immunoglobulins. Amsterdam: Elsevier, 1987.
141. Duncan, A.R., Woof, J.M., Partridge, L.J., Burton, D.R., Winter, G. Nature (London) 332, 563 (1988).
142. Duncan, A.R., Winter, G. Nature (London) 332, 738 (1988).
143. Steplewski, Z., Sun, L.K., Shearman, C.W., Ghrayeb, J., Daddona, P., Koprowski, H. Proc. Natl. Acad. Sci. USA 85, 4852 (1988).
144. Lucisano. V.Y.M., Lachmann, P.J. Clin. Exp. Immunol. 84, 1 (1991).
145. Chaudhary, V.K., Queen, C., Junghans, R.P., Waldmann, T.A., FitzGerald, D.J., Pastan, I. Nature 339, 394 (1989).
146. Chaudhary, V.K., Batra, J.K., Gallo. M.G., Willingham, M.C., FitzGerald, D.J., Pastan, I. Proc. Natl. Acad. Sci. USA 87, 1066 (1990).
147. O'Hare, M., Brown, A.N., Hussain, K., Gebhardt, A., Watson, G., Roberts, L.M., Vitetta, E.S., Thorpe, P.E., Lord, J.M., FEBS Lett. 273, 200 (1990).

148. Chovnick, A., Schneider, W.P., Tso, J.Y., Queen, C., Chang, C.N. Cancer Res. 51, 465 (1991).
149. Seetharam, S., Chaudhary, V.K., FitzGerald, D., Pastan, I. J. Biol. Chem. 266, 17376 (1991).
150. Gould, B.J., Borowitz, M.J., Groves, E.S., Carter, P.W., Anthony, D., Weiner, L.M., Frankel, A.E. J. Natl. Cancer Inst. 81, 775 (1989).
151. Milstein, C., Cuello, A.C. Nature 305, 537 (1983).
152. Better, M., Horwitz, A.H. Methods Enzymol. 178, 476 (1989).
153. Horwitz, A.H., Chang, P., Better, M., Hellstrom, K.E., Robinson, R.R. Proc. Natl. Acad. Sci. 85, 8678 (1988).
154. Benvenuto, E., Ordas, R.J., Tavazza, R., Ancora, G., Biocca, S., Cattaneo, A., Galeffi, P. Plant Mol. Biol. 17, 865 (1991).
155. Bebbington, C.R. Methods 2, 136 (1991).
156. Whittle, N., Adair, J., Lloyd, C., Jenkins, L., Devine, J., Schlom, J., Raubitschek, A., Colcher, D., Bodmer, M. Protein Eng. 1, 499 (1987).
157. Carter, P., Presta, L., Gorman, C.M., Ridgway, J.B.B., Henner, D., Wong, W.C.T., Rowland, A.M., Koch, C., Carver, M., Shepard, H.M. Proc. Natl. Acad. Sci. 89, 4285 (1992).
158. Southern, P., Berg, P. J. Mol. Appl. Genet. 1, 327 (1982).
159. Mulligan, R.C., Berg. P. Science 209, 1422 (1980).
160. Toneguzzo, F., Keating, A., Glynn, S., McDonald, K. Nucleic Acids Res. 16, 5515 (1988).
161. Gillies, S.D., Morrison, S.L., Oi, V.T. Cell 33, 717 (1983).
162. Rice, D., Baltimore, D. Proc. Natl. Acad. Sci. USA 79, 7862 (1983).
163. Potter, H., Weir, L., Leder, P. Proc. Natl. Acad. Sci. USA 81, 7161 (1984).
164. Plon, S.E., Groudine, M. Curr. Biol. 1, 13 (1991).
165. Banerji, J., Olson, L., Schaffner, W. Cell 33, 729 (1983).
166. Picard, D., Schaffner, W. Nature 307, 80 (1984).
167. Neuberger, M.S., Williams, G.T. Nucl. Acids Res. 16, 6713 (1988).
168. Shearman, C.W., Kanzy, E.J., Lawrie, D.K., Li, Y., Thammana, P., Moore, G.P., Kurrle, R.J. J. Immunol. 146, 928 (1991).
169. Weissenhorn, W., Weiss, E., Schwirzke, M., Kahnze, B., Weidle, U.H. Gene 106, 273 (1991).
170. Caulcott, C.A., Boraston, R., Hill, C., Thompson, P.W., Birch, J.R. In Collins (ed.): Complementary Immunoassays. New York: John Wiley.
171. Beidler, C.B., Ludwig, J.R., Cardenas, J., Phelps, J., Papworth, C.G., Melcher, E., Sierzega, M., Myers, L.J., Unger, B.W., Fisher, M., David, G.S., Johnson, M.J. J. Immunol. 141, 4053 (1988).
172. Hendershot, L.M., Ting, J., Lee, A.S. Mol. Cell. Biol. 8, 4250 (1988).
173. Kohler, G. Proc. Natl. Acad. Sci. USA 77, 2197 (1980).
174. Petterson, S., Cook, G.P., Bruggemann, M., Williams, G.T., Neuberger, M.S. Nature 344, 165 (1990).
175. Meyer, K.B., Neuberger, M.S. EMBO J. 8, 1959 (1989).
176. Eccles, S., Sarver, N., Vidal, M., Cox, A., Grosveld, F. New Biologist 2, 801 (1990).

177. Fell, H.P., Yarnold, S., Hellstrom, I., Hellstrom, K.E., Folger, K.R., Proc. Natl. Acad. Sci. USA 86, 8507 (1989).
178. Wood, C.R., Morris, G.E., Alderman, E.M., Fouser, L., Kaufman, R.J. Proc. Natl. Acad. Sci. USA 88, (1991).
179. Baker, M.D., Schulmann, M.J. Mol. Cell. Biol. 8, 4041 (1988).
180. Smith, K.A., Gorman, P.A., Stark, M.B., Groves, R.P., Stark, G.R. Cell 63, 1219 (1990).
181. Kaufman, R.J. Methods Enzymol. 185, 537 (1990).
182. Dorai, H., Moore, G.P. J. Immunol. 139, 4232 (1987).
183. Gillies, S.D., Dorai, H., Wesolowski, J., Majeau, G., Young, D., Boyd, T., Gardner, J., James K. Bio/Technol. 7, 799 (1989).
184. Gillies, S.D., Lo, K-M., Wesolowski, J. J. Immunol. Methods, 125, 191 (1989).
185. Bebbington, C.R., Renner, G., Thomson, S., King, D., Abrams, D., Yarranton, G.T. Bio/Technology 10, 169 (1992).
186. Page, M.J., Sydenham, M.A. Bio/Technology 9, 64 (1991).
187. Hutzell, K., Kashmiri, S., Colcher, D. Cancer Res. 51, 181 (1991).
188. Colcher, D., Milenic, D., Roselli, M. Cancer Res. 49, 1738 (1989).
189. King, D.J., Adair, J.R., Angal, S., Low, D.C., Proudfoot, K.A., Lloyd, C., Bodmer, M.W., Yarranton, G.T. Biochem. J. 281, 317 (1992).
190. King, D.J., Mountain, A., Adair, J.R., Owens, R.J., Harvey, A., Weir, N., Proudfoot, K.A., Phipps, A., Lawson, A., Rhind, S.K., Pedley, B., Boden, J., Boden, R., Begent, R.H.J., Yarranton, G.T. Antibody Immunoconj. Radiopharm. 5, 159 (1992).
191. Owens, R.J., King, D.J., Howat, D., Mountain, A., Harvey, A., Lawson, A., Rhind, S., Pedley, B., Bode, J., Begent, R., Yarranton, G.T. Antibody Immunoconj. Radiopharm. 4, 459 (1991).
192. Wood, C.R., Dorner, A.J., Morris, G.E., Alderman, E.M., Wilson, D., O'Hara, R.M., Kaufman, R.J. J. Immunol. 145, 3011 (1990).
193. Fouser, L.A., Swanberg, S.L., Lin, B-Y., Benedict, M., Kelleher, K., Cumming, D., Riedel, G.E. Bio/Technology 10, 1121 (1992).
194. Cox, A., Emtage, J.S. Nucleic Acids Res. 17, 10439.
195. Riechmann, L., Foote, J., Winter, G. J. Mol. Biol. 203, 825 (1988).
196. Adair, J.R. Immunol. Rev. 130, 5 (1992).
197. Carter, P., Kelley, R.F., Rodrigues, M.L., Snedcor, B., Covarrubias, M., Velligan, M.D., Wong, W.L.T., Rowland, A.M., Kotts, C.E., Carver, M.E., Yang, M., Bourell, J.H., Shepard, H.M., Henner, D. Bio/Technology 10, 163 (1992).
198. Nakaki, T., Deans, R.J., Lee, A.S. Mol. Cell. Biol. 9, 2233 (1989).
199. Liu, D.T-Y. Trends Biotechnol. 10, 114 (1992).
200. Bebbington, C.R. In Zola, H. (ed.): Monoclonal Antibodies: The Next Generation (p. 163). Bios. Scientific, 1994.
201. Manley, K.F., Givens, J.F., Taber, R.L., Ziegel, R.F. J. Gen. Virol. 39, 505 (1978).
202. Stoye, J.P., Coffin, J.M. J. Virol. 61. 2659 (1987).
203. Risser, R., Horowitz, J.M., McCubrey, J. Annu. Rev. Genet. 17, 85 (1983).

CHAPTER 4

MODIFICATION OF ANTIBODIES BY CHEMICAL METHODS

JANIS UPESLACIS AND LOIS HINMAN
Medical Research Division, Lederle Laboratories, American Cyanamid Company, Pearl River, NY 10965-1299

ARNOLD ORONSKY
InterWest Partners, Menlo Park, CA 94025-7112

4.1. INTRODUCTION

Chemical modification of monoclonal antibodies (MoAbs) has been used to turn these proteins into reporter molecules so that the extent of targeting or the quantity of an antigen can be determined. Such manipulation has made it possible to develop MoAbs into successful diagnostic agents. Chemical modification has also been used to rebuild the basic structure of MoAbs, to introduce effector functions, and even to convert MoAbs into cytotoxic agents, but consistent and successful use of such altered MoAbs in human therapy still remains the elusive goal of numerous research groups. This chapter describes the chemistry that has been used for modifying MoAbs. Since most such technologies have been developed for oncology, we have restricted our review of MoAb chemistry to this use.

A MoAb conjugate is made up of three primary elements: the antibody itself, the molecule to be delivered to the target antigen, and the linkage between the two elements. How these elements are designed and assembled is crucial to the success of a construct and forms the primary focus of this chapter. However, the physiology and microenvironment of the tumor target, the vasculature feeding the tumor, the characteristics of the target epitope, and the mode of action and potency of the molecules being delivered are all crucial factors that influence the ultimate success of MoAb targeting and should impact on the strategy used in conjugate design. Before progressing to specifics, some general comments about tumor characteristics and the elements of a conjugate are appropriate.

Monoclonal Antibodies: Principles and Applications, pages 187–230
© 1995 Wiley-Liss, Inc.

4.2. THE BASICS

4.2.1. Tumor Characteristics

First, virtually all human tumors are heterogeneous and may ultimately require treatment with more than one MoAb conjugate ("cocktails") for a complete therapeutic response. Nevertheless, treating tumors as though they are "monoclonal" for the purposes of developing the first useful MoAb conjugate, then applying this technology to the cocktail, has some merit.

Beyond the issue of nonhomogeneity, several other characteristics of human tumors are important for designing MoAb conjugates. Their blood supply is highly heterogeneous and tumor vasculature may be completely different than in normal organs [1]. Tumors recruit their blood supply either from the preexisting network or through an angiogenic response, but such vasculature is disorganized and often restricted in flow, and it may in fact be antigenically district from normal vasculature. Moreover, unlike normal vasculature, that of tumors lacks lymphatic drainage, which can cause the buildup of pressure, as much as 30 mM Hg between the center of a tumor and its periphery [2]. All of these features work against entry into and localization of large molecules within tumors to depths beyond just a few layers of cells near their periphery [3]. A characteristic of tumors favoring MoAb conjugates is that sometimes their blood vessels are leaky and devoid of the barriers to large molecules that normal vasculature possesses. In fact, tumors have been shown to trap large molecules such as immunoglobulins somewhat more effectively than normal tissues [4].

Much of the initial excitement about the potential that MoAbs offered for delivering agents *in vivo* was based on the nude mouse human tumor xenograft model, in which it has been routine to observe localization of 10% or more of the injected dose of a radiolabeled MoAb per gram of tumor. Although the nude mouse model is still routinely used to evaluate candidate MoAbs for their potential as antitumor agents, it is a model that grossly overestimates the amount of conjugate that will localize and exaggerates the potential therapeutic effects of a conjugate. Based on human studies, more realistic MoAb localization values appear to be in the range of 0.002% to 0.032% of the injected dose per gram of tumor in 24 h [5]. This does, however, contrast with a nonspecific immunoglobulin uptake of 0.00011%–0.0002% of the injected dose per gram of tumor.

The absolute amount of material targeted has little relevance in the use of radiolabeled diagnostic MoAbs, since these simply need to show a slight preferential localization into and retention by tumors, long enough to achieve imaging. However, localization and processing of cytotoxic conjugates is key to treatment of solid tumors: If an insufficient amount of cytotoxic agent is delivered, little antitumor effect can be expected.

Two metabolic characteristics that have been used for designing mechanisms to release drugs after targeting are the relatively acid pH and the high levels of peptidases and proteases in the tumor environment. Under the anaerobic conditions that exist in many solid tumors, hydrolysis of adenosine triphosphate (ATP) and the incomplete metabolism of glucose produce larger amounts of lactic acid than in

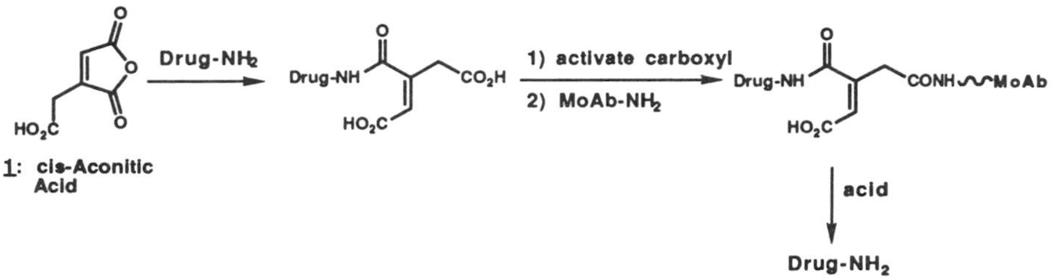

Fig. 4.1. cis-Aconitic acid as a cleavable linker for MoAb conjugates.

normal tissues. This contributes to the maintenance of an acidic extracellular environment in many tumors; a pH range of 5.8–7.6 has been recorded in one study [6] and pH 6.0–6.7 in another [7]. The *cis*-aconitic acid spacer (*1*)* was originally designed to capitalize on tumor acidity to release daunorubicin from polymers [8], but it has also been used to construct drug–MoAb conjugates, as shown in Figure 4.1. This linker is also useful for releasing drugs if the conjugate is routed through cellular lysosomal compartments, where the pH is reported to be 4.5–4.8 [9].

A few attempts to exploit high levels of amino peptidases and lysosomal acid proteases have been made to achieve drug release from MoAb conjugates, by incorporating short peptide sequences between a drug and the MoAb [10]. Some of these studies are mentioned below. This approach also provides for drug release if a conjugate is internalized into a tumor cell, through the action of lysosomal enzymes. Unfortunately, most of the peptide sequences used in this approach so far have been constructed from nonpolar amino acids, which make marginally water-soluble drugs even less soluble and thus further limit the number of molecules conjugated to a MoAb.

4.2.2. The Antibody

Numerous MoAbs are now available that recognize a diversity of antigens [11], though no truly tumor-specific MoAb has as yet been identified. Nevertheless, several classes of MoAbs with sufficient specificity to recognize antigens expressed in larger amounts on tumor cells compared with normal tissues have been produced. Since a MoAb must retain its ability to differentiate among antigens, the chemistry by which it is modified must not significantly alter its immunoreactivity. Thus, the modification of any residue in the antigen-binding domain must be avoided. Of the numerous functional groups that can be modified along the backbone of a MoAb, in reality only four or five are accessible under conditions that do not alter the integrity of the protein; the chemistry associated with modification of these groups will be discussed in greater detail in later sections.

Immunoglobulin G's (IgG's) have been used much more frequently than the other

---

*Italic numerals in parentheses throughout text indicate corresponding chemical structures illustrated in Figures 4.1–4.7.

isotypes for *in vivo* applications because their relatively smaller size (~150,000 MW) makes them better candidates for penetration into tissue targets. The ultimate size of a conjugate is also important for favorable biodistribution and good tumor penetration. Conjugation of a radioiodine to a MoAb causes insignificant changes in the size of the protein, and even conjugation to ten methotrexate residues increases the size of the MoAb by only about 3%. However, conjugation to most enzymes produces constructs substantially larger than the original MoAb with a corresponding decreased ability to penetrate into tumors. Diminishing the size of a construct is possible through use of smaller fragments, generated either by enzymatic digestion of whole MoAb or through genetic engineering [12]. Such manipulation produces $F(ab')_2$ (~100,000 MW) or F(ab') (~50,000 MW) fragments, which are useful for limiting the size of conjugates when large molecules are being delivered. Even smaller fragments, such as single chain Fv's (~27,000 MW) have been generated [13]. These are reported to have good tumor-penetrating characteristics, although they exhibit extremely short circulating half-lives.

The role of MoAb affinity has not been resolved. A high-affinity MoAb, with its correspondingly longer residence time on a target, might appear to be an advantage for targeting. However, theoretical modeling of constructs suggests that high-affinity MoAbs accumulate only in the periphery of a tumor mass, preventing percolation deep into the core [14]. This could be a disadvantage for drug targeting. On the other hand, isotopes rely on a prolonged residence time on the cell surface and cross-fire effects to kill adjacent tumor cells; thus high-affinity antibodies with a long residence time on target cells have theoretical advantages.

Selection of a MoAb for a particular use extends to the nature of the antigen it recognizes. Secreted antigens are ideal for *in vitro* diagnostic purposes; their serum levels, and the ability to quantitate such levels, correlate with the presence and size of several human tumors. For *in vivo* targeting, however, circulating target antigens can complex with a conjugate, not only interfering with its targeting potential, but also contributing to its diminished circulating half-life. As a general rule, MoAbs that recognize nonsecreted, noninternalized antigens are appropriate for targeting of radioisotopes and certain enzymes. For most drugs and toxins that work inside cells, conjugates prepared from MoAbs that are actively internalized are preferable, although the pathway by which the conjugate compartmentalizes inside the cell can be important for retaining drug or enzyme activity.

4.2.3. Molecules Conjugated to MoAbs

Three general types of molecules have been used to prepare MoAb conjugates: radioisotopes, low-molecular-weight compounds, and enzymes. Examples of each type can be found in both the therapeutic and diagnostic fields, but again, since a much larger diversity of chemistry has been applied to the conjugation of therapeutic agents, this chapter concentrates on such applications.

The primary target of virtually all commercial oncolytic drugs is a compartment or structure inside the tumor cell, often nuclear DNA. It would appear obvious to select MoAbs that are internalized for targeting of such drugs [15], yet in many reported studies this has not been the case. The molecular structure of a drug must be preserved for the drug to operate effectively on its cellular target. Even seemingly minor changes in structure can have a profound effect on the activity of a drug. Thus conjugation strategies that make no provisions for release of the parent drug or, worst yet, manipulations that destroy a portion of a drug in an effort to generate a functional group suitable for conjugation (such as oxidation of the daunosamine ring of daunorubicin to generate aldehydes) [16] are doomed to failure.

The two classes of antitumor drugs that have been conjugated to MoAbs most frequently are anthracyclines, exemplified by doxorubicin (*2*) and daunorubicin (*3*), and the dihydrofolate reductase (DHFR) inhibitor methotrexate (*4*). The structures of these drugs are shown in Figure 4.2. All three drugs are used clinically, but achieving better localization and higher concentrations in tumors through targeting by MoAbs seemed appropriate. In the case of the anthracyclines, overcoming their cardiotoxicity was also a compelling reason for conjugation to MoAbs. Although these drugs will be discussed in greater detail in later sections, several lessons for the future of drug targeting in a general sense can be learned from the collective experience with these agents.

Daunorubicin and doxorubicin kill cells predominantly through intercalation and single-stranded cleavage of DNA. Although the daunosamine sugar residue is not directly involved in the cleavage, the full basicity of the C-3′ amine is required for tight binding in the minor groove of DNA [17]. However, many of the methods described for conjugating anthracyclines to MoAbs rely on converting the C-3′ amine to an amide, often without provisions for reconverting the linker back to the free amine. The anthracyclines also have a propensity to bind MoAbs noncovalently: As much as 30% of daunorubicin has been reported bound in a noncovalent form [18]. This could explain the cytotoxicity of some anthracycline–MoAb conjugates. However, an important reason for mixed success with the anthracyclines as MoAb conjugates is their potency. In order to eradicate a tumor, it is necessary to deliver between 10^6 and 10^7 molecules of these drugs to each tumor cell [19]. Successful conjugates prepared from the anthracyclines mandate high loading, exceptionally high levels of antigen expression by target cells, and a good rate of conjugate internalization. Although the last two requirements can be satisfied relatively easily through the routine screening and evaluation of MoAbs, achieving high loading of the anthracyclines has been extremely difficult.

Methotrexate (*4*) is even less potent: To kill a single cell 1.44×10^8 molecules are required [20]. Aminopterin (*5*), a close analogue of methotrexate, is about 10 times more potent in animal models yet has received relatively little attention as far as MoAb conjugation is concerned. Both molecules are unique in that they can tolerate significant modification at the γ-carboxyl group of their glutamic acid residues and still retain full DHFR inhibitory activity. The α-carboxyl group, on the other hand, needs to remain unmodified for binding to an arginine residue of DHFR [21]. Methotrexate–MoAb conjugates are catabolized by cells to smaller peptide and protein fragments, which, if these contain methotrexate, are still capable of inhibiting DHFR [15]. Unfortunately, most methods of activating methotrexate for MoAb conjugation have paid little attention to the regiospecificity of which carboxyl group becomes bound.

On the other end of the potency scale are a number of plant and bacterial

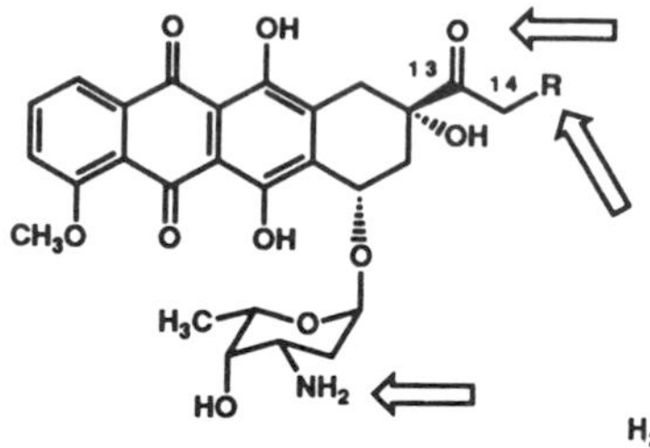

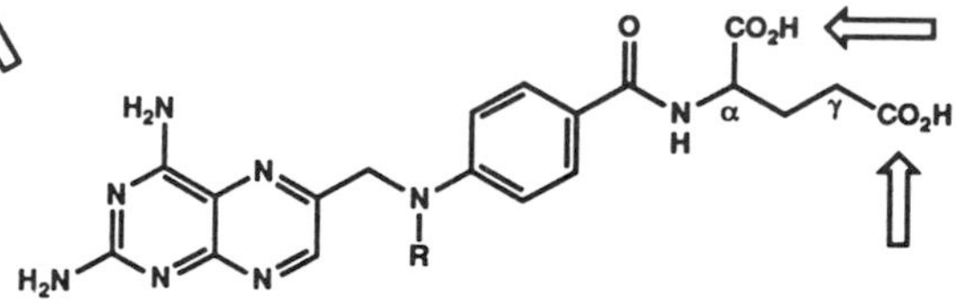

2: Doxorubicin: R = OH
3: Daunorubicin: R = H

4: Methotrexate: R = CH_3
5: Aminopterin: R = H

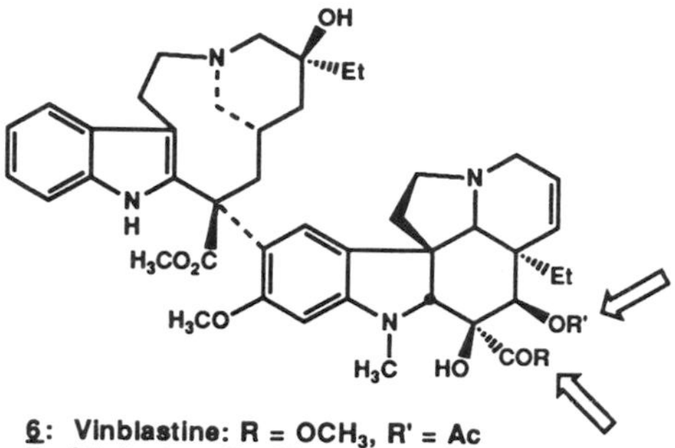

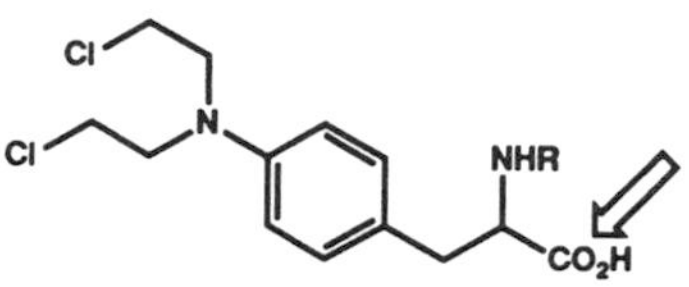

6: Vinblastine: R = OCH_3, R' = Ac
7: Vindesine: R = NH_2, R' = H

8: Melphalan: R = H
9: N-Acetylmelphalan: R = Ac

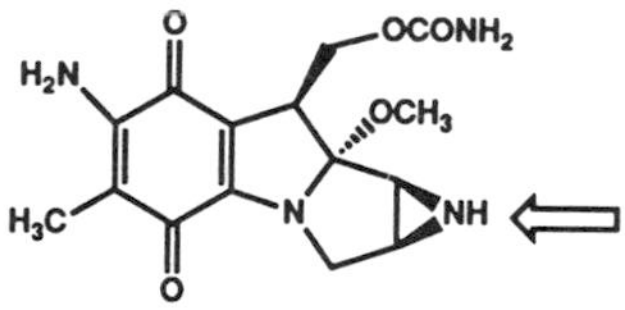

10: Mitomycin C

Fig. 4.2. Structures of the most common anticancer drugs that have been conjugated to MoAbs. The arrows denote the functional groups of the molecules used for conjugation.

enzymes that destroy cells through inhibition of protein synthesis. These enzymes act catalytically rather than stoicheometrically. A single molecule of ricin, for example, is sufficient to destroy a cell. A drawback with using enzymes for constructing MoAb conjugates is their highly immunogenic nature. Nevertheless, numerous conjugates have been prepared from such toxins and these will be discussed in a later section.

Several research groups have also prepared MoAb conjugates from compounds that exhibit potencies intermediate between the classical anticancer agents and the protein toxins. These compounds, many of which are shown in Figure 4.3, are of considerably lower molecular weight than the protein toxins; hence, conjugates prepared from them are expected to be less immunogenic than those prepared from the protein toxins. In addition, by using more potent compounds, the likelihood of

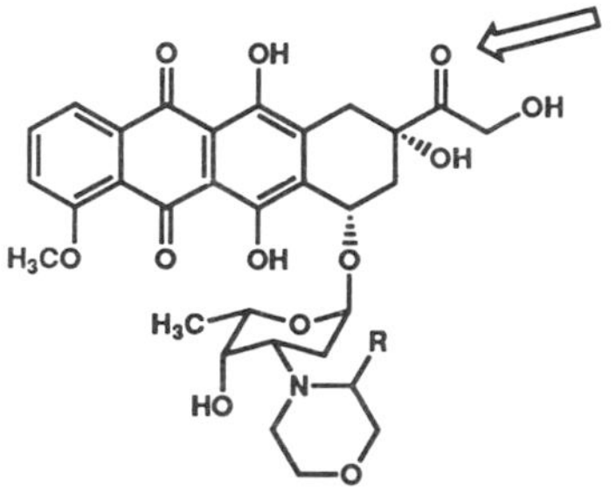

11: morpholinodoxorubicin: R = H
12: cyanomorpholinodoxorubicin: R = CN

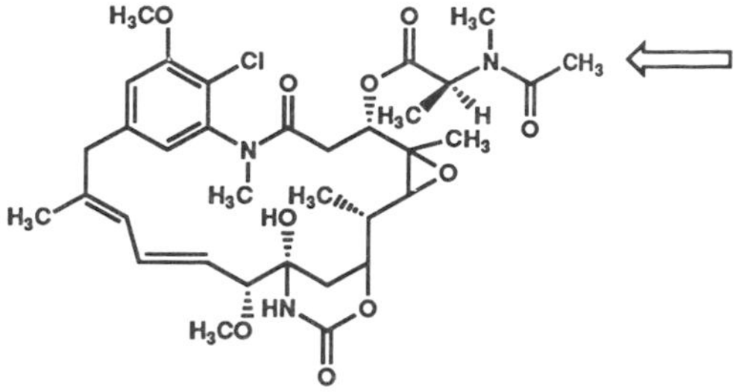

13: Maytansine

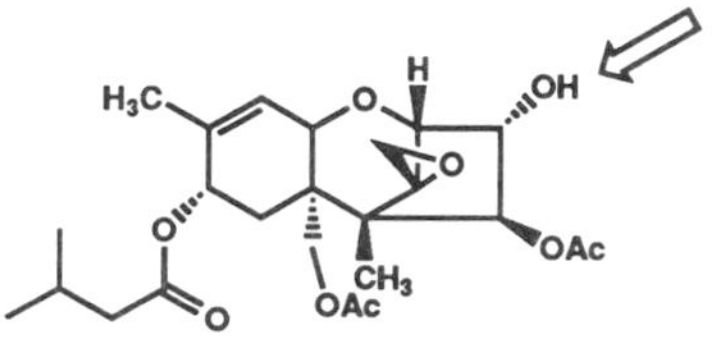

14: Mycotoxin T-2

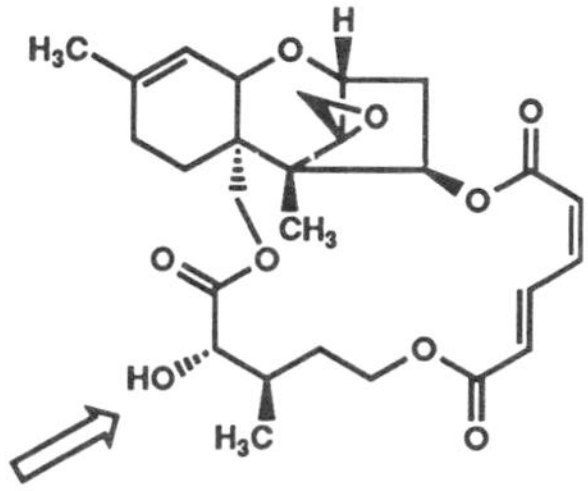

15: Verrucarin A

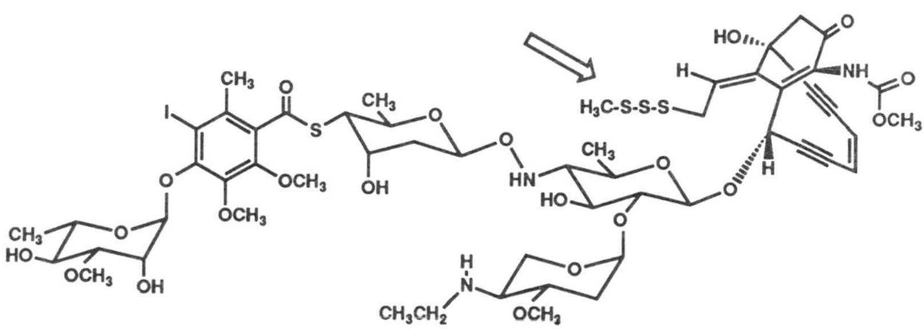

16: Calicheamicin γ_1^{I}

Fig. 4.3. Structures of some of the highly potent small molecules that have been used to prepare MoAb conjugates. Arrows denote the sites of attachment.

delivering enough drug to achieve tumor destruction is increased. These compounds include the morpholino (*11*) and cyanomorpholino (*12*) derivatives of doxorubicin, which are from two to 10,000 times more potent than the parent drug *in vitro*, depending on the target cell line [22]. Other low-molecular-weight drugs that have *in vitro* IC_{50}'s in the 10^{-10} to 10^{-12} M range are also shown in Figure 4.3. Calicheamicin γ_1^{I} (*16*), which destroys cells by causing double-stranded DNA breaks [23], is a member of the ene-diyne family of antitumor antibiotics and as such is related to neocarzinostatin chromophore [24]. However, unlike neocarzinostatin, calicheamicin $\gamma_1{}^I$ does not require a protective protein coat for stability, and it can thus be conjugated to MoAbs directly. Mycotoxin T-2 (*14*) [25] and verrucarin A (*15*) [26] are fungal metabolites of the trichothecene family, and act by inhibiting protein and DNA synthesis. Maytansine (*13*) [27], which kills cells through destabilization of microtubules, and palytoxin (not shown) [28], which is a potent cell-membrane-acting compound, have also been used with MoAbs in an attempt to overcome the potency limitations of classical anticancer agents.

Many radionuclides also improve significantly on the stoicheometry of classical antitumor agents, and thus are good candidates for MoAb targeting. Only three to six transverses of the nucleus by an alpha particle or 1,200–2,400 transverses by a beta particle are sufficient to kill a tumor cell [29]. One interesting source of alpha particles is ^{10}B. When this nonradioactive isotope captures a thermal neutron, it is converted to ^{11}B, and immediately fragments by ejecting an alpha particle to produce ^{7}Li. Unfortunately, only one in 10^8 ^{10}B atoms actually captures a thermal neutron, making it necessary to deliver about 10^9 borons per cell; this corresponds to a loading of 10^3 to 10^4 borons on each MoAb [30]. Given the size of such constructs, there appears to be little chance of success with boron neutron capture therapy as far as MoAb delivery is concerned.

4.2.4. The Linker

Though the conjugation literature is extensive, the number of cross-linking reagents most often used to construct MoAb conjugates is relatively small. The most common reagents are shown in Figure 4.4. All are reactive and sufficiently stable within the pH limits and aqueous conditions required by MoAbs. Symmetrical cross-linking reagents such as glutaraldehyde (*17*), which has been used to join the amines of a MoAb to the amines of an effector molecule, have been largely replaced by more selective bifunctional reagents. SPDP (*18*) and 2-IT (*19*) are commonly used to introduce thiol groups onto MoAbs or amine-containing polymers. MBS (*20*), IAHS (*21*), SMCC (*22*), and SIAB (*23*) react with amino groups to introduce thiol-reactive functionality. DCC (*24*) and the more water-soluble carbodiimide EDCI (*25*) have been used occasionally to cross-link MoAb carboxyl groups with amines, but they are usually used to activate small-molecule carboxyls for conjugation to MoAb lysines. Succinic and glutaric anhydrides have been used to convert amino groups to carboxyls; the resultant hemisuccinate or hemiglutarate is then activated for conjugation to MoAbs. A few variations of these reagents have also been described, such as replacing the hydroxysuccinimide-activating group with α-sulfohydroxysuccinimide to make the materials more water-soluble.

SPDP (*18*) has been a particularly versatile spacer, with applications in constructing all forms of MoAb conjugates. After reaction with amines, the disulfide can be reduced with, for example, dithiothreitol and the resulting thiol can be used to react with a molecule containing a thiol acceptor. Alternatively, the dithiopyridyl group is sufficiently reactive to be displaced by thiols directly to form disulfide bonds. The ultraviolet chromophore of the by-product mercaptopyridine is sufficiently strong to be useful for assessing the number of SPDP molecules introduced onto a MoAb or polymer.

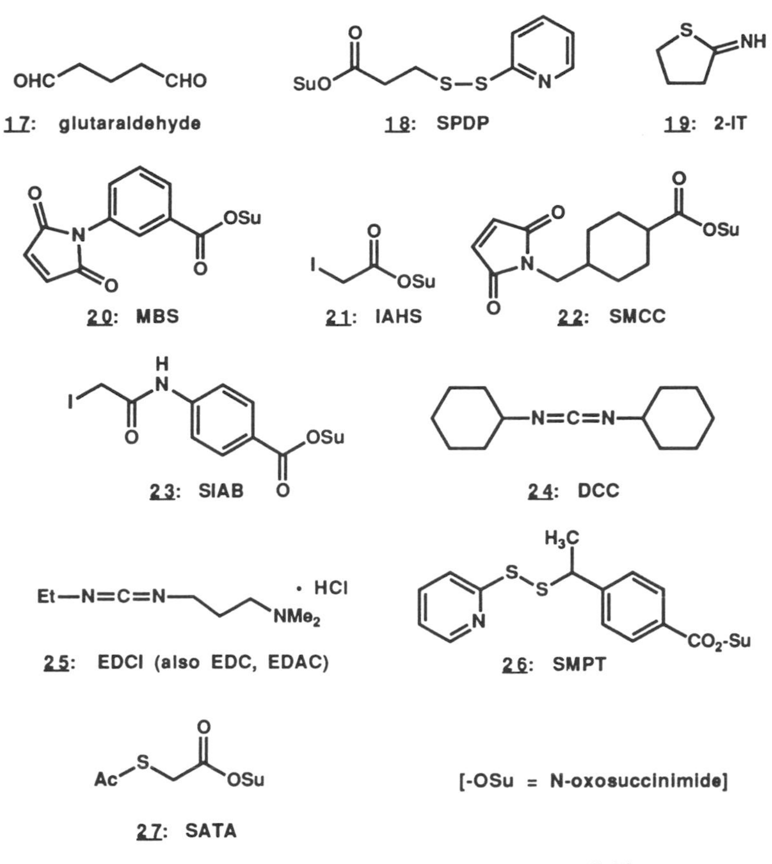

Fig. 4.4. Structure of some commercial protein cross-linking agents.

Many of the linkers and activating reagents mentioned above are adequate for constructing conjugates when release of the effector molecule is not desirable and functionalization of the MoAb or effector molecule with these reagents does not alter their biological functions. For other applications, however, proper design of the linker can be crucial. It can be tailored to ensure stability of a conjugate in serum, designed to increase aqueous solubility or to release a molecule within the tumor, or constructed to encourage rapid metabolic elimination if the conjugate lodges in a normal organ. The following sections exemplify some of the chemical methods and the linkers by which various molecules have been conjugated to MoAbs.

4.3. NONSPECIFIC MoAb CONJUGATION

MoAbs, like all glycoproteins, contain an abundance of side-chain residues along their backbones. Although chemical methods exist for accessing the majority of these side chain functional groups, most such methods are incompatible with preserving the targeting ability of MoAbs and thus are not useful for constructing functional conjugates. Only the ε-amines of lysines and the phenols of tyrosines are sufficiently reactive under conditions that preserve the integrity of MoAbs, and even with that, the latter residues have been used only for radiohalogenations. Moreover, since these residues are dispersed more or less randomly along the backbone of MoAbs, using these to construct conjugates always involves the risk of modifying critical residues in the antigen recognition sites.

In addition to the ε-amines of lysines, all MoAbs also contain four terminal α-amines, one each at the variable end of the heavy and light chains in or near the antigen-binding domains, and these are also susceptible to modification. In fact, these terminal residues have been used selectively for constructing site-specific conjugates (see section 4.3.2). Lysine amines are unprotonated and become reactive at a pH above 8.0 (the terminal amines above pH 7.0), so in fact the reactivity of the α-amines is greater than that of the lysine side chains at a pH between 7 and 8 [31].

Five main reactive groups have been used to functionalize MoAb lysines: activated carboxylates, aldehydes, α-haloesters, isocyanates, and isothiocyanates. All methods of modifying lysines rely on converting an agent to one of these functional groups. This can be achieved either with an existing functional group on the molecule, such as the carboxyls of methotrexate, or after introduction of a spacer—for example, by reacting the amine of daunorubicin with a diacid anhydride—followed by activation of the resultant carboxyl. Three types of activated carboxylates are useful for preparing conjugates: hydroxysuccinimides, iminoesters (from treatment of the carboxyl with a carbodiimide), and mixed or cyclic anhydrides. All such groups react with MoAbs at $pH > 8$, conditions under which the lysines are unprotonated.

4.3.1. Drugs

A representative listing of drugs and the functional groups used to achieve reaction directly to the lysines of MoAbs are given in Table 4.1. The following discussion relates to issues specific to the conjugation of drugs through lysine residues.

A comparison of several bis-electrophilic linkers has been made, using the bromo- and chloroacyl halides of p-isothiocyanatobenzoic acid, p-isothiocyanatophenylpropionic acid, α-bromacetic acid, and p-fluorosulfonylbenzoic acid [58]. The acid chloride ends of the linkers were reacted with the C-3′ amino groups of daunorubicin or doxorubicin, the other ends with MoAb lysines. For conjugation to MoAbs, the p-isothiocyanatophenylpropionyl halides produced the best balance between reactivity, stability of linkage, and cytotoxicity of conjugates.

The hydroxysuccinimide, mixed anhydride (isobutyl chloroformate), and imidate (EDCI) methods have been compared for coupling methotrexate to MoAbs [40]. Incorporation was linear up to a reagent-to-MoAb ratio of 60:1, with the hydroxysuccinimide (OSu) ester producing the highest coupling efficiency and the mixed anhydride the least. Above a 60:1 reagent ratio, EDCI produced the best loading of methotrexate (20 molecules), but it also resulted in a significantly decreased recovery of protein ($< 10\%$) and a severe loss of immunoreactivity ($< 30\%$ compared with unmodified MoAb). Overall, the OSu method produced conjugates with the highest retention of activity, followed by the mixed anhydride, then EDCI.

TABLE 4.1. Examples of Agents That Have Been Conjugated Through MoAb Lysines

Agent (Function Group)	Spacer	Activating group	Loading (molecules/ MoAb)	Reference
Desacetylvindesine				
(*via* 4-OH)	Hemisuccinate	Mixed anhydride (i-BuOCO-)	≤6.2	[32]
(*via* 4-OH)	Hemisuccinate	-OSu	3.6–9	[33]
Desacetylvinblastine (*via* 3-CON_3)	None	Azide	3.6–9	[33]
Daunorbicin				
(*via* C-3′-NH_2)	cis-Asconitate	-OSu	10–40	[34]
(*via* C-3′-NH_2)	cis-Aconitate	Imidate (EDCI)	≤10	[35]
(*via* C-3′-NH_2)	Glutaraldehyde	Aldehyde	0.5–13	[36]
(*via* C-14)	None	Convert to drug-bromide	8.7	[37]
Doxorubicin				
(*via* C-13 carbonyl)	Thiopropinyl hydrazide	SPDP on MoAb, thio-pyridyl on linker	3–8	[38]
(*via* C-3′-NH_2)	-succinyl-Ala-Leu-Ala-Leu	-OSu	≤3.5	[32]
Morpholinodoxorubicin (*via* 13-keto)	=$NNHSO_2$-Ph-CO_2-	-OSu		[22]
Idarubicin (*via* C-14)	None	Convert to drug-bromide	3–5	[39]
Methotrexate				
(*via* α- and γ-carboxyl)	None	-OSU	Up to 18	[40]
(*via* α- and γ-carboxyl)	-Leu-Ala-Leu-Ala-NHNH-	Iodoacetamide	~4.8	[41]
(*via* α- and γ-carboxyl)	-Cys-	SPDP on MoAb	Up to 5	[42]
(*via* α- and γ-carboxyl)	None	-OSu	1.6–1.9	[43]

(*continued*)

TABLE 4.1. *(Continued)*

Agent (Function Group)	Spacer	Activating group	Loading (molecules/ MoAb	Reference
Aminopterin (*via* α- and γ-carboxyl)	None	Imidate (DCC)	~11	[44]
Chlorambucil (*via* CO_2H)	None	Isocyanate	37	[45]
N-Acetylmelphalan (*via* CO_2H)	None	-OSu	15–20	[46]
Melphalan (*via* CO_2H)	None	Imidate (EDCI)	3–5	[47]
Mitomycin C (*via* aziridine)	Hemiglutarate	-OSu	6–8	[48]
5-FUdr (*via* 3-OH)	Hemisuccinate	-OSu	7–9	[49]
Heamatoporphyrin	None	EDCI	15.7	[50]
Chlorin e_6	None	EDCI	~40	[50]
B_9 and B_{10} cage carboranes	Phenyl ring	Isothiocyanate or azide	4 cages	[51]
Ellipticinium acetate (*via* C-10)	None	Oxidize drug to quinone imine, add MoAb	33	[52]
Formyl-Met-Leu-Phe	None	EDCI	Variable	[53]
Misonidazole	Hemisuccinate	Mixed anhydride (i-BuOCO-)	4–6	[54]
Mycotoxin T-2	Hemiglutarate	-OSu		[55]
Verrucarin A	Hemisuccinate	-OSu		[56]
Maytansine	Side chain at C-3 synthetically introduced with terminal thiol functionality	SPDP on MoAb	1–6	[57]

Although OSu esters of drugs are widely used to modify the lysine groups of MoAbs, their reactivity does not appear to be restricted entirely to amines. When the OSu ester of methotrexate was reacted with MoAb MM46 and the resultant conjugate was treated with hydroxylamine in borate buffer at pH 9.0, 41% of the bound methotrexate was released within 24 h. No further release occurred with subsequent treatment [59]. When the reaction was repeated with dextran (a biological polymer that contains no amino functionality), 97% of the bound methotrexate was released by hydroxylamine, implying that a significant amount of the methotrexate bound to MoAb was ester-linked, probably to the carbohydrate residues. Moreover, the hydroxylamine-treated conjugate was 6.3-fold less cytotoxic.

Lysine residues can exist in or near the antigen-binding pockets, and modification of such residues can alter the immunoreactivity of the product significantly. Dinitrofluorobenzene, a reagent that reacts specifically with lysine amines, has been used to assess whether such residues are involved with antigen binding. Of four MoAbs modified with dinitrofluorobenzene, two retained full immunoreactivity, one lost 80% of its immunoreactivity, and the fourth was completely inactivated [60]. However, in this particular case no correlation was found between dinitrofluorobenzene inactivation and the suitability of these same four MoAbs for lysine-based modification.

Dimethylmalonic anhydride has been used to protect the chemically most reactive lysine groups [61] and is a particularly useful reagent if such groups are in or near the antigen-binding sites of the MoAb. The malonamide can be removed by hydroxylamine at pH 8.5–9 after modification of the remaining available lysines with the desired agent. This technique has been used to conjugate the OSu ester of methotrexate to the antimelanoma MoAbs 96.5 and ZME018.

Not all lysines are available for modification; many are buried deep into the backbone of the MoAb and are not accessible. MoAb 9.2.27 contains 52 lysine residues, and only 15 of these can be reacted with doxorubicin before significant loss in immunoreactivity occurs; only 5 residues out of 63 can be modified on MoAb 14.G2a [62]. Another study, in which four additional MoAbs were evaluated, revealed that of the 35–84 lysines in the backbone of these proteins, only 2–30 could be modified with SPDP without precipitation of the product [63].

Incorporation of a disulfide bond within the linker has been common for constructing protein–MoAb conjugates (see section 4.2.3.3). This takes advantage of the reducing environment inside cells to facilitate release of the proteins once targeting and internalization has occurred. Several publications wherein disulfide-based linkers have been used to facilitate release of small molecules have also appeared, and some of these are listed in Table 4.1. For example, attaching the anthracyclines to a MoAb through their 13-keto functionality by using hydrazones facilitates release of the parent drugs within the acidic environment of lysosomal compartments; this technique has been applied to the conjugation of doxorubicin [38] as well as the more potent morpholinodoxorubicin [22]. Examples in which both the disulfide and hydrazone elements are combined within a single linker, to form two "hot spots" for drug release, have also been described [38].

4.3.2. Isotopes

The virtues and failings of diagnostic and therapeutic isotopes as conjugates of proteins in a general sense, and of MoAbs specifically, have been the subject of several recent extensive reviews, so discussion will not be repeated here [29,64]. However, a summary of the chemistry that has been applied to the isotope delivery field may be useful. Unlike the situation with drugs, release of which after targeting is desirable, leakage of even small amounts of isotope from a conjugate can be disastrous. As before, the chemistry used to construct isotope conjugates is important for controlling their stability.

Commercial reagents that are a source of positively charged iodine, such as chloramine T, lodogen, lactoperoxidase, and iodobeads, are used extensively for iodinating MoAbs. Such iodination takes place in a nonspecific manner on the most electronegative sites of amino acid side chains, specifically the tyrosine residues [64e]. The advantage of using these radioiodination reagents is that they involve a single-step procedure and are not encumbered by the decay byproducts that are formed during shipping of the isotope. In practice, iodination of one tyrosine or less per MoAb produces constructs that are sufficiently "hot" even for therapeutic uses; radiolysis of the proteins becomes significantly severe beyond iodination of one tyrosine.

The common iodination reagents chemically damage MoAbs to differing degrees, with lodogen being used most extensively because it produces conjugates with the best balance between high radiochemical incorporation and least damage to proteins. A study using the Iodogen method has been reported in which 13 MoAbs were evaluated for immunoreactivity changes upon iodinations to a specific activity of 0.5–20 mCi/mg and the sites on the protein where such iodination occurred were determined [65]. For two of these MoAbs, iodine was incorporated exclusively in their light chains; in two others, exclusively in their heavy chains. The site of iodination was less specific in the remaining nine, yielding heavy to light chain modification ratios in the range of 2–6.5. Moreover, although this extent of iodination altered binding affinity in six of the MoAbs, this altered affinity did not correlate with a preferential labeling in either of the protein chains.

Metabolic deiodination occurs in mammals when MoAbs are iodinated through tyrosine residues. Therefore, several reagents have been developed that reduce or eliminate this liability (Fig. 4.5). The commercial Bolton–Hunter reagent (*28*) is one such example. Even though the iodine is situated on a phenyl ring, absence of a phenolic hydroxyl group stabilizes the iodine–carbon bond toward enzymatic deiodination. Iodides (*29*) through (*31*) have also been developed with this in mind; all can be produced from their corresponding trialkyltin analogues in 65%–95% radiochemical yield, and show a coupling efficiency of 35%–50% to MoAb lysines [66]. The pyridine ring of tri-n-butyltin precursor to (*31*) makes iodination somewhat more difficult due to the lower nucleophilicity of that compound compared with the corresponding precursor to (*29*). Nevertheless, thyroid uptake of these compounds is reduced to less than 0.1–0.2% of injected dose, or about 100-fold less than tyrosine-based radioiodides. Iodide (*32*), which is prepared by electrophilic dis-

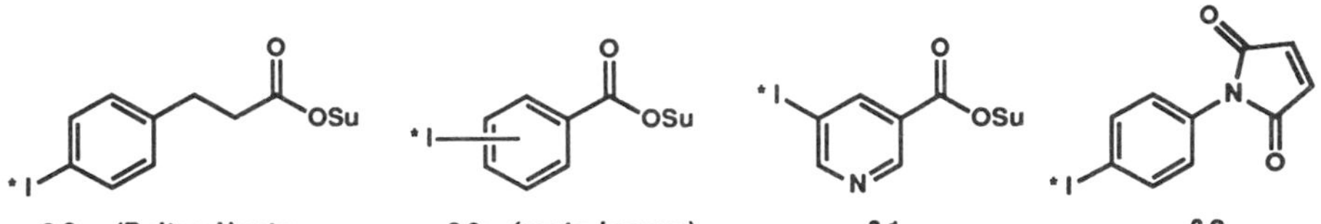

28: (Bolton-Hunter reagent) 29: (*meta* isomer) 30: (*para* isomer) 31 32

Fig. 4.5. MoAb radioiodination reagents useful for labeling lysine residues.

placement of a mercuric acetate by iodonium ion, has been developed for coupling to MoAb thiol groups, particularly those derived from the heterobifunctional cross-linking agents SPDP or 2-IT [67]. A negative aspect of these reagents is that they all involve at least two steps: iodination of the intermediate mercural or stannane derivative, followed by coupling to the MoAb. Compound (*32*) requires one additional step, that of modifying the MoAb with a thiol group.

Most metallic radioisotopes require chelation for their tight binding to proteins. The structures of some chelators used with MoAbs are shown in Figure 4.6. By far the most common chelator used in this regard has been diethylenetriaminepentaacetic acid (DTPA) (*33*). Either an internal cyclic anhydride or one of several mixed anhydrides have been used to conjugate DTPA through a carboxymethyl arm to MoAb lysines. The number of DTPA residues (and other macrocycles as well) that become attached per MoAb is not as critical as in the case of drugs, since for most applications, only a small fraction of MoAbs in a final preparation contain a radiometal. In practice, however, the five or six DTPAs per MoAb normally quoted in publications are necessary to sequester the radiometal rapidly from solution, and also minimize the competition for these sites by the large excess of cold metals that are present in isotope preparations. DTPA (*33*) has been used to conjugate diagnostic and therapeutic isotopes of indium, yttrium, copper, technetium, rhenium, cobalt, bismuth, lead, and astatine, among others. Although the kinetics of radiometal incorporation into DTPA are rapid, it is by no means the ideal chelator. Serum Ca^{+2} and Mg^{+2} ions compete with radiometals for chelation sites, and serum proteins such as albumin compete for chelated copper [68] and transferrin completes for indium [68]. Yttrium, once it is released from its chelate, rapidly accumulates in bone [70]. In part, this instability can be attributed to the use of one of the DTPA arms to form the amide bond to a MoAb, thus preventing full chelation.

A variety of chelators have been synthesized that are designed to hold metals more securely and thereby diminish competition by endogenous ions and proteins. Among open-chain chelators, the benzylisocyanate derivatives of DTPA and EDTA (*34* and *35*, resp.) have been prepared for conjugations through MoAb lysines, and *36* for conjugation through thiols [68,71]. These and other open-chain chelators have been compared as MoAb conjugates for their ability to localize isotopes into xenograft tumors. Chelator (*34*) localized 40% of the injected dose per gram of ^{88}Y into a xenograft, whereas (*35*) localized only 6%–8%. As an indirect measure of *in vivo* stability, chelator (*35*) led to a maximum bone accumulation of 14% injected

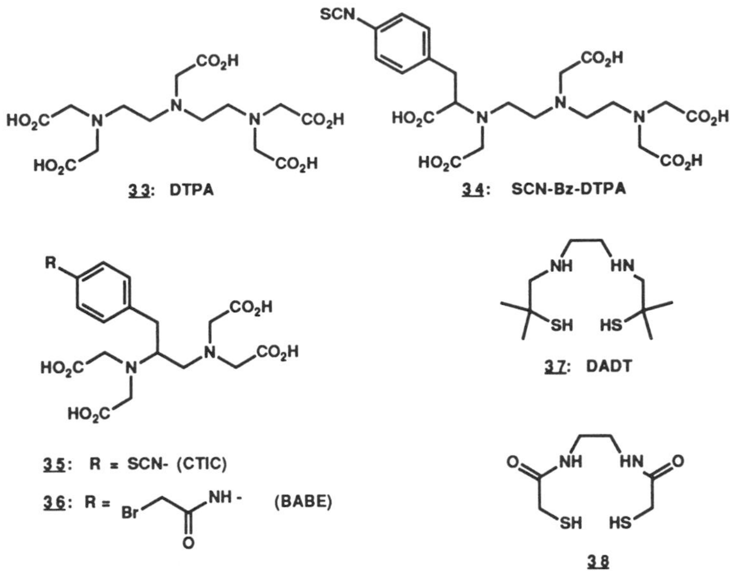

33: DTPA

34: SCN-Bz-DTPA

35: R = SCN- (CTIC)

36: R = Br–CH2–C(=O)–NH- (BABE)

37: DADT

38

Fig. 4.6. Common open chain chelators used to sequester radiometals on MoAb conjugates.

dose per gram, whereas (*34*) showed only 3% bone accumulation with ^{88}Y [70,72]. This contrasted with a 20% bone accumulation of unchelated ^{88}Y-acetate and this material remained there for up to 7 days. On the other hand, the EDTA-based chelator (*35*) held ^{111}In better.

DTPA-derived chelators have also been used for both ^{99m}Tc and ^{186}Re. However, considerable leakage of these metals, particularly ^{186}Re, has been noted [73]. Technetium has also been linked directly to MoAb thiols, generated by mild reduction of disulfide bonds [74]. Several thiol-based chelators have been developed for this pair of isotopes, and two are shown as (*37*) and (*38*). Conjugation of these chelators to MoAbs has been achieved by introducing a linker into the ethylene bridge, or through one of the nitrogen atoms [75].

In order to increase stability of some radiometal conjugates, a number of new macrocyclic chelators have been synthesized (Fig. 4.7). These include tetraaza macrocycles (*39*) through (*42*) and triaza macrocycles (*43*) and (*44*). The primary amino groups at the end of the methylene chains on the macrocycles shown provide the appropriate functionality for further elaboration with spacers. These spacers have included bromacetamide [76], vinyl pyridine [77], isothiocyanate [78], and maleimide groups [79]. The tetraaza macrocycles, particularly those based on the TETA system (*39*), are a necessity for chelation of copper: Open-chain chelators do not hold this metal tightly enough to avoid loss to albumin [80]. Macrocycle (*40*) has been reported to bind both ^{64}Cu and ^{99m}Tc [81], the DOTA macrocycle has been used for chelation of ^{88}Y, ^{90}Y, and ^{111}In [76], and the triaza macrocycles (*43*) and (*44*) have been used with ^{111}In [75].

Although strong chelation of radiometals is essential to avoid kinetic exchange of

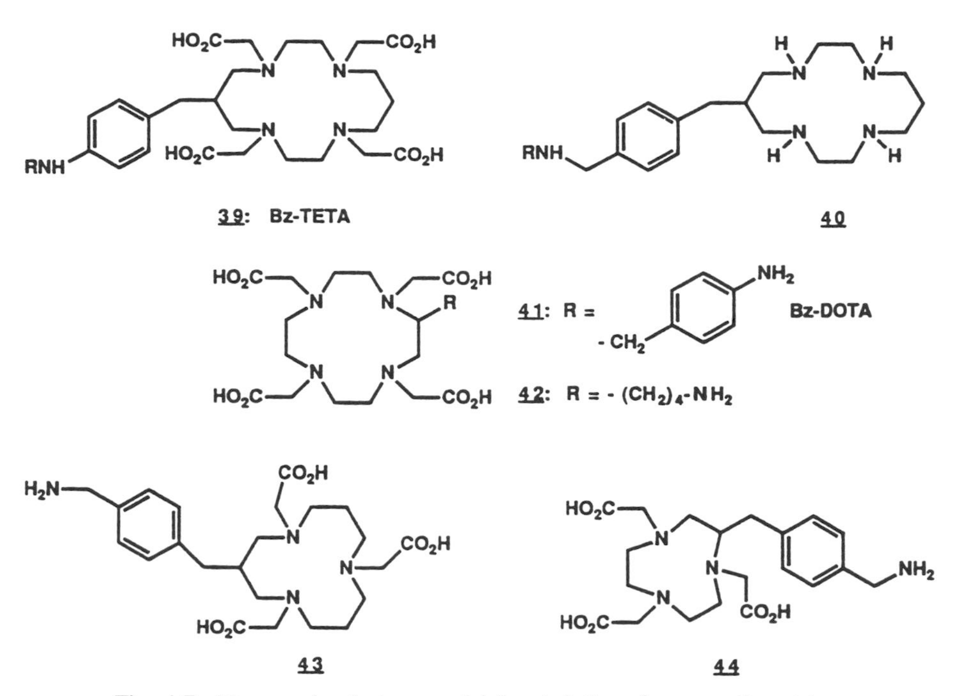

Fig. 4.7. Macrocycles that are useful for chelation of some radiometals.

the isotopes as discussed above, the typical biodistribution of such conjugates results in substantial accumulation in normal organs, particularly in the liver. If accumulation of therapeutic isotopes persists for prolonged periods of time, radiation damage to these organs can occur. Elimination of these complexes can be accomplished by metabolic reactions producing unconjugated chelators that are rapidly eliminated.

For this reason, a number of spacers have been developed to facilitate metabolic release of conjugates containing chelated radiometals by normal organs. In a study in which a CTIC-type chelator was linked to a MoAb through one of four linkers containing a thioether, a peptide, a thiourea, or a disulfide bond, clearance of all constructs was similar ($T_{1/2}$ = 12 h) [69]. Liver uptake, measured as percentage injected dose per gram at 72 h, was 13.4%, 7.6%, 20%, and 2.2%, respectively [82]. Another study concluded that ester bonds within the linker also lead to a better elimination rate from normal organs [83]. Incorporating a cleavable tetrapeptide (–Ala–Leu–Ala–Leu–) between a MoAb and a benzyl-EDTA chelator showed good release of radioisotope from the MoAb *in vitro,* but when the system was tested *in vivo,* approximately the same amount of radioactivity accumulated in the livers of mice regardless of the presence or absence of the tetrapeptide [84].

One additional method of radiolabeling MoAbs deserves mention. Kemptide, a synthetic heptapeptide substrate for protein kinases, was modified at its N terminus with an iodoacetamide group, and this was reacted with the thiols of an SPDP-modified MoAb. The conjugate was labeled with ^{32}P in the presence of a protein kinase and [^{32}P]γ-ATP to a specific activity of over 10 $\mu Ci/\mu g$. This method introduced up to 2.2 molecules of ^{32}P per MoAb, a level sufficiently high to permit the evaluation of this isotope in targeted tumor therapy [85].

4.3.3. Enzyme–MoAb Conjugates

A variety of enzymes have been conjugated to MoAbs in an effort to overcome the stoichiometric limitations of drugs and isotopes. A single enzyme, if it exhibits a high turnover rate, can act on thousands of substrate molecules in a catalytic manner. If the selected enzyme can generate cytotoxic molecules from nontoxic precursors or act on a critical component of the tumor cell, delivery of the enzyme by a MoAb might be more effective than delivery of the cytotoxic species itself. Three tactics have been developed by this approach: delivery of an enzyme that converts an endogenous substance to a cytotoxic species; delivery of an enzyme that converts a nontoxic prodrug to a cytotoxic agent; and delivery of an enzyme that inhibits protein synthesis through the inactivation of ribosomes. Considerable overlap between these tactics exists, and the chemistries used to construct such conjugates are in large part identical.

Enzymes That Generate Cytotoxins From Endogenous Substances. Table 4.2 presents a sampling of enzyme–MoAb conjugates that act by converting endogenous substances to cytotoxic agents or deprive cells of an essential metabolite after binding to target cells. Only the cobra venom factor conjugate has been evaluated *in vivo,* where it was reported to suppress the growth of a human tumor xenograft for an additional 2 weeks relative to controls [90].

Enzymes That Generate Cytotoxins From Prodrugs. Table 4.3 summarizes some of the conjugates prepared for another enzyme-based approach, one that has been referred to as "antibody-directed enzyme prodrug therapy" (ADEPT) [92] or "antibody-directed catalysis" (ADC) [93]. The enzymes used with this approach act by cleaving inactive prodrug forms of anticancer agents into their active counterparts at the tumor site. For studying such systems *in vivo,* the conjugates are injected into animals and allowed to localize and clear from circulation; then the appropriate prodrug is injected [94]. Three potential advantages of this type of therapy are that (a) internalization into cells is not required, (b) drugs are released in the vicinity of tumor cells containing the target antigen as well as those that do not, and (c) a single enzyme localized on a tumor cell can liberate thousands of drug molecules. Such attributes may help overcome limitations due to low drug potency. Moreover, by using enzymes that have low substrate specificity, it may be possible to administer several prodrugs in combinations, and still have all of them liberated within tumors.

To avoid nonspecific toxicity toward normal organs, the prodrugs must be at least 100-fold less toxic than the parent drug; this has been achieved with many of the derivatives mentioned in Table 4.3. However, the size of the MoAb constructs may present a problem. In one study where a whole IgG–carboxypeptidase G_2 conjugate

TABLE 4.2. MoAb–Enzyme Conjugates That Act on Endogenous Substrates

Enzyme	Mode of Action	Method of MoAb Conjugation	Reference
Glucose oxidase	Converts glucose to H_2O_2	Oxidize enzyme carbohydrates; add MoAb; reduce with borohydride	[86]
Glucose oxidase + lactoperoxidase	Second enzyme converts NaI to I_2 in the presence of H_2O_2	Couple individually to MoAb using glutaraldehyde	[87]
C_{3b}	Chemotactic component of complement system, triggers complement cascade	SPDP modification of C_{3b} and MoAb; link through disulfide bond	[88]
Xanthine oxidase	Converts hypoxanthine to uric acid and in the process generates oxygen free radicals	Modify both enzyme and MoAb with SPDP; link through disulfide bond	[89]
Cobra venom factor	Activates human complement cascade	Modify both enzyme and MoAb with SPDP; link through disulfide bond	[90]
Carboxypeptidase G_2	Depletes cellular supplies of folates	Functionalize MoAb with SPDP, enzyme with bromoacetate or MBS; form thioether bond	[91]

was compared with the corresponding $F(ab')_2$ construct, the latter produced threefold higher levels of enzyme within a tumor [95]. Another major concern with this type of therapy, as well as others involving conjugation of natural and synthetic polymers to MoAbs, is the potential immunogenicity of the constructs. This is a particular concern when proteins of nonhuman origin are used, and it remains a problem with the clinical evaluation of such conjugates.

The potency of the released drug has begun to find a place with this technology as well. Palytoxin, a relatively small molecule (MW ~2680), is a compound isolated from soft coral and acts extracellularly by creating a pore in cell membranes. The unmodified toxin exhibits an IC_{50} of about 3×10^{-12}. The molecule contains a single primary amine at one of its ends; conversion of this amine to an N-(4-hydroxyphenylacetyl) derivative diminishes the potency by a factor of 100. *In vitro* use of the latter derivative has been described wherein the prodrug was reconverted to palytoxin by the enzyme penicillin G amidase conjugated to the MoAb L6 [101].

Enzymes That Inhibit Protein Synthesis (The Immunotoxins). By far the most common class of enzymes that have been studied as MoAb conjugates are the

TABLE 4.3. Conjugates Containing Enzymes That Convert Prodrugs to Parent Drugs in Tumor

Enzyme	Mode of Action	Method of MoAb Conjugation	Reference
β-Lactamase	Cleaves β-lactam ring, releasing drug from C-3′ position of cephalosporin nucleus; delivery of desacetylvinblastine described	React enzyme with sulfo derivative of SMCC; react this with Fab′-SH	[93]
Alkaline phosphatase	Cleaves phosphate residues from alcohols; exemplified with phosphate derivatives of etoposide, mitomycin alcohol, adriamycin, and phenol mustard	React MoAb with 2-IT and enzyme with SMCC; link through thioether bond	[94]
Carboxypeptidase G_2	Cleaves glutamic acid derivative of benzoic acid mustard	Use whole MoAb or $F(ab')_2$ fragment; link to enzyme through thioether bond using MBS/SPDP system	[95]
β-Lactamase	Release of phenylenediamine mustard from C-3′ of cephalosporin described	React $F(ab')_2$ with SMCC and enzyme with 2-IT; link through thioether bond	[96]
Carboxypeptidase A	Generates methotrexate from α-alanine derivative of methotrexate	React MoAb with SPDP and enzyme with SMCC; link through thioether bond	[97]
β-Glucuronidase	Cleaves β-glucuronic acid from phenol mustard	Activate both enzyme and MoAb with SPDP; link through disulfide bond	[98]
Cytosine deaminase	Converts 5-fluorocytosine to 5-fluorouricil	Modify enzyme with SMCC and MoAb with 2-IT; link through thioether bond	[99]
Penicillin-V-amidase	Cleaves phenoxyacetamide groups from amines; use with adriamycin and melphalan described	Modify enzyme with SMCC and MoAb with SMCC; link through thioether bond	[100]

phytotoxins. These are a large group of proteins that have been isolated from both plant and bacterial sources; they act by altering ribosomal function and inhibiting protein synthesis in a catalytic manner. Thus, trafficking to the cytosolic compartment of the cell is required [102], preferably by way of internalizing MoAbs [103]. Research with the phytotoxins has produced several MoAb-linked immunotoxins, or ITs, that have been taken into Phase I clinical trials for the treatment of lymphomas, leukemias, solid tumors, and graft-versus-host disease [104]. Recently, ITs have been designed to inhibit HIV replication as well [105]. The general review of phytotoxins [106] as well as the preclinical literature on ITs is covered in detail elsewhere [107].

Three structural variants of the phytotoxins have been used for constructing ITs. The first, typified by ricin and abrin, contain both an A, or active, domain as well as a B, or binding, domain. The natural function of the B chain in this type of phytotoxin is to bind galactose residues on the surface of cells in a nonspecific manner. The two domains are joined by a disulfide bond that can be cleaved by reducing agents such as dithiothreitol or mercaptoethanol. After cleavage, both chains contain an exposed sulfhydryl group that can be used as a handle for further modification or be directly linked to MoAbs containing activated thiol groups. Replacing the B chain by a MoAb confers specificity to these molecules, but it also eliminates the second function of the B chain—that of assisting the translocation of the holotoxin from endosomes to their site of action, the ribosomes. Numerous constructs have been prepared from these molecules, most from only the A chain, but some containing the B chain as well.

The second group of phytotoxins used includes the bacterial holotoxins such as diphtheria toxin (DT) and *Pseudomonas exotoxin* (PE). These are similar to ricin and abrin by containing the A and B domains, but they are linked together through a T or translocating domain. PE40, a genetically engineered construct of PE missing the B domain, has also been widely used in IT construction [108].

Numerous single chain ribosomal inactivating proteins (RIPs) have also been described; they constitute the third group of enzymes used to construct immunotoxins. They are typified by proteins such as gelonin, saporin, barley toxin, and pokeweed antiviral protein but differ from the holotoxins in that since they do not contain a B chain, they do not enter whole cells readily on their own.

The potency, size, and mechanism of cellular destruction of the toxins impart some unique problems to IT construction. The holotoxins, which contain both binding and catalytic domains, are extremely toxic on their own. Ricin, by far the most studied of the toxins, is reported to be 30,000 times more potent on a molar basis than methotrexate or doxorubicin. A single molecule of ricin, for example, can inactivate all 2 million elongation factor molecules associated with ribosomes per day, which is sufficient to kill a cell [109]. Such potency makes the issue of maximizing loading less relevant for ITs than in drug conjugates, though linker stability is a serious concern. In IT construction, the general aim is to achieve a 1:1 ratio of toxin to MoAb.

Two fundamental types of linkages have been used to construct ITs; both rely on the nucleophilicity of thiols and incorporate a heterobifunctional cross-linker onto the MoAb, the toxin, or both. These pieces are then assembled to contain either a reducible disulfide bond or the more stable thioether linkage. The A chains of ricin and abrin, if these are generated by reductive cleavage of the holotoxins, already contain a thiol group, as do Fab′ fragments of MoAbs; these are useful for generating constructs without further modification. The single chain toxins, however, require modification for introduction of an active thiol.

The heterobifunctional linkers most commonly used to generate ITs are SPDP

(*18*), 2-iminothiolane (2-IT) (*19*), or substituted and improved versions thereof, as shown in Figure 4.4. These reagents react with lysine amines and incorporate a free thiol onto either conjugation partner, as in the case of 2-IT, or an activated disulfide, as in the case of SPDP. The SPDP method is described repeatedly in the literature and has been used to introduce 3-(2-pyridyldithio)propionyl groups onto one or both of the proteins to be coupled [110]. If both conjugation partners are modified with SPDP, one of the 2-pyridyldisulfides is reduced with dithiothreitol or mercaptoethanol, and the two derivatized proteins are reacted at neutral pH to form a disulfide through displacement of 2-mercaptopyridine [111]. Numerous conjugates have been prepared by the SPDP–disulfide linking method, yielding 1:1 constructs in 20%–40% yield, and most of these have shown excellent *in vitro* and *in vivo* activity. When the intact holotoxins are used and both conjugation partners are reacted with SPDP, the free thiol is generated from the MoAb half to avoid reductive cleavage of the A and B chains.

2-IT (*19*) reacts with lysine amines to form a stable amidinium derivative. The resulting sulfhydryl group can be used to displace mercaptopyride from an SPDP-modified protein to generate a disulfide linkage, or it can be activated itself by the addition of 5′5′-dithiobis(2-nitrobenzoic acid), also known as Ellman's reagent [112]. The latter reagent has been particularly useful for activating the thiol of Fab′ fragments, then reacting them with freshly reduced ricin A chain to form disulfide-linked ITs [113]. Conjugates have also been prepared in which the toxins and the MoAbs are both modified with 2-IT [114]. Interest in increasing the stability of 2-IT-linked conjugates has prompted the synthesis and testing of hindered iminothiolanes [115].

Two factors have contributed to rapid *in vivo* clearance of disulfide-based ITs: low stability of the disulfide bonds in circulation, and rapid clearance of the ITs by galactose receptors in the liver, owing to the high galactose content of some of the toxins. Cross-linking reagents designed to introduce sterically hindered disulfide groups have helped address the first problem, and oxidative deglycosylation of the toxin carbohydrates has dramatically improved circulation half-life of certain ITs [116]. In conjugates prepared from intact toxins such as ricin, blocking or eliminating nonspecific binding of the ricin B chain component to mannose residues on cells has been accomplished by a variety of chemical methods or through genetic reengineering of the toxins [108]. The SMPT (*26*) cross-linker has been particularly useful in producing conjugates with improved serum stability, yet retaining the cleavability of a disulfide once targeting has occurred. Several deglycosylated, SMPT-based ricin A chain ITs are being developed for the clinic [117].

Other efforts to improve disulfide stability have been made. Thioacetylsuccinic anhydride produces a linkage with a carboxymethyl group adjacent to the sulfur, but conjugates prepared with this linker have been no more stable than the SPDP- or 2-iminothiolane-linked ITs [[118]. A more promising new thiol-containing cross-linker, N-[3(acetylthio)-3-methylbutyryl]-β-alanine (ATMBA), in which the carbon adjacent to the thiol is substituted with two methyl groups, shows improved stability and activity *in vitro* [118]. Another disulfide linking method has involved coupling cystamine to the toxin with EDCI. After reduction, the toxin was reacted with an SPDP-modified MoAb, and the resultant conjugate showed antitumor activity comparable to that of the more stable disulfide conjugates [120].

Direct comparisons of thioether with disulfide-linked ricin conjugates has shown that the latter linkage produces significantly more cytotoxic products, presumably because the toxic moiety must be released from the MoAb after targeting to realize full potency. However, some promising antitumor activity has been seen with conjugates of the bacterial toxins prepared with thioether bonds, and this has led to improvements in thioether-type linkages. The basic strategy in preparing thioether-linked conjugates involves modification of lysine amines of a toxin or MoAb with a thiol acceptor such as MBS, then reacting this with a thiol donor [121].

SMCC (*22*), another maleimide-containing linker, has been used extensively in constructing PE conjugates [122]. Poor reactivity of PE40 with SMCC was overcome by the preparation of a recombinant LysPE40 containing a lysine near the amino terminus that reacted with SMCC more efficiently. The derivatized toxin was then reacted with an iminothiolane-modified MoAb to form a thioether linkage [123].

Immunogenicity of several maleimide-based cross-linkers has been evaluated. The more flexible 6-(N-maleimido)-n-hexanoate induced significantly less linker-specific immunogenicity than the more constrained maleimide linkers MBS (*20*) or SMCC (*22*) [124].

Another strategy for producing thioether linkages involves the addition of an iodoacetyl group to the toxin by using a reagent such as IAHS (*21*), followed by reaction with a thiol-modified MoAb. Alternatively, N-succinimidyl S-acetylthioacetate (SATA) (*27*) has been used to introduce a thiol group onto whole ricin. Deacylation of the thiol was achieved by reacting the product with hydroxylamine, thus avoiding the reductive conditions used to produce thiols out of SPDP-modified materials. This method also avoided the possibility of cleaving the two chains of ricin [125].

Several reports have described the construction of immunotoxins that contain cleavage points other than disulfide bonds within their linkers. Intact ricin conjugates have been prepared with a modified insulin B chain as a linker for enzymatic cleavability and are reported to show a tenfold increase in specificity *in vitro*, compared with conjugates that do not contain the spacer [126]. Acid-cleavable, rather than reducible, linkers have also been used in preparing DT conjugates [127].

4.4. SITE-SPECIFIC MoAb MODIFICATIONS

A limited number of methods have been developed for modifying MoAbs either at specific amino acids, or on their carbohydrate residues. Such methodology offers the potential of not altering the immunoreactivity of the parent MoAb, since most such modifications are done at sites well removed from the antigen recognition domains. In practice, however, these methods have their own set of limitations.

4.4.1. Conjugation Through Carbohydrate Residues

Carbohydrate residues are located on both CH_2 domains of the MoAb Fc region, attached through N-linkages to asparagine residues in the amino acid sequence Asn-X-Ser/Thr [128]. Carbohydrate residues help mediate effector functions, such as complement protein binding and ADCC. In addition, their depletion or alteration can result in increased circulation half-time for a MoAb compared with the unmodified counterpart [129]. The oligosaccharide structure of human IgG molecules is fairly consistent, and is composed of the core sequence Siaα2 → Galβ1 → 4GlcNAcβ1 → 2Manα1 → 6(Siaα2 → Galβ1 → 4-GlcNAcβ1 → 2Manα1 → 3)(GlcNAcβ1 → 4)Manβ1 → 4GlcNAcβ1 → 4(Fucα1 → 6)GlcNAc– [130]. The posttranslational processing of MoAb carbohydrate residues has been inhibited by growing hybridomas in the presence of 1-deoxynojirimycin, an inhibitor of mannosidase 1, or tunicamycin. Such treatment produces MoAbs with a high mannose content [131].

Since the carbohydrate residues are located in a region remote from the antigen recognition sites, they are good candidates for use in conjugations without altering MoAb immunoreactivity. Detracting from their usefulness, however, is that (a) the carbohydrates constitute less than 5% of the total MoAb IgG structure, (b) a few MoAbs have oligosaccharides linked in the V_L or V_H domains [132], and (c) the conditions used to oxidize the carbohydrate residues for conjugation can also result in oxidation of certain amino acid residues, particularly methionines. If such residues occur in or near the antigen recognition sites, immunoreactivity can be altered or lost. Nevertheless, a number of examples wherein the carbohydrate residues are used for constructing successful conjugates have been reported, and all are based on the oxidation of vicinal hydroxyl groups in the carbohydrate backbone to aldehydes. Mild oxidation generates aldehydes at only the C-7 position of the sialic acid residues; the more common conditions, involving 5–10 mM sodium periodate at pH 5–6, also generate aldehydes deeper in the structure [133]. Oxidation at pH 4 is reported to suppress concurrent oxidation of Tyr, Trp, Cys, and Met residues [124]. Several enzymatic methods for achieving oxidation have also been developed. A combination of neuraminidase and galactose oxidase has been used, either in a solution phase [133] or immobilized on Eupergit C beads [135] in an attempt to suppress oxidation of sensitive amino acids. The immobilized system, which produces two aldehydes per MoAb, contrasts with chemical oxidation with periodate, which produces as many as 22 aldehydes per MoAb [136].

The steric environment near the newly generated aldehydes influences their accessibility. Different methods of measuring the numbers of aldehydes that are generated through oxidation can produce differing results. For example, when the identical samples of oxidized MoAb 96.5 was reacted with NaB^3H_4 or treated with fluorescein thiosemicarbazide (both reagents react with aldehydes, but differ in size and reactivity), the apparent numbers of aldehydes generated was 25 and 6.8, respectively [137]. Such steric effects presumably extend to the introduction of other molecules as well.

Since the carbohydrate-derived aldehydes can cross-link with MoAb amines, they must be generated and used in acidic buffers—conditions under which the lysine residues are protonated. Hydrazides have generally been used as cross-linking functionality, since their pKa of ~2.6 is well below the pKa of 8–9 for MoAb amines [133]; analine amino groups have also been used for such purposes. The Schiff's bases formed as a result of the reaction can be reduced with sodium cyanoborohydride or sodium borohydride, although complete reduction is difficult to achieve without severely compromising the immunoreactivity of the MoAb. Sufficient stability for the Schiff's bases has been noted at neutral pH such that for some applications (as, for example, immobilization on a solid support) reduction is not required [138]. However, under lysosomotropic conditions, 70%–85% of unreduced methotrexate-γ-hydrazide was released [139] and 30% of desacetylvinblastine carbohydrazide was released at pH 5.2 in 7 days from their respective conjugates [140].

Table 4.4 presents a sampling of the different types of agents that have been conjugated to MoAbs through their carbohydrate residues. Calicheamicin $\gamma_1{}^I$ (structure (*16*) in Fig. 3) is an example of a potent, low-molecular-weight drug that has been conjugated by this technique. Hydrazine-containing linkers were introduced into the molecule by displacing the methyltrisulfide moiety by 3-mercaptopropionyl hydrazide as well as other thiols, to yield more stable disulfide derivatives. These were then reacted with oxidized MoAb. High potency and specificity were noted when conjugates were prepared from internalizing MoAbs, and overall stability could be adjusted by varying the steric and electronic environment adjacent to the disulfide group in the linker [141].

One final application of aldehydes generated from MoAb carbohydrates has been to react these with ethylenediamine to increase the number of amines available for amine-based conjugation. By this technique, an average of 25 daunorubicins were loaded through a *cis*-aconitic acid spacer [34].

4.4.2. Conjugation Through Terminal Amino Acids

One group has described the use of the N-terminal amino acids for modifying a MoAb [149]. The heavy chain N termini of the MoAb Lym-1 are glutamic acids. The α-amino groups of these residues are involved in cyclic pyroglutamate structures, so they are not available for modification. However, the light chains contain aspartic acid residues at their N termini. These residues have been modified with the CTIC chelator (*35*) (Fig. 4.6) at pH 7. Immunoreactivity of the MoAb was not significantly altered, implying that the modified residues were not part of the antigen recognition site, nor were any of the 86 lysine residues of Lym-1 modified. This method, however, will probably not find general use because the chemistry operates so near the antigen recognition end of the MoAb.

The carboxy-terminal amino acids of MoAb chains have also been used for preparing conjugates. Lysine endopeptidase was used to conjugate carbohydrazide to the C-terminal carboxyl groups of an $F(ab')_2$ fragment derived from chimeric B72.3 MoAb. This hydrazine-containing construct was then conjugated to aldehyde-modified desferrioxamine. Immunoreactivity of the product was essen-

TABLE 4.4. MoAb Conjugates Prepared by Linking Through Carbohydrate Residues

Agent (Functional Group)	Spacer	Activating Group	Loading (molecules/ MoAb)	Reference
Methotrexate (γ-CO_2H)	Hydrazine		4.5	[142]
Methotrexate (α- or γ-CO_2H)	Hydrazine		4.5–8.5	[136]
Desacetyl vinblastine (3-CO_2H)	Hydrazine		2–6	[140]
Doxorubicin (13-keto)	Adipic dihydrazide			[143]
Calicheamicin γ_1^I (-$SSSCH_3$)	3-Mercaptopropionyl hydrazides, displacing methyltrisulfide		1.5–2.5	[141]
Chlorin e_6	Dextran hydrazide		1.7–19	[144]
Chlorin e_6	HPMA copolymer	Nitrophenyl esters, reacted with ethylenediamine derivative of drug	1.7–5.1	[145]
^{111}In	Aminoaniline-DTPA			[146]
^{111}In	H·Gly-Tyr-Lys-DTPA			[147]
^{125}I (label on Tyr of spacer)	H·Gly-Tyr-Gly-Gly-Arg·OH			[148]
Affi-Gel 10 or 15	Hydrazine			[138]
Horseradish peroxidase				[134]

tially unchanged. Modification of the light chain was less efficient when this system was used, and amounted to ≤ 10% of that of the heavy chain [150].

4.5. MISCELLANEOUS CONJUGATES

4.5.1. Conjugation via Polymers

Attempts to achieve higher delivery rates by conjugating larger amounts of a reagent directly onto the surface of a MoAb have generally met with failure, particularly where conjugation has been done through lysine residues. Immunoreactivity of such conjugates has been inversely proportional to the amount of agent loaded. Thus, a number of natural and synthetic polymers have been investigated as carriers in an effort to load more cytotoxic agent, predominantly drugs, per MoAb. The rationale for using polymers has been to select one that is highly water-soluble, load this as heavily as possible with drugs, and then conjugate one or more of these units to a MoAb. Attachment to MoAbs has been made through both lysine and carbohydrate residues, and a representative listing of such constructs is given in Table 4.5.

Such constructs, however, have not solved the drug efficacy problem. In practice, drug loading has not been increased by the anticipated one or two orders of magnitude necessary to produce the needed activity, mainly because the polymers themselves have a finite loading capacity. The experience with *cis*-platinum is illustrative. Two of the ligands of *cis*-diaminodichloro platinum (*cis*-DDP) and *cis*-diaquodichloro platinum (*cis*-aq) can be replaced by carboxyl ligands, and the molecules still retain strong antitumor activity. Coordination of *cis*-DDP and *cis*-aq directly through MoAb aspartic and glutamic acid residues was attempted, but the MoAbs held the platinum too tightly. A series of polymers was then investigated as carriers of platinum, including several of the polyamino acids and dextrans listed in Table 4.5. Although a few polymers were identified that released platinum well and in a controlled manner, maximum loading on all the polymers was one platinum for every five to seven of the repeating units [151]. This study typifies the experience with other drugs, except as molecular weight and lipophilicity increase, loadings decrease.

Size and charge of the ultimate constructs have also been a problem. To achieve a loading of 23–36 doxorubicins per MoAb required the conjugation of 8–10 molecules of T-10 dextran (MW = 10,000, a small size compared with many of the polymers used) in one study [19] and yielded a final construct with MW ~270,000. Polyamines such as poly-L-lysine are "sticky" molecules as far as cell surfaces are concerned, and it is difficult to separate the "stickiness" of the carrier from the targeting conferred by the MoAb, which translates to an overall decrease in antibody specificity [167]. Another study showed a marked difference in biological half-life based on the charge of the carrier polymer. Dextran that was modified with carboxyl groups, loaded with mitomycin C, and then conjugated to a MoAb exhibited a long serum half-life in biodistribution studies. When the same carboxyls were converted to amines with ethylenediamine, the conjugates were rapidly removed from circulation [164].

TABLE 4.5. MoAb Conjugates Prepared With Polymeric Carriers

Agent (Functional Group)	Polymer	MoAb Attachment	Loading (molecules/ MoAb)	Reference
Bleomycin	Oxidized dextran, direct conj.	Through aldehyde; reduce	57.5	[152]
Adriamycin				
(*via* 3′-NH_2; EDCl)	Poly(ethyleneglycol)-block-poly(aspartic acid)	SPDP on MoAb and polymer	22	[153]
(*via* 3′-NH_2)	Oxidized dextran	Through polymer aldehydes; reduce	~30	[19]
(*via* 13-keto)	Poly(glutamic acid), modified with adipic dihydrazide	Oxidized MoAb	110–120	[154]
(*via* 3′-NH_2; EDCl)	Poly(glutamic acid), modified with SPDP	MBS	7–28	[155]
(*via* 13-keto)	Carboxymethyldextran hydrazide	Glutaraldehyde		[156]
Daunomycin				
(*via* 3′-NH_2)	N-(2-hydroxypropyl)methacrylamide, containing -G-P-L-G- spacer, activated as 4-nitrophenyl ester	4-Nitrophenyl group of polymer		[157]
(*via* 13-keto)	Carboxymethyldextran hydrazide	Oxidized MoAb		[158]
cis Platinum				
(replace CI ligands)	Carboxymethyldextran	React polymer with MoAb in the presence of EDCl, then add Pt	40–60	[159]

Methotrexate				
(α- and γ-CO_2H)	Dextran using EDCl	Oxidized MoAb; reduce	30–40	[160]
(*via* -OSu or EDCl)	Oxidized dextran subst. with 1,3-diamino-2-hydroxypropane, reduced with borohydride	Oxidized MoAb; reduce with cyanoborohydride	23	[161]
(*via* -OSu)	HSA; protect -SH with iodoacetamide	4-Maleimidomethylcyclohexylcarboxylate (SMCC); link to mercaptosuccinic group on HSA	23	[162]
Mitomycin C				
(*via* aziridine; DCC)	Poly-glutamic acid; introduce protected thiol	SMCC	10–48	[163]
(*via* aziridine; EDCl)	Dextran modified with 6-bromohexanoid acid	SPDP on both polymer and MoAb	40	[164]
(*via* aziridine)	Oxidized dextran; add MMC after polymer linked to MoAb	Add MoAb to polymer prior to add'n of MMC; reduce	20	[165]
ARA-C	Oxidized dextran	Through polymer aldehydes; reduce	10–30/IgG 80–130/IgM	[166]
FUR				
(oxidized)	Poly-L-lysine; reduce	Oxidized MoAb, succinylate to reduce cationic charge	45–100	[167]
(oxidized)	Aminodextran; reduce	Oxidized MoAb; reduce	25–35	[168]
^{99m}Tc	Metallothionein	SMCC or sulfo-SMCC		[169]

Even the use of an endogenous protein such as human serum albumin (HSA) as a carrier has met with difficulties. A construct between methotrexate, HSA, and the MoAb 791T/36 produced a conjugate with a loading of 30 methotrexates per MoAb. When this material was assessed in humans, its was removed unusually rapidly from circulation, showing particularly fast clearance through the liver. The investigators concluded that this was probably due to the high net negative charge of the construct [170].

Efforts to increase radioisotope loadings, even for therapeutic purposes, have not been necessary. In fact, the radiolysis potential of most isotopes makes loadings of more than one isotope per MoAb counterproductive. However, polymers have been investigated as carriers for cold metals, for applications in nuclear magnetic resonance imaging. Polylysine has been modified with EDTA or DTPA, the remaining amines have been blocked by succinylation, and these constructs have been conjugated to MoAbs. The conjugates were capable of holding 75 molecules of cold indium or 60 molecules of cold gadolinium, but such loadings were not sufficiently high to image tumors by nuclear magnetic resonance [171].

4.5.2. Chemical Modifications to Minimize Immunogenicity

A human anti-mouse response (HAMA) has been observed in essentially all human clinical trials in which mouse-derived MoAbs have been used. Efforts to overcome the HAMA problem are mainly being addressed through genetic engineering, by converting mouse-derived MoAbs to their "humanized" counterparts while retaining the antigen recognition properties of the mouse material. However, immunogenicity of the targeted molecule or the linkers attached to the MoAb may present a special set of problems even if the immune response to the MoAb itself is minimized.

Several attempts have been made to mask the immunogenicity of mouse MoAb conjugates. Polyethylene glycol (PEG), a polymer that has been covalently bonded to a number of proteins to increase their circulation half-lives and make them less immunogenic, has also been used in several MoAb studies. In one case, conjugation of 31 molecules of PEG to a MoAb suppressed 95% of its immunogenicity for over 300 days, though the suppression was dependant on dosing schedule [172]. A neocarzinostatin (NCS)-A7 conjugate produced a HAMA response in all patients treated. In an attempt to make this conjugate more tolerated and thereby increase its circulation half-time, the conjugate was also reacted with PEG. Although modification with PEG doubled the circulation half-life in mice to 48 h, no additional amounts of conjugate were localized into tumors compared to NCS-A7 itself. Conjugation of dextran to the same material decreased circulation half-time [173]. In another study, conjugation of eight molecules of 10,000-MW PEG to a MoAb reduced anti-MoAb titer by a factor of only four [174]. However, the consequences of such modifications have not been evaluated in human trials, where merely diminishing rather than eliminating the HAMA response may well show some beneficial effects.

4.6. SUMMARY AND CONCLUSIONS

Our purpose in writing this chapter has been to review various chemical methods by which MoAb conjugates have been constructed. It should be evident that even though the proteinaceous nature of MoAbs severely limits the scope of reagents and conditions that can be used, a surprising diversity of constructs have been prepared. This diversity includes numerous isotopes, enzymes, classical anticancer agents, and potent toxins. To achieve such diversity, however, most of the chemical innovation has been applied to the assembly and modification of molecules prior to their attachment to MoAbs.

We have resisted making biological efficacy comparisons of conjugates because, given the large number of variables inherent in such constructs, as well as the diversity of biological test systems used to evaluate these products, such comparisons are meaningless. Instead, we have concentrated on defining the key factors that need to be considered for designing a MoAb conjugate, such as the nature of the target, the potency of the delivered agent, the structure of the linker, the chemistry that is selected to assemble the product, and so forth. Each type of agent delivered by the MoAb requires emphasizing different design factors. Sufficient information about most of these factors is now available to make the process of designing a conjugate one that is rational and predictable.

This is not to imply that the antigen-specific delivery of agents by MoAbs is a mature field. So far, modified MoAbs have only secured a place in science as research tools. Thus, even though MoAbs are the basis of numerous *in vitro* diagnostic kits, the first MoAb based *in vivo* imaging agent, OncoScint CR/OV, was approved for human use by the FDA only on December 29, 1992 [175]. Human therapeutic applications, whether they are based on the delivery of enzymes, isotopes, or drugs, are still a number of years away from approval.

The development of therapeutic MoAbs is an extremely complex undertaking that requires diverse scientific skills. Molecular biologists need to engineer the sequences and adjust the size of MoAbs to optimize their biodistribution, penetrability, and immunogenicity. Chemists and biochemists need to select the appropriate antigens and MoAbs, then design and synthesize better linkers to balance serum stability with cleavability (if desired) at the target. Pharmacologists need to develop the appropriate dosing schedules, as well as define additives to help conjugates achieve the highest levels of targeting and therapeutic effectiveness. And finally, oncologists need to consider clinical regimens that by present standards may seem unorthodox. Together, these disciplines may in fact make the therapeutic use of MoAbs a reality.

4.7. REFERENCES

1. Weinstein, J.N., Van Osdol, W. The macroscopic and microscopic pharmacology of monoclonal antibodies. Int. J. Immunopharmacol. 14, 457 (1992).

2. **a:** Jain, R.K., Baxter, L.T. Mechanisms of heterogeneous distribution of monoclonal antibodies and other macromolecules in tumors: Significance of elevated interstitial pressure. Cancer Res. 48, 7022 (1988). **b:** Jain, R.K. Delivery of novel therapeutic agents in tumors: Physiological barriers and strategies. J. Natl. Cancer Inst. 81, 570 (1989).
3. Shockley, T.R., Lin, K., Nagy, J.A., Tompkins, R.G., Yarmush, M.L., Dvorak, H.F. Spatial distribution of tumor-specific monoclonal antibodies in human melanoma xenografts. Cancer Res. 52, 367 (1992).
4. Maeda, H., Matsumura, Y. Tumoritropic and lymphotropic principles of macromolecular drugs. Crit. Rev. Ther. Drug Carrier Syst. 6, 193 (1989).
5. Epenetos, A.A., Snook, D., Durbin, H., Johnson, P.H., Taylor-Papodimetrius, J. Limitations of radiolabeled monoclonal antibodies for localization of human neoplasms. Cancer Res. 46, 3183 (1986).
6. Tannock, I.F., Rotin, D. Acid pH in tumors and its potential for therapeutic exploitation. Cancer Res. 49, 4373 (1989).
7. Lavie, E., Hirschberg, D.L., Schreiber, G., Thor, K., Hill, L., Hellstrom, I., Hellstrom, K.-E. Monoclonal antibody L6-daunomycin conjugates constructed to release free drug at the lower pH of tumor tissue. Cancer Immunol. Immunother. 33, 223 (1991).
8. Shen, W.-C., Ryser, H. J.-P. cis-Aconityl spacer between daunomycin and macromolecular carriers: A model of pH-sensitive linkage releasing drug from a lysosomotropic conjugate. Biochem. Biophys. Res. Commun. 102, 1048 (1981).
9. Ohkuma, S., Poole, B. Fluorescence probe measurement of the intralysosomal pH in living cells and the perturbation of pH by various agents. Proc. Natl. Acad. Sci. USA 75, 3327 (1978).
10. Baurain, R., Masquelier, M., Deprez-De Campeneere, D., Trouet, A. Amino acid and dipeptide derivatives of daunorubicin. 2. Cellular pharmacology and antitumor activity on L1210 leukemic cells *in vitro* and *in vivo*. J. Med. Chem. 23, 1171 (1980).
11. **a:** Griffin, T.W., Rybak, M.E., Giordano, L. Monoclonal antibodies as new antitumor agents. Adv. Biotechnol. Processes 11, 335 (1989). **b:** Carlsson, R., Glad, C., Borrebaeck, C.K.A. Monoclonal antibodies into the '90's: The all-purpose tool. Bio/Technology 7, 567 (1989).
12. Lyons, A., King, D.J., Owens, R.J., Yarranton, G.T., Millican, A., Whittle, N.R., Adair, J.R. Protein-specific attachment of recombinant antibodies via introduced surface cysteine residues. Protein Eng. 3, 703 (1990).
13. Yokota, T., Milenic, D.E., Whitlow, M., Schlom, J. Rapid tumor penetration of a single-chain Fv and comparison with other immunoglobulin forms. Cancer Res. 52, 3402 (1992).
14. **a:** Fujimori, K., Covell, D.G., Fletcher, J.E., Weinstein, J.N. Modeling analysis of the global and microscopic distribution of immunoglobulin G, $F(ab')_2$, and Fab in tumors. Cancer Res. 49, 5656 (1989). **b:** Weinstein, J.N., Eger, R.R., Covell, D.G., Black, C.D.V., Mulshine, J., Carrasquillo, J.A., Larson, S.M., Keenan, A.M. The pharmacology of monoclonal antibodies. Ann. N.Y. Acad. Sci. 507, 199 (1987).
15. Uadia, P., Blair, A.H., Ghose, T., Ferrone, S. Uptake of methotrexate linked to polyclonal and monoclonal antimelanoma antibodies by a human melanoma cell line. J. Natl. Cancer Inst. 74, 29 (1985).

16. Arnon, R., Sela, M. *In vitro* and *in vivo* efficacy of conjugates of daunomycin with anti-tumor antibodies. Immunol. Res. 62, 5 (1982).

17. DiMarco, A., Arcamone, F. DNA complexing antibiotics: Daunomycin, adriamycin and their derivatives. Arzneim Forsch. 25, 368 (1975).

18. Ogden, J.R., Leung, K., Kunda, S.A., Telander, M.W., Avner, B.P., Liao, S.-K., Thurman, G.B., Oldham, R.K. Immunoconjugates of doxorubicin and murine anti-tumor breast carcinoma monoclonal antibodies prepared via an N-hydroxysuccinimide active ester intermediate of *cis*-aconityl-doxorubicin: Preparation and *in vitro* cytotoxicity. Mol. Biother. 1, 170 (1989).

19. Biddle, W.C., Haruta, Y., Seon, B.K., Henderson, E.S., Sarcione, E.J. *In vitro* and *in vivo* cytotoxic activity of anti-human leukemia monoclonal antibodies SN5c and SN6 daunomycin conjugates. Leuk. Res. 13, 699 (1989).

20. Katzenellenbogen, J.A., Zablocki, J.A. Cytotoxic oestrogens and antioestrogens: Concepts, progress and evaluation. In Furr, B.J., Wakeling, A.E. (eds.): Pharmacology and Clinical Uses of Inhibitors of Hormone Secretion and Action (chap. 3). London: Bailliere Tindall, 1987.

21. Kralovec, J., Spencer, G., Blair, A.H., Mammen, M., Singh, M., Ghose, T. Synthesis of methotrexate-antibody conjugates by regiospecific coupling and assessment of drug and antibody activities. J. Med. Chem. 32, 2426 (1989).

22. Mueller, B.M., Wrasidlo, W.A., Reisfeld, R.A. Antibody conjugates with morpholinodoxorubicin and acid-cleavable linkers. Bioconjugate Chem. 1, 325 (1990).

23. **a:** Lee, M.D., Manning, J.K., Williams, D.R., Kuck, N.A., Testa, R.T., Borders, D.B. Calicheamicins, a novel family of antitumor antibiotics. 3. Isolation, purification and characterization of calicheamicins β_1^{Br}, γ_1^{Br}, α_3^{I}, β_1^{I}, γ_1^{I} and δ_1^{I}. J. Antibiot. 42, 1070 (1989). **b:** Zein, N., Sinha, A.M., McGahren, W.J., Ellestald, G.A. Calicheamicin γ_1^{I}: An antitumor antibiotic that cleaves double-stranded DNA site-specifically. Science 240, 1198 (1988).

24. Edo, K., Akiyama, Y., Saito, K., Mizugaki, M., Koide, Y., Ishida, N. Absolute configuration of the amino sugar moiety of the neocarzinostatin chromophore. J. Antibiot. 39, 1615 (1986).

25. Bamburg, J.R., Riggs, N.V., Strong, F.M. The structures of toxins from two strains of *Fusarium tricinctum*. Tetrahedron 24, 3329 (1968).

26. Härri, E., Loeffler, W., Sigg, H.P., Stähelin, H., Stoll, C., Tamm, C., Wiesinger, D. Über die Verrucarin and Roridine, eine Gruppe von cytostatisch hochvirksamen Antibiotica aus Myrothedium-Arten. Helv. Chim. Acta 45, 839 (1962).

27. Kupchan, S.M., Komoda, Y., Court, W.A., Thomas, G.J., Smith, R.M., Karim, A., Gilmore, C.J., Haltiwanger, R.C., Bryan, R.F. Maytansine, a novel antileukemic ansa macrolide from *Maytenus ovatus*. J. Am. Chem. Soc. 94, 1354 (1972).

28. Moore, R.E., Bartolini, G. Structure of Palytoxin. J. Am. Chem. Soc. 103, 2491 (1981).

29. Humm, J.L. Dosimetric aspects of radiolabeled antibodies for tumor therapy. J. Nuclear Med. 27, 1490 (1986).

30. Alam, F., Soloway, A.H., Barth, R.F., Johnson, C.W., Carey, W.E., Knoth, W.H. Boronation of polyclonal and monoclonal antibodies for neutron capture. Brookhaven Natl. Lab., [Rep.] BNL 229 (1983).

31. Brinkley, M. A Brief survey of methods for preparing protein conjugates with dyes, haptens and cross-linking reagents. Bioconjugate Chem. 3, 2 (1992).
32. Aboud-Pirak, E., Lesur, B., Bhushana Rao, K.S.P., Baurain, R., Trouet, A., Schneider, Y.-J. Cytotoxic activity of daunorubicin or vindesine conjugated to a monoclonal antibody on cultured MCF-7 breast carcinoma cells. Biochem. Pharmacol. 38, 641 (1989).
33. Rowland, G.F., Axton, C.A., Baldwin, R.W., Brown, J.P., Corvalan, J.R.F., Embleton, M.J., Gore, V.A., Hellström, I., Hellström, K.E., Jacobs, E., Marsden, C.H., Pimm, M.V., Simmonds, R.G., Smith, W. Antitumor properties of vindesine-monoclonal antibody conjugates. Cancer Immunol. Immunother. 19, 1 (1985).
34. Dillman, R.O., Johnson, D.F., Shawler, D.L., Koziol, J.A. Superiority of an acid-labile daunorubicin-monoclonal antibody immunoconjugate compared to free drug. Cancer Res. 48, 6097 (1988).
35. Hudecz, F., Ross, H., Price, M.R., Baldwin, R.W. Immunoconjugate design: A predictive approach for coupling of daunomycin to monoclonal antibodies. Bioconjugate Chem. 1, 197 (1990).
36. Page, M., Thibeault, D., Noël, C., Dumas, L. Coupling a preactivated daunorubicin derivative to antibody. A new approach. Anticancer Res. 10, 353 (1990).
37. Pimm, M.V., Paul, M.A., Ogumuyiwa, Y., Baldwin, R.W. Biodistribution and tumor localization of a daunomycin-monoclonal antibody conjugate in nude mice with human tumor xenografts. Cancer Immunol. Immunother. 27, 267 (1988).
38. Kaneko, T., Willner, D., Monkovic, I., Knipe, J.O., Braslawsky, G.R., Greenfield, R.S., Vyas, D.M. New hydrazone derivatives of adriamycin and their immunoconjugates—A correlation between acid stability and cytotoxicity. Bioconjugate Chem. 2, 133 (1991).
39. Page, M., Thibeault, D. Coupling anthracyclines to antibodies without polymerization. Tumor Biol. 8, 365 (1988).
40. Ghosh, M.K., Kildsig, D.O., Mitra, A.K. Preparation and characterization of methotrexate-immunoglobulin conjugates. Drug Des. Delivery 4, 13 (1989).
41. Umemoto, N., Kato, Y., Endo, N., Takeda, Y., Hara, T. Preparation and *in vitro* cytotoxicity of a methotrexate-anti-MM46 monoclonal antibody conjugate *via* an oligopeptide spacer. Int. J. Cancer 43, 677 (1989).
42. Umemoto, N., Kato, Y., Hara, T. Cytotoxicities of two disulfide-bond-linked conjugates of methotrexate with monoclonal anti-MM46 antibody. Cancer Immunol. Immunother. 28, 9 (1989).
43. Pimm, M.V., Clegg, J.A., Garnett, M.D., Baldwin, R.W. Biodistribution and tumor localization of a methotrexate-monoclonal antibody 791T/36 conjugate in nude mice with human tumor xenografts. Int. J. Cancer 41, 886 (1988).
44. Kanellos, J., Pietersz, G.A., Cunningham, Z., McKenzie, I.F.C. Anti-tumor activity of aminopterin-monoclonal antibody conjugates: *In vitro* and *in vivo* comparison with methotrexate-monoclonal antibody conjugates. Immunol. Cell Biol. 65, 483 (1987).
45. Bernier, L.G., Page, M., Dumas, L., Gauthier, C., Gaudreault, R., Joly, L.P. Monoclonal anti-CEA antibodies for targeting chlorambucil to human colon carcinoma cells *in vitro*. Peptides Biol. Fluids 31, 787 (1983).
46. Pietersz, G.A., Smyth, M.J., Tsandra, J.J., McKenzie, I.F.C. Pre-clinical and clinical studies with N-acetyl melphalan immunoconjugates and tumor necrosis factor α. Antibody Immunoconj., Radiopharm. 2, 47 (1989).

47. Sugita, N., Niimura, K., Fujii, M., Matsunaga, K., Fujii, T., Furusho, T., Yoshikumi, C., Kawai, Y., Taguchi, T. *In vivo* antineoplastic activity of mouse or human IgG conjugated with melphalan. In Vivo 1, 205 (1987).
48. Kato, Y., Tsukada, Y., Hara, T., Hirai, H. Enhanced antitumor activity of mitomycin C conjugated with anti-α-fetoprotein antibody by a novel method of conjugation. J. Appl. Biochem. 5, 313 (1983).
49. Krauer, K.G., McKenzie, I.F.C., Pietersz, G.A. Antitumor effect of 2′-deoxy-5-fluorouridine conjugates against a murine thymoma and colon carcinoma xenografts. Cancer Res. 52, 132 (1992).
50. He, D., Taniuchi, S., Sun, C.H., Berns, M.W., Cardiff, R.D. The monoclonal antibody UCD/AB 6.01 conjugated to photosensitizers attaches to and photoinactivates epithelial tumors cells. Antibody Immunoconj. Radiopharm. 3, 199 (1990).
51. Varadarajan, A., Sharkey, R.M., Goldenberg, D.M., Hawthorne, M.F. Conjugation of phenyl isothiocyanate derivatives of carborane to antitumor antibody and in vivo localization of conjugates in nude mice. Bioconjugate Chem. 2, 102 (1991).
52. Alberici, G.F., Pallardy, M., Manil, L., Dessaux, J.-J., Fournier, J., Mondesir, J.-M., Bohuon, C., Gros, P. Conjugates of elliptinium acetate with mouse monoclonal anti-α-fetoprotein antibodies or Fab fragments: *In vitro* cytotoxic effects upon human hepatoma cell lines. Int. J. Cancer 41, 309 (1988).
53. Obrist, R., Schmidli, J., Obrecht, J.P. Chemotactic monoclonal antibody conjugates: A comparison of four different f-Met-Peptide-Conjugates. Biochem. Biophys. Res. Commun. 155, 1139 (1988).
54. Pierce, D.L., Heindel, N.D., Schray, K.J., Jetter, M.M., Emrich, J.G., Woo, D.V. Misonidazole conjugates of the colorectal tumor associated monoclonal antibody 17-1A. Bioconjugate Chem. 1, 314 (1990).
55. Ohtani, K., Murakami, H., Shibuya, O., Kawamura, O., Ohi, K., Chiba, J., Otokawa, M., Ueno, Y. Antibody activity of T-2 toxin-conjugated monoclonal antibody to murine thymoma. Jpn. J. Exp. Med. 60, 57 (1990).
56. Pavanasasivam, G. Trichothecene antibody conjugates. U.S. Pat. No. 4,744,981, issued May 17, 1988.
57. Chari, R.V.J., Martell, B.A., Gross, J.L., Cook, S.B., Shah, S.A., Blättler, W.A., McKenzie, S.J., Goldmacher, V.S. Immunoconjugates containing novel maytansinoids: Promising anticancer drugs. Cancer Res. 52, 127 (1992).
58. Rosik, L.O., Sweet, F. Electrophilic analogs of daunorubicin and doxorubicin. Bioconjugate Chem. 1, 251 (1990).
59. Endo, N., Takeda, Y., Umemoto, N., Kishida, K., Watanabe, K., Saito, M., Kato, Y., Hara, T. Nature of linkage and mode of action of methotrexate conjugated with antitumor antibodies: Implications for future preparation of conjugates. Cancer Res. 48, 3330 (1988).
60. Jeanson, A., Cloes, J.-M., Bouchet, M., Rentier, B. Comparison of conjugation procedures for the preparation of monoclonal antibody-enzyme conjugates. J. Immunol. Methods 111, 261 (1988).
61. Endo, N., Umemoto, N., Kato, Y., Takeda, Y., Hara, T. A novel covalent modification of antibodies at their amino groups with retention of antigen-binding activity. J. Immunol. Methods 104, 253 (1987).
62. Reisfeld, R.A., Yang, H.M., Muller, B., Wargalla, U.C., Schrappe, M., Wrasidlo, W.

Promises, problems, and prospects of monoclonal antibody-drug conjugates for cancer therapy. Antibody Immunoconjugates Radiopharm. 2, 217 (1989).

63. Mueller, B.M., Wrasidlo, W.A., Reisfeld, R.A. Determination of the number of ϵ-amino groups available for conjugation of effector molecules to monoclonal antibodies. Hybridoma 7, 453 (1988).

64. **a:** Kozak, R.W., Waldmann, T.A., Atcher, R.W., Gansow, O.A. Radionuclide-conjugated monoclonal antibodies: A synthesis of immunology, inorganic chemistry and nuclear medicine. Trends Biotechol. 4, 259 (1986). **b:** Wolf, W., Shani, J. Criteria for the selection of the most desirable radionuclide for radiolabeling monoclonal antibodies. Nucl. Med. Biol. 13, 319 (1986). **c:** Volkert, W.A., Goeckeler, W.F., Ehrhardt, G.J., Ketring, A.R. Therapeutic radionuclides: Production and decay property considerations. J. Nucl. Med. 32, 174 (1991). **d:** Wheldon, T.E., O'Donoghue, J.A. The radiobiology of targeted radiotherapy. Int. J. Radiat. Biol. 58, 1 (1990). **e:** Wilbur, D.S. Radiohalogenation of proteins: An overview of radionuclides, labeling methods, and reagents for conjugate labeling. Bioconjugate Chem. 3, 434 (1992).

65. Matzku, S., Kirchgebner, H., Nissen, M. Iodination of monoclonal IgG antibodies at a sub-stoichiometric level: Immunoreactivity changes related to the site of iodine incorporation. Nucl. Med. Biol. 14, 451 (1987).

66. **a:** Zalutsky, M.R., Narula, A.S. Radiohalogenation of a monoclonal antibody using an N-succinimidyl 3-(tri-*n*-butylstannyl)benzoate intermediate. Cancer Res. 48, 1446 (1988). **b:** Wilbur, D.S., Jones, D.S., Fritzberg, A.R. Synthesis and radioiodinations of some aryltin compounds for radiolabeling of monoclonal antibodies. J. Labelled Compd. Radiopharm. 23, 1304 (1986). **c:** Garg, S., Garg, P.K., Zalutsky, M.R. N-Succinimidyl 5-(trialkylstannyl)-3-pyridinecarboxylates: A new class of reagents for protein radioiodination. Bioconjugate Chem. 2, 50 (1991).

67. Srivastava, P.C., Buchsbaum, D.J., Allred, J.F., Brubaker, P.G., Hanna, D.E., Spicker, J.K. A new conjugating agent for radioiodination of proteins: Low in vivo deiodination of a radiolabeled antibody in a tumor model. BioTechniques 8, 536 (1990).

68. Meares, C.F. Chelating agents for the binding of metal ions to antibodies. Nucl. Med. Biol. 13, 311 (1986).

69. Meares, C.F., McCall, M.J., Deshpande, S.V., DeNardo, S.J., Goodwin, D.A. Chelate radiochemistry: Cleavable linkers lead to altered levels of radioactivity in the liver. Int. J. Cancer Suppl. 2, 99 (1988).

70. Roselli, M., Schlom, J., Gansow, O.A., Raubitschek, A., Mirzadeh, S., Brechbiel, M.W., Colcher, D. Comparative biodistributions of yttrium- and indium-labeled monoclonal antibody B72.3 in athymic mice bearing human colon carcinoma xenografts. J. Nucl. Med. 30, 672 (1989).

71. Brechbiel, M.W., Gansow, O.A., Atcher, R.W., Schlom, J., Esteban, J., Simpson, D.E., Colcher, D. Synthesis of 1-(p-isothiocyanatobenzyl) derivatives of DTPA and EDTA. Antibody labeling and tumor-imaging studies. Inorg. Chem. 25, 2772 (1986).

72. **a:** Brandt, K.D., Johnson, D.K. Structure-function relationships in indium-111 radioimmunoconjugates. Bioconjugate Chem. 3, 118 (1992). **b:** Kline, S.J., Betebenner, D.A., Johnson, D.K. Carboxymethyl-substituted bifunctional chelators: Preparation of aryl isothiocyanate derivatives of 3-(carboxymethyl)-3-azapentanedioic acid, 3,12-bis(carboxymethyl)-6,9-dioxa-3,12-diazatetradecanedioic acid, and 1,4,7,10-tetraazacyclododecane-N,N′,N″,N‴-tetraacetic acid for use as protein labels. Bioconjugate Chem. 2, 26 (1991).

73. Quadri, S.D., Wessels, B.W. Radiolabeled biomolecules with ^{186}Re: Potential of radioimmunotherapy. Nucl. Med. Biol. 13, 447 (1986).

74. Mather, S.J., Ellison, D. Reduction-mediated technetium-99m labeling of monoclonal antibodies. J. Nucl. Med. 31, 692 (1990).

75. **a:** Fritzberg, A.R., Abrams, P.G., Beaumier, P.L., Kasina, S., Morgan, Jr., A.C., Rao, T.N., Reno, J.M., Sanderson, J.A., Srinivasan, A., Wilbur, D.S., Vanderheyden, J.-L. Specific and stable labeling of antibodies with technetium-99m with a diamide dithiolate chelating agent. Proc. Natl. Acad. Sci. USA 85, 4025 (1988). **b:** Liang, F.H., Virzi, F., Hnatowich, D.J. Serum stability and non-specific binding of technetium-99m labeled diaminodithiol for protein labeling. Nucl Med. Biol. 14, 555 (1987).

76. Deshpande, S.V., DeNardo, S.J., Kukis, D.L., Moi, M.K., McCall, M.J., DeNardo, G.L., Meares, C.F. Yttrium-90-labeled monoclonal antibody for therapy: Labeling by a new macrocyclic bifunctional chelating agent. J. Nucl. Med. 31, 473 (1990).

77. Cox, J.P.L., Jankowski, K.J., Kataky, R., Parker, D., Beeley, N.R.A., Boyce, B.A., Eaton, M.A.W., Millar, K., Millican, A.T., Harrison, A., Walker, C. Synthesis of a kinetically stable yttrium-90 labelled macrocycle-antibody conjugate. J. Chem. Soc. Chem. Commun., p. 797 (1989).

78. McMurry, T.J., Brechbiel, M., Kumar, K., Gansow, O.A. Convenient synthesis of bifunctional tetraaza macrocycles. Bioconjugate Chem. 3, 108 (1992).

79. Craig, A.S., Helps, I.M., Jankowski, K.J., Parker, D., Beeley, N.R.A., Boyce, B.A., Eaton, M.A.W., Millican, A.T., Millar, K., Phipps, A., Rhind, S.K., Harrison, A., Walker, C. Towards tumor imaging with indium-111 labelled macrocycle-antibody conjugates. J. Chem. Soc. Chem. Commun., p. 794 (1989).

80. **a:** Cole, W.C., DeNardo, S.J., Meares, C.F., McCall, M.J., DeNardo, G.L., Epstein, A.L., O'Brien, H.A., Moi, M.K. Serum stability of ^{67}Cu chelates: Comparison with ^{111}In and ^{57}Co. Nucl. Med. Biol. 13, 363 (1986). **b:** Deshpande, S.V., DeNardo, S.J., Meares, C.F., McCall, M.J., Adams, G.P., Moi, M.K., DeNardo, G.L. Copper-67-labeled monoclonal antibody Lym-1, a potential radiopharmaceutical for cancer therapy: Labeling and biodistribution in RAJI tumored mice. J. Nucl. Med. 29, 217 (1988).

81. Morphy, J.R., Parker, D., Alexander, R., Bains, A., Carne, A.F., Eaton, M.A.W., Harrison, A., Millican, A., Phipps, A., Rhind, S.K., Titmas, R., Weatherby, D. Antibody labelling with functionalized cyclam macrocycle. J. Chem. Soc. Chem. Commun., p. 156 (1988).

82. Deshpande, S.V., DeNardo, S.J., Meares, C.F., McCall, M.J., Adams, G.P., DeNardo, G.L. Effect of different linkages between chelates and monoclonal antibodies on levels of radioactivity in the liver. Nucl. Med. Biol. 16, 587 (1989).

83. Paik, C.H., Yokoyama, K., Reynolds, J.C., Quadri, S.M., Min, C.Y., Shin, S.Y., Maloney, P.J., Larson, S.M., Reba, R.C. Reduction of background activities by introduction of a diester linkage between antibody and a chelate in radioimmunodetection of tumor. J. Nucl. Med. 30, 1693 (1989).

84. **a:** Studer, M., Meares, C.F. A convenient and flexible approach for introducing linkers in bifunctional chelating agents. Bioconjugate Chem. 3, 420 (1992). **b:** Studer, M., Kroger, L.A., DeNardo, S.J., Kukis, D.L., Meares, C.F. Influence of a peptide linker on biodistribution and metabolism of antibody-conjugated benzyl-EDTA. Comparison of enzymatic digestion in vitro and in vivo. Bioconjugate Chem. 3, 424 (1992).

85. Foxwell, B.M.J., Band, H.A., Long, J., Jeffery, W.A., Snook, D., Thorpe, P.E., Watson, G., Parker, P.J., Epenetos, A.A., Creighton, A.M. Conjugation of monoclo-

nal antibodies to a synthetic peptide substrate for protein kinase: A method for labelling antibodies with ^{32}P. Br. J. Cancer 57, 489 (1988).

86. Order, S.E. Analysis, results, and future prospective of the therapeutic use of radiolabelled antibody in cancer therapy. In Monoclonal Antibodies for Cancer Detection and Therapy (p. 303). London: Academic Press, 1985.
87. Stanislawski, M., Rousseau, V., Goavec, M., Ito, H.-O. Immunotoxins containing glucose oxidase and lactoperoxidase with tumoricidal properties: *In vitro* killing effectiveness in a mouse plasmacytoma cell model. Cancer Res. 49, 5497 (1989).
88. Reiter, Y., Fishelson, Z. Targeting of complement to tumor cells by heteroconjugates composed of antibodies and of the complement component C3b. J. Immunol. 142, 2771 (1989).
89. Battelli, M.G., Abbondanza, A., Tazzari, P.L., Dinota, A., Rizzi, S., Grassi, G., Gobbi, M., Stirpe, F. Selective cytotoxicity of an oxygen-radical-generating enzyme conjugated to a monoclonal antibody. Clin. Exp. Immunol. 73, 128 (1988).
90. Vogel, C.-W., Wilkie, S.D., Morgan, A.C. *In vivo* studies with covalent conjugates of cobra venom factor and monoclonal antibodies to human tumors. Haematol. Blood Transfus. 29, 514 (1985).
91. Searle, F., Bier, C., Buckley, R.G., Newman, S., Pedley, R.B., Bagshawe, K.D., Melton, R.G., Alwan, S.M., Sherwood, R.F. The potential of carboxypeptidase G_2-antibody conjugates as anti-tumor agents. I. Preparation of anti-human chorionic gonadotrophic-carboxypeptidase G_2 and cytotoxicity of the conjugate against JAR choriocarcinoma cells *in vitro*. Br. J. Cancer 53, 377 (1986).
92. **a:** Bagshawe, K.D. Towards generating cytotoxic agents at cancer sites. Br. J. Cancer 60, 275 (1989). **b:** Senter, P.D., Wallace, P.M., Svensson, H.P., Vrudhula, V.M., Kerr, D.E., Hellström, I., Hellström, K.E. Generation of cytotoxic agents by targeted enzymes. Bioconjugate Chem. 4, 3 (1993).
93. Meyer, D.L., Jungheim, L.N., Mkolajczyk, S.D., Shepherd, T.A., Starling, J.J., Ahlem, C.N. Preparation and characterization of a β-lactamase-Fab′ conjugate for the site-specific activation of oncolytic agents. Bioconjugate Chem. 3, 42 (1992).
94. Senter, P.D., Wallace, P.M., Svensson, H.P., Kerr, D.E., Hellström, I., Hellström, K.E. Activation of prodrugs by antibody-enzyme conjugates. Adv. Exp. Med. Biol. 303, 97 (1991).
95. **a:** Melton, R.G., Searle, F., Bier, C., Boden, J.A., Pedley, R.B., Green, A.J., Bagshawe, K.D., Sherwood, R.F. Antibody-carboxypeptidase G_2 conjugates as potential tumor imaging agents. NATO ASI Ser. A 152, 377 (1988). **b:** Bagshawe, K.D., Sharma, S.K., Springer, C.J., Antoniw, P., Rogers, G.T., Burke, P.J., Melton, R., Sherwood, R. Antibody-enzyme conjugates can generate cytotoxic drugs from inactive precursors at tumor site. Antibody Immunoconj. Radiopharm. 4, 915 (1991).
96. Svensson, H.P., Kadow, J.F., Vrudhula, V.M., Wallace, P.M., Senter, P.D. Monoclonal antibody-β-lactamase conjugates for the activation of a cephalosporin mustard prodrug. Bioconjugate Chem. 3, 176 (1992).
97. **a:** Haenseler, E., Esswein, A., Vitols, K.S., Montejano, Y., Mueller, B.M., Reisfeld, R.A., Huennekens, F.M. Activation of methotrexate-α-alanine by carboxypeptidase A–monoclonal antibody conjugate. Biochemistry 31, 891 (1992). **b:** Vitols, K.S., Haenseler, E., Montejano, Y., Baer, T., Huennekens, F.M. Activation of methotrexate prodrugs by enzyme/monoclonal antibody conjugates. Pteridines 3, 125 (1992).

98. Roffler, S.R., Wang, S.-M., Chern, J.-W., Yeh, M.-Y., Tung, E. Anti-neoplastic glucuronide prodrug treatment of human tumor cells targeted with a monoclonal antibody-enzyme conjugate. Biochem. Pharmacol. 42, 2062 (1991).
99. Senter, P.D., Su, P.C.D., Katsuragi, T., Sakai, T., Cosand, W.L., Hellström, I., Hellström, K.E. Generation of 5-fluorouricil from 5-fluorocytosine by monoclonal antibody-cytosine deaminase conjugates. Bioconjugate Chem. 2, 447 (1991).
100. Kerr, D.E., Senter, P.D., Burnett, W.V., Hirschberg, D.L., Hellström, I., Hellström, K.E. Antibody-penicillin-V-amidase conjugates kill antigen-positive tumor cells when combined with doxorubicin phenoxyacetamide. Cancer Immunol. Immunother. 31, 202 (1990).
101. Bignani, G.S., Senter, P.D., Grothaus, P.G., Fisher, K.I., Humphreys, T., Wallace, P.M. N-(4′-Hydroxphenylacetyl)palytoxin: A palytoxin prodrug that can be activated by a monoclonal antibody-penicillin G amidase conjugate. Cancer Res. 52, 5759 (1992).
102. Spooner, R.A., Lord, J.M. Immunotoxins: Status and prospects. Trends Biotechnol. 8, 189 (1990).
103. Preijers, F.W.M.B., Tax, W.J.M., DeWitte, T., Janssen, A., Heijden, H.V.D., Vidal, H., Wessels, J.M.C., Capel, P.J.A. Relationship between internalization and cytotoxicity of ricin A-chain immunotoxins. Br. J. Haematol. 70, 289 (1988).
104. Vitetta, E.S., Thorpe, P.E. Immunotoxins. p. 482 in DeVita, V., Hellman, S., Rosenberg, S. (eds.): Biologic Therapy of Cancer: Principles and Practice (p. 482). Philadelphia: J.B. Lippincott, 1991.
105. Zarling, J.M., Moran, P.A., Haffar, O., Sias, J., Richman, D.D., Spina, C.A., Myers, D.E., Kuebelbeck, V., Ledbetter, J.A., Uckun, F.M. Inhibition of HIV replication by pokeweed antiviral protein targeted to CD4+ cells by monoclonal antibodies. Nature 347, 92 (1990).
106. Barbieri, L., Stirpe, F. Ribosome-inactivating proteins from plants: Properties and possible uses. Cancer Surv. 1, 489 (1982).
107. **a:** FitzGerald, D., Pastan, I. Targeted toxin therapy for the treatment of cancer. J. Natl. Cancer Inst. 81, 1455 (1989). **b:** Blakey, D.C., Wawrzynczak, E.J., Wallace, P.M., Thorpe, P.E. Antibody toxin conjugates: A perspective. Prog. Allergy 45, 50 (1988). **c:** Lord, J.M., Spooner, R.A., Hussain, H., Roberts, L.M. Immunotoxins: Properties, applications and current limitations. Adv. Drug Delivery Rev. 2, 297 (1988).
108. Morgan, Jr., A.C., Manger, R., Pearson, J.W., Longo, D., Abrams, P., Bjorn, M., Sivam, G. Holotoxin immunoconjugates of Pseudomonas exotoxin A: Evaluation in mice, monkeys, and man. Antibody Immunoconj. Radiopharm. 2, 165 (1989).
109. Collier, R.J., Kaplan, D.A. Immunotoxins. Sci. Am. 251, 56 (1984).
110. Cumber, A.J., Forrester, J.A., Foxwell, B.M.J., Ross, W.C.J., Thorpe, P.E. Preparation of antibody-toxin conjugates. Methods Enzymol. 112, 207 (1985).
111. Knowles, P.P., Thorpe, P.E. Purification of immunotoxins containing ricin A-chain and abrin A-chain using blue Sepharose CL-6B. Anal. Biochem. 160, 440 (1987).
112. FitzGerald, D.J.P. Construction of immunotoxins using *Pseudomonas* exotoxin A. Methods Enzymol. 151, 139 (1987).
113. Engert, A., Burrows, F., Jung, W., Tazzari, P.L., Stein, H., Pfreundschuh, M., Diehl, V., Thorpe, P. Evaluation of ricin A chain-containing immunotoxins directed against

the CD30 antigen as potential reagents for the treatment of Hodgkin's disease. Cancer Res. 50, 84 (1990).

114. Kozak, R.W., FitzGerald, D.P., Atcher, R.W., Goldman, C.K., Nelson, D.L., Gansow, O.A., Pastan, I., Waldmann, T.A. Selective elimination in vitro of alloresponsive T cells to human transplantation antigens by toxin or radionuclide conjugated anti-IL-2 receptor (Tac) monoclonal antibody. J. Immunol. 144, 3417 (1990).

115. **a:** Goff, D.A., Carroll, S.F. Substituted 2-iminothiolanes: Reagents for the preparation of disulfide cross-linked conjugates with increased stability. Bioconjugate Chem. 1, 381 (1990). **b:** Carroll, S.F., Goff, D.A. Hindered linking agents derived from 2-iminothiolanes and methods. U.S. Pat. No. 5,093,475, issued Mar. 3, 1992.

116. **a:** Thorpe, P.E., Wallace, P.M., Knowles, P.P., Relf, M.G., Brown, A.N.F., Watson, G.J., Knyba, P.E., Wawrzynczak, E.J., Blakey, D.C. New coupling agents for the synthesis of immunotoxins containing a hindered disulfide bond with improved stability *in vivo*. Cancer Res. 47, 5924 (1987). **b:** Worrell, N.R., Cumber, A.J., Parnell, G.D., Mirza, A., Forrester, J.A., Ross, W.C.J. Effect of linkage variation on pharmacokinetics of ricin A chain-antibody conjugates in normal rats. Anti-Cancer Drug Design 1, 170 (1986).

117. **a:** Thorpe, P.E., Wallace, P.M., Knowles, P.P., Relf, M.G., Brown, A.N.F., Watson, G.J., Blakey, D.C., Newell, D.R. Improved antitumor effects of immunotoxins prepared with deglycosylated ricin A-chain and hindered disulfide linkages. Cancer Res. 48, 6396 (1988). **b:** Engert, A., Martin, G., Amlot, P., Wudenes, J., Diehl, V., Thorpe, P. Immunotoxins constructed with anti-CD25 monoclonal antibodies and deglycosylated ricin A-chain have potent anti-tumor effects against human Hodgkin cells *in vitro* and solid Hodgkin tumors in mice. Int. J. Cancer 49, 450 (1991).

118. Kanellos, J., McKenzie, I.F.C., Pietersz, G.A. Intratumor therapy of solid tumors with ricin-antibody conjugates. Immunol. Cell Biol. 67, 89 (1989).

119. Greenfield, L., Bloch, W., Moreland, M. Thiol-containing cross-linking agent with enhanced steric hinderance. Bioconjugate Chem. 1, 400 (1990).

120. Ovadia, M., Hager, C.C., Oeltmann, T.N. An antimelanoma-barley ribosome inactivating protein conjugate is cytotoxic to melanoma cells *in vitro*. Anticancer Res. 10, 671 (1990).

121. **a:** Colombatti, M., Dell'Arciprete, L., Rappuoli, R., Tridente, G. Selective immunotoxins prepared with mutant diphtheria toxins coupled to monoclonal antibodies. Methods Enzymol. 178, 404 (1989). **b:** Myers, D.E., Uckun, F.M., Swaim, S.E., Vallera, D.A. The effects of aromatic and aliphatic maleimide crosslinkers on anti-CD5 ricin immunotoxins. J. Immunol. Methods 121, 129 (1989).

122. Morgan, Jr., A.C., Sivam, G., Beaumier, P., McIntyre, R., Bjorn, M., Abrams, P.G. Immunotoxins of Pseudomonas exotoxin A (PE): Effect of linkage on conjugate yield, potency, selectivity and toxicity. Mol. Immunol. 27, 273 (1990).

123. Batra, J.K., Jinno, Y., Chaudhary, V.K., Kondo, T., Willingham, M.C., FitzGerald, D.J., Pastan, I. Antitumor activity in mice of an immunotoxin made with anti-transferrin receptor and a recombinant form of *Pseudomonas* exotoxin. Proc. Natl. Acad. Sci. USA 86, 8545 (1989).

124. Peeters, J.M., Hazendonk, T.G., Beuvery, E.C., Tesser, G.I. Comparison of four bifunctional reagents for coupling peptides to proteins and the effect of the three moieties on the immunogenicity of the conjugates. J. Immunol. Methods 120, 133 (1989).

125. Cattel, L., Delprino, L., Brusa, P., Dosio, F., Comoglio, P.M., Prat, M. Comparison of blocked and non-blocked ricin-antibody immunotoxins against human gastric carcinoma and colorectal adenocarcinoma cell lines. Cancer Immunol. Immunother. 27, 233 (1988).
126. Marsh, J.W., Neville, Jr., D.M. A flexible peptide spacer increases the efficacy of holoricin anti-T Cell immunotoxins. J. Immunol. 140, 3674 (1988).
127. Neville, Jr., D.M., Srinivasachar, K., Stone, R., Scharff, J. Enhancement of immunotoxin efficacy by acid-cleavable cross-linking agents utilizing diphtheria toxin and toxin mutants. J. Biol. Chem. 264, 14653 (1989).
128. Eisen, H.N. Immunoglobulins and Immunoglobulin Genes. In Davis, B.D., Dulbecco, R., Eisen, H.N., Ginsberg, H.A. (eds.): Microbiology (4th ed., p. 291). Philadelphia: J.B. Lippincott, 1990.
129. Heyman, B., Nose, M., Weigle, W.D. Carbohydrate chains on IgG2b: A requirement for efficient feedback immunosuppression. J. Immunol. 134, 4018 (1985).
130. Mizuochi, T., Taniguchi, T., Shimizu, A., Kobata, A. Structural and numerical variations of the carbohydrate moiety of immunoglobulin G. J. Immunol. 129, 2016 (1986).
131. Gallagher, J.T. The cell-surface membrane in malignancy. In Farmer, P.B., Walker, J.M. (eds.): The Molecular Basis of Cancer (p. 37). New York: John Wiley & Sons, 1985.
132. Hymes, A.J., Mullinax, G.L., Mullinax, F. Immunoglobulin carbohydrate requirement for formation of an IgG-IgG complex. J. Biol. Chem. 254, 3148 (1979).
133. O'Shannessy, D.J., Quarles, R.H. Labeling of the oligosaccharide moieties of immunoglobulins. J. Immunol. Methods 99, 153 (1987).
134. Murayama, A., Shimada, K., Yamamoto, T. Modification of immunoglobulin G using specific reactivity of sugar moiety. Immunochemistry 15, 523 (1978).
135. Solomon, B., Koppel, R., Schwartz, F., Fleminger, G. Enzymatic oxidation of monoclonal antibodies by soluble and immobilized bifunctional enzyme complexes. J. Chromatogr. 510, 321 (1990).
136. Kralovec, J., Singh, M., Mammem, M., Blair, A.H., Ghose, T. Synthesis of site-specific methotrexate-IgG conjugates. Cancer Immunol. Immunother. 29, 293 (1989).
137. Abraham, R., Moller, D., Gabel, D., Senter, P., Hellström, I., Hellström, K.E. The influence of periodate oxidation on monoclonal antibody avidity and immunoreactivity. J. Immunol. Methods 144, 77 (1991).
138. O'Shannessy, D.J., Hoffman, W.L. Site-directed immobilization of glycoproteins on hydrazide-containing solid supports. Biotechnol. Appl. Biochem. 9, 488 (1987).
139. Kralovec, J., Singh, M., Mammen, M., Blair, A.H., Ghose, T. Comparison of the stability and antitumor activity of MTX-IgG conjugates prepared by amide and hydrazone linkages. Proc. Am. Assoc. Cancer Res. 29, 288 (1988).
140. Laguzza, B.C., Nichols, C.L., Briggs, S.L., Cullinan, G.J., Johnson, D.A., Starling, J.J., Baker, A.L., Bumol, T.F., Corvalan, J.R.F. New antitumor monoclonal antibody-vinca conjugates LY203725 and related compounds: Design, preparation, and representative *in vivo* activity. J. Med. Chem. 32, 548 (1989).
141. Hinman, L.M., Wallace, R., Hamann, P.R., Durr, F.E., Upeslacis, J. Calicheamicin immunoconjugates: Influence of analog and linker modification on activity *in vivo*. Chemoimmunoconjugates (Abst. No. 84), 59 (1990).
142. Lopes, A.D., Radcliffe, R.D., Caughlin, D.J., Lee, L.S., McKearn, T.J., Rodwell,

J.D. Site-selective methotrexate-antibody conjugates yield a superior therapeutic effect in tumor xenograft-bearing nude mice. Int. Congr. Ser.–Excerpta Med. 718, 303 (1987).

143. King, D.H., Coughlin, D.J., Rodwell, J.D., Lopes, A.D., Radcliffe, R.D. Amine derivatives of anthracycline antibiotics and antibody conjugates thereof. U.S. Pat. No. 4,950,738, issued Aug. 21, 1990.

144. Rakestraw, S.L., Tompkins, R.G., Yarmush, M.L. Preparation and characterization of immunoconjugates for antibody-targeted photolysis. Bioconjugate Chem. 1, 212 (1990).

145. Krinick, N.L., Rihova, B., Ulbrich, K., Strohalm, J., Kopecek, J. Targetable photoactivatable drugs. 2: Synthesis of N-(2-hydroxypropyl)methacrylamide copolymer-anti-Thy 1.2 antibody-chlorin e6 conjugates and a preliminary study of their photodynamic effect on mouse splenocytes *in vitro*. Makromol. Chem. 191, 839 (1990).

146. Simonson, R.B., Utlee, M.E., Long, C.G., Gillette, R.W., McKearn, T.J., Rodwell, J.D. Inhibition of mannosidase in hybridomas yields monoclonal antibodies with greater capacity for carbohydrate labeling. Clin. Chem. (Winston-Salem, NC) 34, 1713 (1988).

147. Alvarez, V.L., Wen, M.-L., Lee, C., Lopes, D., Rodwell, J.D., McKearn, T.J. Site-specifically modified ^{111}In labelled antibodies give low liver backgrounds and improved radioimmunoscintigraphy. Nucl. Med. Biol. 13, 347 (1986).

148. Rodwell, J.D., Alvarez, V.L., Lee, C., Lopes, A.D., Goers, J.W.F., King, H.D., Powsner, H.J., McKcarn, T.J. Site-specific covalent modification of monoclonal antibodies: *In vitro* and *in vivo* evaluations. Proc. Natl. Acad. Sci. USA 83, 2632 (1986).

149. Rana, T.M., Meares, C.F. N-terminal modification of immunoglobulin polypeptide chains tagged with isothiocyanato chelates. Bioconjugate Chem. 1, 357 (1990).

150. Fisch, I., Künzi, G., Rose, K., Offord, R.E. Site-specific modification of a fragment of a chimeric monoclonal antibody using reverse proteolysis. Bioconjugate Chem. 3, 147 (1992).

151. Schechter, B., Neumann, A., Wilchek, M., Arnon, R. Soluble polymers as carriers of *cis*-platinum. J. Controlled Release 10, 75 (1989).

152. Manabe, Y., Tsubota, T., Haruta, Y., Okazaki, M., Haisa, S., Nakamura, K., Kimura, I. Production of a monoclonal antibody-bleomycin conjugate utilizing dextran T-40 and the antigen-targeting cytotoxicity of the conjugate. Biochem. Biophys. Res. Commun. 115, 1009 (1983).

153. Yokoyama, M., Inoue, S., Kataoka, K., Yui, N., Okano, T., Sakurai, Y. Molecular design for missile drug: Synthesis of adriamycin conjugated with immunoglobulin G using poly(ethylene glycol)-block-poly(aspartic acid) as intermediate carrier. Makromol. Chem. 190, 2041 (1989).

154. Galun, E., Shouval, D., Adler, R., Shahaar, M., Wilchek, M., Hurwitz, E., Sela, M. The effect of anti-α-fetoprotein-adriamycin conjugate on a human hepatoma. Hepatology 11, 578 (1990).

155. Tsukada, Y., Kato, Y., Umemoto, N., Takeda, Y., Hara, T., Hirai, H. An anti-α-fetoprotein antibody-daunorubicin conjugate with a novel poly-L-glutamic acid derivative as intermediate drug carrier. J. Natl. Cancer Inst. 73, 721 (1984).

156. Hurwitz, E., Arnon, R., Sahar, E., Danon, Y. A conjugate of adriamycin and monoclonal antibodies to Thy-1 antigen inhibits human neuroblastoma cells *in vitro*. Ann. N.Y. Acad. Sci. 417, 125 (1983).

157. Rihova, B., Kopecek, J. Biological properties of targetable poly[N-(2-hydroxypropyl)-methacrylamide]-antibody conjugates. J. Controlled Release 2, 289 (1985).
158. Hurwitz, E., Wilchek, M., Pitha, J. Soluble macromolecules as carriers of daunorubicin. J. Appl. Biochem. 2, 25 (1980).
159. Schechter, B., Pauzner, R., Arnon, R., Haimovich, J., Wilchek, M. Selective cytotoxicity against tumor cells by cis-platin complexed to antitumor antibodies *via* carboxymethyl dextran. Cancer Immunol. Immunother. 25, 225 (1987).
160. Shih, L.B., Goldenberg, D.M. Effects of methotrexate-carcinoembryonic-antigen-antibody immunoconjugates on GW-39 human tumors in nude mice. Cancer Immunol. Immunother. 31, 197 (1990).
161. Shih, L.B., Sharkey, R.M., Primus, F.J., Goldenberg, D.M. Site-specific linkage of methotrexate to monoclonal antibodies using an intermediate carrier. Int. J. Cancer 41, 832 (1988).
162. Pietersz, G.A., Cunningham, Z., McKenzie, I.F.C. Specific *in vitro* antitumor activity of methotrexate-monoclonal antibody conjugates prepared using human serum albumin as an intermediate. Immunol. Cell Biol. 66, 43 (1988).
163. Greenfield, R.S., Senter, P.D. Daues, A.T., Fitzgerald, K.A., Gawlak, S., Manger, W., Braslawsky, G.R. *In vitro* evaluation of immunoconjugates prepared by linking mitomycin C to monoclonal antibodies *via* polyglutamic acid carriers. Antibody Immunoconj. Radiopharm. 2, 201 (1989).
164. Naguchi, A., Takahashi, T., Yamaguchi, T., Kitamura, K., Takakura, Y., Hashida, M., Sezaki, H. Preparation and properties of the immunoconjugate composed of anti-human colon cancer monoclonal antibody and mitomycin C-dextran conjugate. Bioconjugate Chem. 3, 132 (1992).
165. Li, S., Zhang, X.-Y., Zhang, S.-Y., Chen, X.-T., Chen, L.-J., Sho, Y.-H., Zhang, J.-L., Fan, D.-M. Preparation of antigastric cancer monoclonal antibody MGb_2-mitomycin C conjugate with improved antitumor activity. Bioconjugate Chem. 1, 245 (1990).
166. Shouval, D., Adler, R., Wands, J.R., Hurwitz, E. Conjugates between monoclonal antibodies to HBs Ag and cytosine arabinoside. J. Hepatol. 3, S87 (1986).
167. Hurwitz, E., Stancovski, I., Wilchek, M., Shouval, D., Takahashi, H., Wands, J.R., Sela, M. A conjugate of 5-fluorouridine-poly(L-lysine) and an antibody reactive with human colon carcinoma. Bioconjugate Chem. 1, 285 (1990).
168. Shih, L.B., Xuan, H., Sharkey, R.M., Goldenberg, D.M. A fluorouridine-anti-CEA immunoconjugate is therapeutically effective in a human colon cancer xenograft model. Int. J. Cancer 46, 1101 (1990).
169. Brown, B.A., Drozynski, C.A., Dearborn, C.B., Hadjian, R.A., Liberatore, F.A., Tulip, T.H., Tolman, G.L., Haber, S.B. Conjugation of metallothionein to a murine monoclonal antibody. Anal. Biochem. 172, 22 (1988).
170. Pimm, M.V., Clegg, J.A., Caten, J.E., Ballantyne, K.D., Perkins, A.C., Garnett, M.C., Baldwin, R.W., Biodistribution of methotrexate-monoclonal antibody conjugates and complexes: Experimental and clinical studies. Cancer Treat Rev. 14, 411 (1987).
171. Torchilin, V.P., Klibanov, A.L., Nossiff, N.D., Slinkin, M.A., Strauss, H.W., Haber, E., Smirnov, V.N., Khan, B.A. Monoclonal antibody modification with chelate-linked high-molecular-weight polymers: Major increases in polyvalent cation binding without loss of antigen binding. Hybridoma 6, 229 (1987).

172. Maiti, P.K., Lang, G.M., Sehon, A.H., Tolerogenic conjugates of xenogeneic monoclonal antibodies with monomethoxypolyethylene glycol. I. Induction of long-lasting tolerance to xenogeneic monoclonal antibodies. Int. J. Cancer Suppl. 3, 17 (1988).

173. Takashina, K.-I., Kitamura, K., Yamaguchi, T., Noguchi, A., Noguchi, A., Tsurumi, H., Takahashi, T. Comparative pharmacokinetic properties of murine monoclonal antibody A7 modified with neocarzinostatin, dextran and polyethylene glycol. Jpn. J. Cancer Res. 82, 1145 (1991).

174. Shih, L.B., Sharkey, R.M., Primus, F.J., Goldenberg, D.M. Modification of immunotoxins for improved cancer therapy. Proc. Am. Assoc. Cancer Res. 28, 383 (1987).

175. Clark, S.A. Monoclonals and the FDA. Curr. Opin. Invest. Drugs 2, 165 (1993).

CHAPTER 5

THE PRODUCTION OF MONOCLONAL ANTIBODIES

J.R. BIRCH, J. BONNERJEA, S. FLATMAN, AND S. VRANCH
Celltech Biologics, Slough, Berkshire, SL1 4EN, UK

5.1. INTRODUCTION

Earlier chapters have dealt with the uses to which monoclonal antibodies are being put and applications that may arise in the future. Product application obviously has a great bearing on choice of manufacturing process both with respect to scale and quality requirements. While diagnostic antibodies may be required in tens or hundreds of grams per year, some therapeutic antibodies may be required in tens or even hundreds of kilograms per year. In addition the purity requirements for a therapeutic antibody will be far more exacting than for an *in vitro* diagnostic antibody. Consequently in selecting a manufacturing process it is essential at the outset to choose a system that is capable of delivering product in appropriate quantities, to an acceptable quality specification, and at an acceptable price. It is important to develop processes with the ultimate manufacturing process in mind, particularly for a therapeutic product, because it may be difficult or impossible, for regulatory reasons, to make significant process changes once the product is in a late stage of clinical development or on the market. Particularly for therapeutic products great attention must be paid not just to the quality of the product but also to the control of the production process, including validation and monitoring of the process. This is because, for a complex biological molecule, uncontrolled process changes may have unexpected effects on the properties of the product. Consequently there is a major emphasis on the development of analytical methods for monitoring both product and process.

5.2. MANUFACTURING AND REGULATORY ISSUES

Regulatory standards are applied to the manufacture of antibodies for use in the pharmaceutical industry, and for the design and construction of facilities and equip-

Monoclonal Antibodies: Principles and Applications, pages 231–265
© 1995 Wiley-Liss, Inc.

ment used in their manufacture. These quality standards for manufacturing are described in current good manufacturing practice (cGMP) regulations [1,2]. Facility requirements for factories making antibodies are also described in regulations [2,3]. These require that the process be contained, that cross-contamination is avoided, that pure water is used in the manufacturing process, and that the air is filtered and conditioned. Environmental monitoring takes place to ensure that hygienic conditions are maintained. The design of equipment is the key to successful manufacture, particularly the ability to clean and sterilize fermenters and purification equipment. Equipment and facilities must be validated and regularly maintained. cGMP regulations also require that staff be carefully selected and trained. The flow of materials throughout the facility must be controlled, including the removal of waste products. Often, regulatory authorities will comment on flow plans and designs before a facility is built. High-quality services are needed to ensure batch consistency. Hence, water-for-injection is normally used for both fermentation and purification steps. Where facilities are used for making different products, special attention is paid to operating procedures, segregation of operations, such as cleaning, and the avoidance of cross-contamination [4].

Compared with most chemical pharmaceuticals, biopharmaceuticals are complex molecules, often showing microheterogeneity. Consequently, there is considerable emphasis on controlling the process itself rather than relying on an analysis of final product [5]. Process validation is addressed during the development phase, in which impurities and contaminants are identified, and clearance studies are carried out. Process validation is used to stress the process and demonstrate its robustness and reproducibility. Pilot-scale batches provide information and material for stability studies and for the establishment of reference standards for the product. Before production begins, manufacturing directions, often over 1,000 pages for a single batch, have to be written and approved. At least three batches have to be made at full scale to demonstrate consistency. Changes in the process, such as scale-up, can affect product quality and contaminant profile, and this requires a large investment in process validation before changes can be authorized. The function of in-process controls is crucial, and quality control procedures must also be validated. Documents relating to the control of operations comprise the manufacturing and control data package, which is an integral part of the quality control system.

5.3. ANTIBODY MANUFACTURE WITH MAMMALIAN CELLS

5.3.1. Cell Types

The most common type of cell used to produce monoclonal antibodies is the mouse hybridoma. A variety of mouse myeloma cells have been used to create hybridomas, most of them descended from the parental MOPC 21 cell line [6]. Rat hybridomas have also been a useful source of monoclonal antibodies [7].

Production of human antibodies has proved more problematic and a variety of strategies have been adopted that use human hybridomas, human–mouse "heterohybridomas," and human lymphoid cells immortalized with Epstein-Barr virus. The limitations and potential of these approaches are reviewed by Boyd and James [8]. While these types of cell culture have been successful in some cases, they have not provided a consistently reliable route to producing the human antibodies that are preferred for human therapeutic applications. To overcome these limitations, increasing use has been made of recombinant DNA technology to "humanize" rodent antibodies.

Choice of cell line and genetic expression system are dictated by the productivity achieved, cell line stability, absence of undesirable adventitious agents, and the ability of the cell to provide appropriate posttranslational processing, especially glycosylation. In practice engineered antibodies have most commonly been expressed in Chinese hamster ovary (CHO) cells [9–11] and mouse myeloma cells [12,13] by means of a variety of genetic expression systems. Both cell types have been widely studied in industry and lend themselves to large-scale manufacturing processes because they grow in suspension culture. The use of highly efficient genetic expression systems has led to high productivity, often exceeding that achieved with hybridomas. Page and Sydenham [11], for example, report a yield of 200 mg antibody per liter in batch culture of DHFR-amplified CHO cells. With mouse myeloma cells and the glutamine synthetase expression system, yields as high as 1 g/L have been achieved in optimized batch culture [14] (Fig. 5.2). Hybridomas typically produce tens to hundreds of milligrams per liter in batch culture [15]. In some cases it may be useful to use recombinant DNA technology to "rescue" antibody genes from unsuitable host cells. This may be to overcome problems of low productivity, cell line instability, or the presence of adventitious agents. Gillies et al. [13] and Lewis et al. [16], for example, have described the rescue of immunoglobulin genes from human cell lines into mouse and rat myeloma cell lines, respectively.

Other cell types, such as insect cells [17,18], have been used to a lesser extent, but they do not appear to offer any advantage (for example, in productivity) compared with the more commonly used mammalian cell types. A disadvantage of insect cells at least for therapeutic applications is that glycosylation of proteins is different from that achieved with mammalian cells. Insect cells may be useful, however, for the rapid expression of research quantities of antibody using the baculovirus system.

5.3.2. Cell Banks

For the commercial production of monoclonal antibodies the manufacturing process for each production cycle starts with cells taken from a highly characterized cell bank. In the case of mammalian cells the banks consist of aliquots of cells taken from a homogeneous pool that are cryopreserved in liquid nitrogen. It is usual to establish a master cell bank (MCB) of perhaps 100–200 ampoules and a manufacturer's working cell bank (MWCB) of similar size prepared by expansion of cells from an ampoule of the MCB. Ampoules from the MWCB are used to initiate production lots and additional MWCBs can be generated as required from the MCB.

The use of cell banks prepared under cGMP conditions is a key element in providing a consistent process and in contributing to confidence in the safety of the eventual product. Cell banks are highly characterized with respect to adventitious agents, genetic characteristics (especially for recombinant cell lines), and product characteristics. Particular attention is paid to the stability of cell and product characteristics over the number of generations required to complete the manufacturing process. In order to provide a reference point and source of material for characterization it is usual to preserve aliquots of cells that are representative of those found at the end of the production process. A detailed account of the strategies employed in preparing and characterizing cell banks has been given by Wiebe and May [19]. The requirements for establishing and characterizing cell banks are laid down in guidelines provided by regulatory authorities (Chapter 6).

5.3.3. Culture Systems for Animal Cell Culture

Many monoclonal antibodies are produced by growing hybridoma cells as ascites tumors in mice or rats. Typically antibody accumulates to concentrations of about 10 g/L in the ascites fluid. However, for therapeutic antibodies most manufacturers have now adopted *in vitro* cell culture techniques. This approach reduces the chance of introducing adventitious agents and irrelevant antibodies from the animal host. It is probable that ethical pressures to reduce the unnecessary use of animals in research and manufacture will result in a further shift to *in vitro* methodology, even for nontherapeutic applications.

Turning to *in vitro* methods, many types of cell culture systems have been described. For small-scale applications requiring milligrams or grams of product, a variety of methods are used that range from simple roller bottles and spinner flasks to hollow-fiber perfusion culture systems. These small-scale systems cannot usually be scaled up and for larger scale (hundreds of grams to multikilograms) a variety of reactors have been developed or are in development (Table 5.1). Since antibody-producing cells can be grown in suspension, it is possible to use conventional batch reactors with mixing provided either by stirring [22] or by airlift pump [15]. These

TABLE 5.1. Examples of Animal Cell Reactors for Monoclonal Antibody Production

Reactor	Mode of Operation	Reference
Stirred tank	Batch	Reuveny and Lazar [20]
	Continuous cytostat	Fazekas de St. Groth [21]
	Perfused	Shevitz et al. [22]
	Perfused microcapsules	Tyler [23]
	Perfused alginate microbeads	Sinacore et al. [24]
Airlift	Fed batch	Birch et al. [15]
	Continuous with cell recycle	Hulscher et al. [25]
Fluidized bed	Perfused porous carriers	Runstadler et al. [26]
Ceramic matrix	Perfused	Applegate and Stephanopoulos [27]
Hollow fiber	Perfused	Altshuler et al. [28]

Fig. 5.1. A 1,000-liter airlift fermenter for MAb production. courtesy of Celltech Ltd.

types of reactors are used in industry at scales up to 10,000 liters to make a range of proteins including antibodies from animal cells and they are the systems most commonly used. Conventional batch culture has the advantage of relative simplicity and ease of scale-up, leading to relatively rapid process development. The operation of a batch system also helps to give straightforward definition of a product "lot." For very large scale operation some workers argue that there is an economic advantage in operating a continuous process, and a variety of continuous systems have been described, mostly at laboratory scale, although some continuous perfused systems have been described that have the potential for producing multikilogram quantities [26].

An example of a production scale airlift fermenter (1,000 liters) for monoclonal antibody production is shown in Figure 5.1. Figure 5.2 shows growth and product accumulation profiles for a mouse NS0 myeloma cell line making a genetically engineered IgG in batch culture. For this cell line yields of 1 g/L can be achieved in a 10-day growth cycle.

5.3.4. Process Optimization and Characterization

At production scale it is usual to optimize not just the culture medium (see below) but also key physicochemical parameters such as pH, dissolved oxygen concentration, and temperature of the culture. More detailed discussions of the issues involved in process optimization, as well as the reactors used, are provided by Mizrahi [29] and Lubiniecki [30].

In a manufacturing process (particularly for a therapeutic product) it is important to characterize the process thoroughly and to have determined the sensitivity of the process to changes in key process parameters. It is usual to monitor the reproducibility of the process rigorously over a number of production runs, not just with respect to fermentation parameters but also with respect to the properties of the product. This is particularly important because changes in fermentation conditions

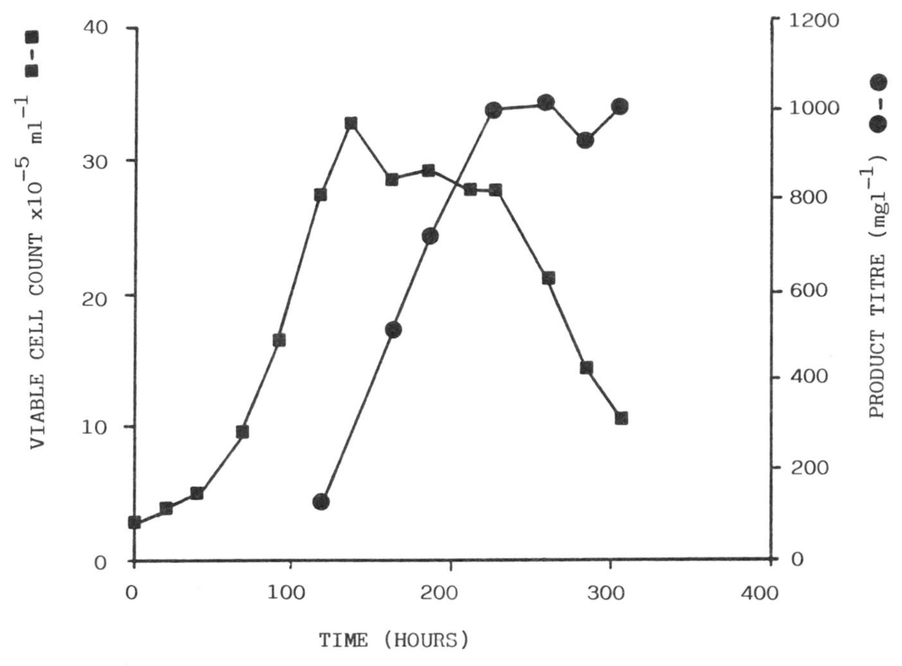

Fig. 5.2. Production of recombinant MAb from the GS-NS0 cell line in serum-free fermentation (courtesy of Celltech Ltd.)

may subtly alter the structure of the product [31,32]. For example, Goochee and Monica [31] have reviewed the effects that environmental conditions can have on one aspect of product structure, namely, glycosylation. They describe the altered glycosylation of a secreted antibody light chain resulting from glucose starvation.

5.3.5. Culture Media for Animal Cell Culture

While media for microbial culture are relatively simple, those used for animal cell culture are complex mixtures containing amino acids, vitamins, salts, and other nutrients. It is common to supplement media with animal serum, although in recent years it has become more usual to replace serum with purified blood proteins such as albumin and transferrin. Murakami [33] has reviewed the status of serum-free media for hybridoma cell culture. Serum-free media are also being used to produce recombinant antibodies in rodent myeloma cell lines [14]. Frequently the protein supplements prove to be acting as carriers of low-molecular-weight nutrients such as lipids and iron [34]. By careful design of the basal medium it has proved possible to grow a number of hybridoma cell lines in completely protein-free media (for example see Kovar and Franek [34], Schneider [35], and Darfler [36]). The use of serum-free and protein-free media has distinct advantages for large-scale production; the levels of contaminating protein (including immunoglobulin derived from serum) to be removed during purification (and measured in the product) are reduced; the risk of introducing adventitious agents via animal-sourced materials is removed, and costs are reduced. The trend towards protein-free media is likely to increase significantly in the next few years.

In addition to the development of serum-free media significant progress has been made in optimizing culture media for growth and productivity based on a quantitative understanding of nutritional requirements. For example, Brown et al. [14] described how iterative improvements to the culture medium improved productivity of a myeloma cell line making a recombinant antibody by a factor of 10, to approximately 600 mg/L. In some cases it may be possible advantageously to modify the nutritional requirements of a cell line by genetic manipulation. For example, the enzyme glutamine synthetase has been used as a selectable marker for isolating cell lines transfected with genes expressing recombinant antibodies [9]. The enzyme confers glutamine independence on the myeloma host cell line, permitting growth in glutamine-free medium. This is an advantage because glutamine is an unstable nutrient that can accumulate to toxic levels in the culture medium (for example, see Murakami [33, p. 125]). Bell et al. [37] described the modification of hybridoma cell metabolism by introduction of the gene for glutamine synthetase.

5.4. MICROBIAL PRODUCTION SYSTEMS

The use of *Escherichia coli* as a host cell has greatly facilitated the manipulation of antibody genes and the screening of gene products (Chapter 3). Because *E. coli* grows rapidly in inexpensive media in conventional ferments it is also attractive as a production vehicle. The genetically modified strain of the bacterium is mostly used for the production of antibody fragments (Fab′, $F(ab')_2$, Fv, single chain antibodies), rather than complete immunoglobulin molecules, since the latter, in common with other large, complex, multichain proteins are difficult to express in active form and at high levels in noneukaryotic cells. An example of fragment production is given by Carter et al. [38], who describe the manufacture of soluble Fab′ fragments in the periplasmic space of *E. coli*. Yields of 1–2 g/L were achieved in a stirred 10-liter reactor following induction by phosphate starvation. $F(ab')_2$ fragments were regenerated *in vitro* from the Fab′ fragments. The high yields were in part due to the hundredfold increase in cell densities that could be achieved under fermenter conditions. In some cases antibody molecules have been expressed as insoluble inclusion bodies in *E. coli* which have then been renatured. For example, Buchner and Rudolph [39] describe such a system for making a Fab fragment. The product represented 40% of total cell protein. Interest has also been expressed in the use of yeast as a production system and Davis et al. [40], for example, have described the expression of a single chain antibody (heavy and light chain variable regions joined by a polypeptide linker) in the yeast *Schizosaccharomyces pombe*.

5.5. OTHER PRODUCTION SYSTEMS

A number of other systems with potential for manufacturing recombinant antibodies have been described. For example, antibody genes have been expressed in the leaves of plants [41] and insect larvae [17]. Whether such alternatives will be used for large-scale production remains to be seen.

5.6. METHODS FOR ANTIBODY PURIFICATION

Most monoclonal antibodies of commercial interest are required to meet very high purity specifications. In order to ensure their safety, antibody preparations intended for therapeutic uses need to carry a very low risk of containing potentially harmful substances such as pyrogens, viruses, fragments of DNA, and immunogenic protein contaminants [42]. Even antibodies used in diagnostic kits generally need to be sufficiently pure to provide a stable, reproducible preparation free of unwanted side reactions.

Despite the very wide variety of production systems now available for the manufacture of monoclonal antibodies (e.g., ascites, bacterial and mammalian cell culture, batch and perfusion systems), a common feature of virtually all these systems is that the antibody is only a minor component of the complex mixture of molecules that is produced. Depending on the production system chosen, various quantities of cells, cell debris, lipids, DNA, proteolytic enzymes, pyrogens, and viruses need to be removed. Often one of the most challenging tasks is the removal of modified versions of the antibody, e.g., proteolytically cleaved, oxidized, or aggregated antibody. Since such molecules may differ only slightly from the "authentic" molecule in terms of chemical composition and behavior, their removal can be a very difficult and challenging task.

The removal of these many different impurities and contaminants must be achieved while the integrity and biological activity of the product molecule are maintained. The purification process also needs to give a high recovery of the antibody and it should not introduce any extraneous substances. In addition, for commercial production the very important economic issues of cost and throughput must be taken into consideration.

In order to meet these requirements, virtually all antibody purification processes rely on one or more column chromatography steps. There are various forms of column chromatography that can be used for the purification of antibodies, for example, ion exchange, affinity, hydrophobic interaction, and gel permeation chromatography. The operation of the individual chromatography steps, their number and their sequence is generally tailored to the specific antibody and the specific application during the development of the purification process.

In order to be able to purify antibodies by column chromatography, the antibody must be in a soluble form and largely free of particulate matter, and, depending on the type of chromatography, it may also need to be in a solution at a certain pH and ionic strength. Primary recovery operations are therefore required to prepare the crude antibody preparation for chromatography. These operations are typically filtration and centrifugation, and as with the chromatography steps, their sequence and number are tailored to the specific antibody and application.

An alternative approach to the use of conventional packed chromatography beds is the use of fluidized or expanded beds [75]. The main advantage of these systems is that feedstreams containing solids can be applied to the column without prior filtration. Thus the clarification steps can be omitted, although adjustment of the pH and conductivity may still be necessary to promote binding of the product to the matrix.

The most elegant approach is the use of an affinity matrix in a fluidized bed, which can allow a fermenter to be coupled directly to a fluidized bed chromatography column without the need for any conditioning steps. However, the hardware and the matrices for this type of operation have only recently become available commercially, and therefore most antibody production relies on clarification and conventional packed bed chromatography.

5.6.1. Primary Recovery Operations

Most chromatography columns become clogged by particulate material; therefore it is standard practice to clarify all solutions before they are applied to a chromatography column. For this reason, and also to maintain a high level of process hygiene, in-line sterilizing-grade filters are used to filter the buffer solutions used in the chromatography operation. Crude antibody preparations often contain a high content of solids derived from cells and cannot be put through a sterilizing filter without prior clarification.

Crude antibody preparations derived from ascites fluid are difficult materials to filter. Ascites fluid contains cells, cell debris, fibrin clots, and considerable quantities of lipid [43]. Most of these substances block membrane filters and therefore centrifugation is used for clarification. High-speed centrifugation at greater than 25,000 *g* is very effective at clarifying ascites fluid. Whole cells, cell debris, and other submicron particles are sedimented, while lipids form a distinct layer at the top of the centrifuge tube. Only small volumes of liquid can be centrifuged at these speeds due to the limited capacity of ultracentrifuge rotors, but in many cases this is not a significant limitation, since only a few milliliters of ascites fluid are available from each animal.

Crude antibody preparations derived from bacterial fermentations are also difficult to filter due to their viscosity and high content of solids. Both centrifugation and filtration have been used for the clarification of bacterial fermentation broths. With small-scale fermentations of up to approximately 10 liters, centrifugation in a conventional batch centrifuge is a simple and effective means of clarification. For larger fermentations, continuous-flow centrifuges are available in which the fermentation broth can be continuously passed through the centrifuge so that the volume of the feedstream is no longer limited to the volume of the centrifuge bowl. With these types of centrifuge the solids build up in the bowl, and when the bowl is full, the centrifuge must be stopped and the bowl emptied. For very large scale industrial processes with fermentation volumes of many hundreds or thousands of liters, intermittent discharge centrifuges are available that periodically eject the solids accumulated in the centrifuge bowl without requiring the machine to stop, which allows truly continuous operation [44].

With most continuous or semicontinuous centrifuges, the limited relative centrifugal force generated in the centrifuge bowl (generally <10,000 *g*) results in incomplete clarification. Very small particles are not sedimented and consequently the supernatant cannot be applied directly to a sterilizing membrane filter without blockage. Further clarification, often by means of depth filters, is required.

Clarification by filtration, either with conventional dead-end filters or with cross-flow filters, is an alternative to centrifugation. Cross-flow filtration, in which the liquid is continuously recirculated through a channel whose walls are composed of filtration membrane, has the advantage that the solids fraction (i.e., the cells and any cell debris) can be retained, and that the membrane can be reused. In contrast, with dead-end filtration the solids cannot be removed from the filter, which must be replaced after a single use. Also, a series of dead-end filters with decreasing pore size is normally required, with high-solids-capacity filters at the front end and sterilizing-grade membrane filters at the tail end.

Crude antibody preparations derived from mammalian cell culture are generally easier to clarify than bacterial fermentation broths owing to a much lower cell density and lower solid, lipid, and DNA content. Centrifugation, cross-flow filtration, and dead-end filtration can all be used for clarification of mammalian cell culture fluids.

5.6.2. Antibody Recovery by Precipitation

Antibody precipitation is a long-established purification method that remains popular as the first step of laboratory-scale processes. Ammonium sulfate, polyethylene glycol, and small organic molecules such as caprylic acid [54] have all been used to precipitate antibodies. For high and reproducible recovery of the antibody, "ageing" of the precipitate may be necessary before centrifugation and resuspension of the pellet. Since soluble proteins inevitably become entrapped in the precipitate, washing of the pellet can be used to improve the purity of the antibody. However, even with repeated washing of the pellet, the degree of purification achieved by precipitation operations is generally much lower than that for column chromatography operations.

Although antibody precipitation can be performed on a large scale, it is difficult to do so in a contained and sanitary manner. Furthermore, disposal or regeneration of large quantities of precipitate can be difficult and expensive; therefore precipitation of antibodies is seldom used for the large-scale production of commercial antibodies.

5.6.3. Chromatography Operations

Antibodies can be purified by chromatographic techniques similar to those used for other proteins [45]. Examples of methods used to purify antibody are summarized in Table 5.2 and discussed in greater detail in Kenney [55].

The purification of antibodies by column chromatography relies on the partitioning of the antibody and impurities between two phases: a mobile phase, generally

TABLE 5.2. Examples of Purification Methods for Monoclonal Antibodies

Purification Method	Antibody Type	Reference
Affinity chromatography	IgG	Lee et al. [46]
Ion exchange chromatography	Fab	Gavit et al. [47]
	IgG	Scott et al. [48]
Size exclusion chromatography	IgM	Jehanli and Hough [49]
Hydrophobic interaction chromatography	IgG	Goheen and Matson [50]
Hydroxylapatite	IgG	Bukovsky and Kennett [51]; Stanker et al. [52]
Mixed-mode chromatography	IgG, IgM	Nau [53]
Precipitation	IgG	Temponi et al. [54]

composed of an aqueous buffer solution (although organic solvents are occasionally used), and a stationary phase or matrix, composed of a packed bed of a porous, inert support material, such as agarose or silica with functional groups attached to it. Ion exchange matrices based on cellulose support material derivatized with diethylaminoethyl (DEAE) and carboxymethyl (CM) functional groups were among the first matrices to be developed for protein purification, and they are still in use today [56].

There are now many different matrices available for antibody purification. One can choose between inert support materials composed of cross-linked polymers, such as agarose, dextran, cellulose, polyacrylamide, and polymethacrylate, and also inorganic materials such as silica and controlled pore glass. These materials vary with respect to properties such as rigidity, pore size, and bead size. Sepharose, for example, is a relatively soft matrix, whereas glass is very rigid and more easily fabricated into small beads. Different applications require different properties. Smaller beads give better resolution between antibody and impurities but with the disadvantage that higher pressures are required to force the mobile phase through the packed bed. For production-scale chromatography of antibodies, beads with diameters of approximately 50–100 μm are generally used, giving rise to back pressures of only 1–3 bar. On a laboratory scale, 2- to 25-μm-diameter beads may be used in high-performance liquid chromatography (HPLC) systems, but these require high pressures to drive liquid through the column and therefore are not generally used for large-scale production of proteins.

The matrices can be derivatized with functional groups to provide the particular separation mechanism required. For example, positively and negatively charged groups are introduced to generate anion and cation exchange matrices, respectively. Underivatized matrices, which do not interact with or bind to proteins, can be suitable for size exclusion chromatography. In this type of chromatography, also known as gel permeation chromatography, separation of product from impurities is based solely on the size of the molecules concerned. Large molecules are excluded from more of the pores in the matrix than small molecules, and they therefore travel faster through the chromatography column and elute before the smaller molecules that are able to enter more pores.

The ideal support material for column chromatography is porous and permeable even to large molecules, yet is also rigid and physically stable. In addition, it should be inert to proteins and chemically stable yet easily derivatized. No single support meets these contradictory requirements; a compromise is always necessary.

5.6.4. Affinity Chromatography

Affinity chromatography using immobilized protein A or protein G is a very popular technique for the purification of antibodies. This application is one of the best examples of the advantages of affinity chromatography: Extremely high purities can be achieved in a single step with virtually full recovery of the antibody, even from dilute feedstreams that contain high concentrations of protein impurities. Protein A/G matrices are commercially available in large quantities with the ligand linked to cross-linked agarose, silica, or controlled pore glass beads. The matrices have a relatively high capacity for many different subclasses of IgG from many different species. For intact human or humanized antibodies, the dynamic capacity is typically between 10–20 mg of antibody per milliliter of packed matrix. Although these matrices have the disadvantage that they cannot be sanitized and depyrogenated with sodium hydroxide solutions, which are typically used with other chromatography media, other cleaning agents such as solutions of chaotropic salts can be used.

Because of their very specific mode of action, the quality of the column packing and the exact operating conditions, such as ionic strength and flow rate, are less critical for the operation of affinity chromatography compared, for example, with an ion exchange process. Automation is simple, since fraction collection is limited and gradient elution or complex washing regimes are usually not required. The protein A or G ligand is also surprisingly robust and can withstand high concentrations of salts and metal ions; it is also relatively insensitive to cleavage by proteases particularly when immobilized.

Two potential disadvantages of affinity chromatography with protein A/G are the cost of the matrix and ligand leakage. Protein A and G matrices are at least an order of magnitude more expensive than, for example, ion exchange matrices based on the same support. However, given the robust nature of the matrix and the ease of automation, their high cost can be overcome by repeated cycling of a relatively small column. The leakage of the ligand from the support is also a surmountable problem. Very sensitive assays have been developed for the measurement of trace amounts of protein A and G even in the presence of a high concentration of antibody. Thus the clearance of the leached ligand through subsequent chromatography steps can normally be demonstrated to low or undetectable levels. In fact the leakage is often very low, no more than a few parts per million of antibody [57], and far too low to have any measureable effect on matrix capacity. Following affinity chromatography, the antibody preparation is often >95% pure, as measured by SDS-polyacrylamide gel electrophoresis. Of the remaining impurities, most are antibody-related—that is, they represent antibody molecules or fragments that have been cleaved, cross-linked, or modified in some other way, yet still bind to protein A. These molecules can be removed by other column chromatography methods.

5.6.5. Ion Exchange Chromatography

Ion exchange chromatography is one of the oldest and most versatile methods of protein purification. The separation mechanism in ion exchange chromatography is very simple, relying on the binding of negatively charged molecules to a positively charged matrix and vice versa. It can be used as a positive or a negative purification step; that is, either the antibody or the impurities can be made to bind to the matrix, depending on the operating conditions and properties of the antibody. The pH, ionic strength, and chemical composition of the buffers can all be altered during the chromatography operation and these changes can be made either stepwise or with a gradient to adjust the binding of impurities and product. The challenge during the development of an antibody purification process is to use such techniques to develop a robust ion exchange step that gives enhanced antibody purity with high recovery.

Anion exchange chromatography is particularly useful in the purification of therapeutic-grade antibodies, since both DNA and endotoxins bind very strongly to such matrices. The binding is strong even in relatively high-ionic-strength solutions, so a simple negative purification operation involving a single pass through an anion exchanger can give a good clearance of DNA and endotoxins.

5.6.6. Gel Permeation Chromatography

Gel permeation or size exclusion chromatography is the only technique available for the separation of antibodies from impurities on the basis of molecular size. Its principle role is in the separation of antibody aggregates from monomer, although it can also be used as a general purification step, provided that there is at least a twofold difference between the molecular weights of the antibody and the impurities. It is particularly useful in the purification of IgM molecules that have molecular weights of approximately 900,000 daltons.

Because there is no chemical interaction between the matrix and the antibody or impurities, and separation is achieved solely by a "sieving" mechanism that retards small molecules, gel permeation chromatography is unique in that the length of the gel bed is an important factor in the degree of purification obtained. Also for the same reason, the capacity of size exclusion chromatography columns is based on the volume of the sample to be purified, and not the amount of protein present in the sample. In general, the sample volume may be between 2% and 10% of the column volume and the bed length approximately 60–100 cm.

5.6.7. Hydrophobic Interaction Chromatography

Hydrophobic interaction chromatography (HIC) is a close relative of reverse phase chromatography but utilizes aqueous solutions instead of organic solvents. HIC matrices are derivatized with various hydrocarbons such as butyl, phenyl, or octyl functional groups and are often available with different degrees of substitution. The mechanism of separation relies on the interaction of hydrophobic "patches" on the surface of proteins with the functional groups of the matrix. This interaction is promoted by high concentrations of salts, such as ammonium sulfate. The salt concentration should be high enough to promote binding of the antibody to the matrix, but not so high as to precipitate the antibody. Once the antibody has bound to the matrix, the salt concentration is reduced, either stepwise or with a linear gradient, until the antibody is eluted.

Intact antibodies are generally more hydrophobic than most proteins present in fermentation media and therefore they elute from HIC matrices at low salt concentrations. In fact it is often necessary to select low-substitution, relatively polar matrices to ensure that antibodies do not bind irreversibly to the matrix. A difficulty often encountered with HIC is that of achieving good separation of antibodies from albumin, particularly when the albumin is present at high concentrations. Since albumin is often present in fermentation media, some pretreatment is often required to reduce the albumin level before hydrophobic interaction can be used.

5.6.8. Other Purification Methods

Hydroxylapatite chromatography [51,52] is an unusual type of absorption chromatography that does not fit into any of the previously mentioned forms of chromatography, but nonetheless has been used successfully for antibody purification. Hydroxylapatite is a crystalline form of calcium phosphate to which some proteins bind reversibly. In low concentrations of phosphate buffer at pH values close to neutral, serum proteins such as albumin and transferrin do not bind to hydroxylapatite, while most antibodies do. They can then be eluted by increasing the phosphate concentration. Sodium chloride and other ions do not generally interfere with antibody binding and elution; therefore buffer exchange operations may not be necessary prior to hydroxylapatite chromatography. The limited stability of hydroxylapatite has restricted its use, but recently novel matrices combining hydroxylapatite with polymers such as agarose have been introduced and have improved physical and chemical stability.

Another unusual form of chromatography that has been used for antibody purification is "mixed mode" chromatography. As the name implies, this form of chromatography relies on several different modes of chromatography in the matrix, for example, both anion and cation exchange, or ion exchange and hydrophobic interaction. One mixed-mode chromatography matrix that is particularly useful for antibody purification is Bakerbond ABx [53]. This derivatized silica matrix is used as a cation exchanger but gives better purification of IgG and IgM antibodies than most conventional cation exchange matrices.

5.6.9. Process Integration

The strategy normally adopted for the purification of antibodies is to achieve as much purification as possible in the first chromatography step, so that later steps can be optimized for the removal of small quantities of a few specific impurities. Hence, affinity chromatography is an ideal first step provided that the relatively expensive affinity matrix is not fouled or damaged by proteases, lipids, or other components of the cell culture medium. Where affinity chromatography cannot be used—for example, because a suitable immobilized ligand does not exist—it is usual to invest a great deal of time and effort in the optimization of the first step, since this will greatly simplify the later operations. Ion exchange is an obvious first step, since it has a high capacity and the operating conditions can be tailored to remove as many impurities as possible. However, the clarified cell culture medium may require a buffer exchange step or dilution and pH adjustment prior to ion exchange chromatography, since most culture media have a high ionic strength and defined pH that may not be appropriate for optimal separation on the first chromatography step. Care must be taken not to precipitate the antibody during adjustment of pH and ionic strength, although precipitation of the impurities may be a positive advantage.

Irrespective of whether affinity or ion exchange chromatography is used as the first step, the antibody should ideally be eluted by a change in pH or ionic strength in a buffer suitable for the next step. Hydrophobic interaction or ion exchange chromatography is frequently specified as a second step. If necessary, the antibody can be buffer-exchanged between the two chromatography operations either by diafiltration or by a desalting step. Depending on the purity specification that the antibody is required to meet, and also the ease with which the antibody can be separated from the impurities, a small number of chromatography operations are generally performed in sequence. For IgG molecules used *in vitro,* a single chromatography operation may be adequate, particularly where affinity chromatography can be used. For therapeutic applications, three or sometimes even more chromatography operations may be required.

Affinity and ion exchange operations have sometimes been regarded as competing techniques, but they are complementary and a great many purification schemes, particularly those for therapeutic antibodies, use both types of chromatography. Where the final product specification calls for an extremely low level of antibody aggregates, gel filtration is frequently used as a final step. It then performs as an aggregate removal step and a buffer exchange step to transfer the antibody into the final formulation buffer.

Figure 5.3 shows two typical processes for the purification of therapeutic grade

<u>Affinity Chromatography-based process</u>:

protein A affinity chromatography → anion exchange chromatography → gel filtration chromatography

<u>Ion exchange Chromatography-based process</u>:

cation exchange chromatography → anion exchange chromatography → gel filtration chromatography

Fig. 5.3. Typical purification processes for therapeutic grade monoclonal antibodies.

antibodies. The protein A-based process is a generic process that will successfully purify most antibodies that bind to protein A. It is attractive in that three different separation mechanisms are used (based on biological structure, charge, and size); therefore most impurities are easily separated from the product. The affinity chromatography step clears the bulk of the impurities, allowing the ion exchange and gel filtration steps to remove trace quantities of DNA, endotoxins, and leached protein A.

With the ion exchange-based process, the antibody is generally less pure after the first step than with affinity-based processes, and subsequent steps therefore must be able to remove significant quantities of impurities. An antibody with an isoelectric point well away from that of most of the impurities can greatly aid the development of ion exchange-based processes.

5.6.10. Influence of Scale

Most (but certainly not all) chromatography operations are linearly scaleable and can be used equally effectively at a laboratory workbench scale for the production of milligram quantities of antibody, and also on a factory scale for the production of kilograms of antibody. However, a few basic requirements should be met:

For the large-scale production of antibodies, it is important to select rigid or at least highly cross-linked chromatographic matrices that can withstand high pressures and high flow rates. Large beads are generally specified to enable high flow rates to be achieved without the need for very high-pressure equipment. The purification process itself must be "robust," that is, insensitive to small alterations in the feedstream, the quality of the column packing, or the chromatographic operating conditions. In practice this means that processes relying on fraction collection or gradient elution to provide the required purity of the product should be avoided if possible.

Column geometry also needs careful consideration for production-scale chromatography. On a laboratory scale the aspect ratio (height to diameter) of a chromatography column is rarely given much thought, but as the scale of operation increases, it has a very significant impact on the throughput of the operation. For most types of chromatography (except size exclusion), short and fat columns are preferred for large-scale applications, since this gives rise to reduced processing time. The greater the surface area of the column, the greater the volumetric flow rate for a given linear flow rate. However, the chromatographic bed must not be so short that the antibody or impurities pass straight through the column without binding to the matrix. In practice, column diameter-to-height ratios are approximately 3–5 for minimum processing time without the risk of reduced recovery or purity.

5.6.11. Influence of Antibody Type

The term *antibody* can be used to cover a very wide variety of molecules from genetically engineered fragments of antibodies, with molecular weights of only a few thousand daltons, to large IgM molecules with molecular weights of 900,000 daltons. In addition to size, other characteristics such as charge, hydrophobicity, solubility, and stability can vary greatly from one type of antibody to another. Even antibodies of the same type and subtype, and from the same species, can have quite different properties. For example, mouse IgG_1 antibodies can have isoelectric points that range between 4.0 and 9.0; therefore purification processes generally need to be specifically tailored to each individual antibody.

Protein A and protein G affinity chromatography are very powerful purification techniques that have been used successfully on both laboratory and production scales for a variety of different antibody types. Most human IgG antibodies bind to protein A or G and many murine IgG molecules can be made to bind by manipulation of the binding conditions, such as pH and salt concentration. Other classes of antibodies and IgG antibodies from other species may not bind to protein A or G and these molecules may require different purification methods.

If affinity chromatography cannot be used, ion exchange chromatography is often the method of choice.

Some IgG_3 and IgM antibodies can be particularly difficult to purify due to their reduced solubility in low-conductivity solutions. Consequently it can be difficult to bind these antibodies to ion exchange matrices without the risk of precipitation. Since most IgM and human IgG_3 antibodies do not bind to protein A, there is often no easy alternative to ion exchange-based processes. Under these circumstances manipulation of pH or the addition of stabilizers may be used to increase the solubility of the antibody [58].

Antibody fragments are also generally purified by ion exchange processes [47]. Some fragments, despite their lack of an Fc moiety, still bind to protein A or G through a secondary binding site [59]. In these cases, the antibody fragments can be purified in the same way as an intact antibody.

5.6.12. Purification Process Validation

For the production of antibodies for clinical use, documented evidence is required to demonstrate that the production process performs as intended and that it will do so consistently. It must be demonstrated that the purification process will remove known impurities such as DNA derived from the cell line, and also potential contaminants such as viruses derived from host cell lines [60].

Validation data is normally gathered by analysis of full-scale manufacturing runs and also by "challenge" studies on scaled-down versions of the full-scale process. The detailed analysis of a small number of full-scale manufacturing runs can provide direct evidence that the purification process can consistently remove known impurities for which sensitive assays exist. For contaminants that do not normally enter the process stream, or where no sensitive assay exists, small-scale challenge studies can be performed. Laboratory-scale replicas of the full-scale process can be challenged with a high titer of contaminant (for example, virus or endotoxin spiked into the feedstream) and their clearance through each step of the process calculated.

A similar approach using both small-scale challenge studies and analysis of full-scale production runs must be used to demonstrate the effect of reuse of chromatography matrices. It is clearly desirable to reuse matrices to minimize column packing and unpacking operations and to reduce consumables costs, but documented evidence is required to demonstrate that antibody purity remains acceptable within the limits set for matrix reuse.

For the production of therapeutic-grade antibodies, process hygiene is a major issue of concern. While most purification operations are not claimed to be sterile operations, nonetheless a very low level of bioburden is required throughout the production process to ensure that, for example, bacterial proteases and endotoxins do not contaminate the product. Both the process itself and the equipment need to be designed to achieve a very high degree of cleanliness. For example, threaded connections where bacterial deposits could accumulate should be avoided and highly polished stainless steels should be used for vessels and piping to minimize bacterial adhesion.

Regular monitoring of raw materials, for example, water systems, and of both the manufacturing environment and the process stream is required to ensure that very high standards of hygiene are maintained.

5.7. PRODUCT CHARACTERIZATION METHODS

Monoclonal antibodies are a group of relatively complex proteins that share many similar structural characteristics but by their nature represent unique entities. The objective of product characterization is to monitor the integrity of the structure and demonstrate elements particular to each individual antibody. In addition the purity of the product is of utmost importance especially in therapeutic applications. A range of analytical methods are required to satisfy the above demands for monitoring the manufacturing process, as quality control tests on product, as stability measuring tests, and as support to process development. Guidelines for the quality control of monoclonal antibodies for therapeutic use are laid down by regulatory authorities (see references in Chapter 6).

Analytical techniques generally fall into two categories: those that are relatively simple to perform and interpret and those more complex techniques that provide quite detailed information. A summary of these methods is presented in Table 5.3. As a general guide the simpler set of assays can be used for routine testing of a manufacturing process, while the more complex are employed for structural characterization, for example, defining reference standards. The following sections describe each of the assays in more detail in terms of the methodologies involved and the information to be gained from their application. The tests will be described for the analysis of IgGs; however, the principles are generally applicable to all types of monoclonal antibody with minor technical changes to the method in some cases.

5.7.1. Product Assays for Routine Testing

Sodium Dodecyl Sulfate Polyacrylamide Gel Electrophoresis. The method for sodium dodecyl sulfate polyacrylamide gel electrophoresis (SDS-PAGE)

TABLE 5.3. Methods for Analysis of Monoclonal Antibodies

Method	Characteristic
SDS-PAGE	Purity, proteolysis, dissociation, cross-linking, contaminants
Isoelectric focusing	Charge isoforms, deamidation, deglycosylation, oxidation
Gel permeation HPLC	Aggregation, dissociation, proteolysis
Absorbance (280 nm)	Protein concentration
Amino acid analysis	Amino acid composition, extinction coefficient
N-terminal sequencing	Identity
Peptide mapping	Purity/identity—deamidation, substitution, oxidation, proteolysis, cross-linking, and primary structure
Mass spectrometry	Molecular mass (LDMS/ESMS) Primary structure (FAB-MS), glycosylation
Biophysical methods Nuclear magnetic resonance Circular dichroism Intrinsic fluorescence	Secondary and tertiary structure, unfolding, denaturation
Immunoassays	Antigen-binding activity, specific contaminants, isotyping
DNA hybridization	DNA contamination
In Vitro cell-based assay	Potency—antigen-binding activity, neutralization activity
Western blotting/antisera probes	Specific contaminants (host cell proteins), product degradation profile (antiproduct probes)
High performance ion exchange–pulsed amperometric detection	Carbohydrate profiling

described by Laemmli [61] is generally that applied to monoclonal antibodies as a test for purity. IgG antibodies are composed of two heavy (H) chain and two light (L) chain polypeptides held together by disulfide bonds to form the intact whole molecule. Analysis of samples in the presence and absence of a reducing agent allows detection of dissociation products, proteolytic fragments, covalently linked (nonthiol) aggregates and protein impurities.

Reduced samples typically give a banding profile consisting of two components; the H chain with a molecular weight of around 50 kDa and the L chain at around 25 kDa. Nonreduced samples should appear as a single band at around 150 kDa (Fig. 5.4). In practice posttranslational modification events, for example, glycosylation, can affect the apparent molecular weights of these species disproportionately and produce multiple banding patterns, particularly of nonreduced samples.

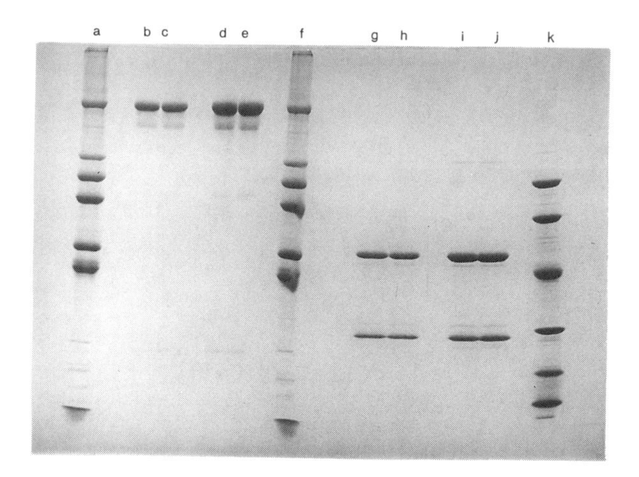

Fig. 5.4. Sodium dodecyl sulphate polyacrylamide gel electrophoresis of monoclonal antibodies. SDS-PAGE was performed under standard Laemmli conditions over a 5%–15% (w/v) polyacrylamide gradient. Samples of a humanized IgG_4 were analyzed in the reduced and nonreduced form at two separate loadings of 10 μg per lane. a,f: High-molecular-weight standards. b,c: 10-μg nonreduced sample. d,e: 25-μg nonreduced sample. g,h: 10-μg reduced sample. i,j: 25-μg reduced sample. k: Low-molecular-weight standards.

Two staining procedures are commonly used to visualize polypeptide bands on polyacrylamide gels. Coomassie blue staining has proved particularly useful for purity measurements of antibodies. Binding of the dye to the H and L chains of antibodies is very similar on a mass basis and therefore can be expected to be equivalent for aggregates or fragments arising from either subunit. In practice an increase in the measured amount of a degradation product is accompanied by a proportional decrease in the H and/or L chain bands, confirming this hypothesis. Purity estimates are usually determined by densitometry with either a red laser or a monochromatic light source of an optimal wavelength. The absorbance units recorded for the H and L chains are summed and expressed relative to the total absorbance units to derive a purity value for the intact antibody. It is important to confirm the linear working range and limit of detection of the assay. Accuracy can be evaluated by spiking with potential protein contaminants over a suitable range (e.g., 0.5%–5.0%, w/w). The limit of detection for Coomassie blue staining is around 50 ng per stained protein band. The approximate limits of linearity of stain for monoclonal antibodies is 25 μg and 10 μg for samples analyzed under reducing and nonreducing conditions, respectively. This equates to limits of detection for contaminant bands of 0.2% (w/w) and 0.5% (w/w) under the respective conditions.

Silver staining of gels is used to increase sensitivity of detection. Amounts of less than 1 ng of a protein band can be detected, equivalent to less than 0.004% (w/w) of a 25-μg antibody load. However, the efficiency of stain between gels and the staining intensity of different proteins can vary markedly. Therefore the method is not appropriate as an accurate estimate of purity. Individual proteins do stain in a proportional way over a defined mass range. Therefore the levels of a particular process contaminant can be determined by applying the relevant standards and controls to each gel.

The main advantages of SDS-PAGE are that it is a well proven technique that is simple to set up and is relatively inexpensive. Gels can be modified to provide optimum resolution (± 0.5%–1 kDa) over a desired molecular weight range. However, the limits of detection and range of linearity of available staining techniques are inferior to those of other techniques. A major problem regarding specificity of the method is the detection of contaminants that have the same molecular size as heavy and light chains and hence comigrate.

Gel Permeation HPLC. Gel permeation (or size exclusion) chromatography is employed as a nondestructive method for monitoring aggregates of native monoclonal antibodies that result from covalent or noncovalent interactions. The chosen method should be able to resolve a dimer of the antibody from the monomer and more highly aggregated species. Dissociation and fragmentation products may also be measured, although the level of resolution achieved between these species is very much inferior to that of SDS-PAGE.

Elution of proteins from gel permeation columns is normally monitored with conventional spectrophotometric detectors set at selected wavelengths (e.g., 215 nm and 280 nm). The linear range for monomeric antibody response is around 0.1–500 μg at 280 nm. Aggregates have a response factor similar to that of their respective monomers; therefore, using a 250-μg load of total antibody, the limit of detection of aggregates is around 0.04% (w/w). In practice a limit of detection of around 0.1% is acceptable.

The major problem encountered with this method is interaction of the antibody molecules with the column matrix, resulting in nonideal chromatography that is observed as peak broadening and elution at apparently lower molecular sizes. Resolution of monomer from other species is greatly affected under such conditions. Interactions occur primarily through two mechanisms involving hydrophobic and/or ionic forces. Hydrophobic interactions can be minimized by the inclusion of small amounts of detergents in the mobile phase, while ionic interactions are principally controlled by altering the pH or ionic strength. Due to the weakly acidic nature of many column support materials basic-hydrophobic antibodies present the main problems. The addition of a charged (basic) chaotropic agent can help minimize both types of forces. In this respect amino acids have been found to be most suitable (e.g., lysine and arginine).

The performance of the assay operated on a routine basis should be checked with calibrated sets of appropriate molecular weight standards. A resolution factor for two adjacent standards should be calculated and remain within predefined limits as a control of assay performance. A number of columns are available commercially. Those designed for high-pressure systems are used most frequently for maximum speed of sample processing. To date the two most commonly used columns for IgG analysis are the Zorbax GF-250 (Dupont) and the TSKGW3000SWXL (Toya soda). Recently the Phenomenex G3000 has appeared and offers at least equivalent performance.

Isoelectric Focusing. Isoelectric focusing (IEF) is a technique that offers high resolution of different charged isoforms of proteins and it is universally used to characterize antibodies. Under optimal conditions it is possible to resolve two species with a net charge (pI) difference equivalent to 0.01 pH units.

There are two main approaches to analytical gel-based systems for IEF that involve the use of either carrier ampholytes or immobilines to form pH gradients. The gel support matrix is composed of either polyacrylamide or agarose. Both gel supports have provided good results for antibodies on a routine basis; however, in the authors' experience problematical antibodies (hydrophobic, precipitating) have focused more readily on the agarose gels. Gels are available that cover broad pH gradients (pH 3–10) and narrow gradients (e.g., pH 3–7, pH 5.5–8.5, pH 8–10) for specific uses. The pH gradient can be altered by addition of extra ampholytes of the required pH range in order to optimize resolution. Exogenous protein stabilizers (detergents, urea) can also be added to minimize interaction of the antibody with the gel support and, for example, to prevent precipitation. Immobiline gels are more difficult to prepare and require longer run times (6–18 h). Suitable precast gels are not readily available from commercial sources. However, they have the advantage of more stable pH gradients and therefore can potentially provide better resolution of isoforms, as gradient profiles can be precisely modified to fit requirements.

Antibodies tend to resolve as a number of bands and this microheterogeneity can be caused by a variety of chemical changes, leading to a net charge difference (e.g., oxidation, deamidation, altered glycosylation). Banding patterns can vary according to the expression system and manufacturing process, but they are generally characteristic for a specific antibody and therefore are widely used for identification purposes. Figure 5.5 shows the results obtained for three complementarity-determining region-grafted human–mouse IgG hybrids. Each contains identical human constant regions (>90% of amino acid residues) but different mouse variable regions. Although there is a relatively high degree of homology between the antibodies, their banding patterns are quite distinct. Four separate batches of one of the antibodies were assayed. Three represent different development-scale process batches, while the fourth is from a full-scale manufacturing run. The banding pattern obtained for all four batches is very consistent and demonstrates the level of results that should be achieved in taking a product from development into manufacturing.

Potency Assays. Potency assays are required to allow the determination of specific activity of the antibody. The potency of antibodies can normally be defined as their ability to bind their respective antigens, although this does not always confirm *in vivo* biological activity.

Cell-based *in vitro* assays are used frequently to assess antibody binding to antigen. They take many forms, including (1) monitoring the direct effect of an antibody that binds to cell surface antigen on cell function or growth; and (2) the neutralization of the effect of antigen on cells. They have the advantage of being more relevant to the *in vivo* situation than non-cell-based assays but generally suffer from lower precision.

Enzyme-linked immunosorbent assays (ELISAs) and radioimmunoassays (RIAs) are the preferred option for obtaining high-precision, reliable, and rapid evaluation

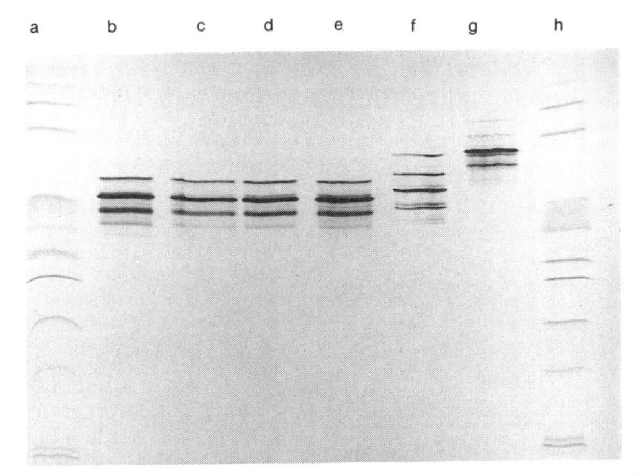

Fig. 5.5. Isoelectric focusing of monoclonal antibodies. Isoelectric focusing of three different humanized IgG_4 monoclonal antibodies on agarose preformed gels (pH 3.5–9.5, Isogel). b–e: Four separate purified batches of antibody A. b: 8L fermentation. c: 100L fermentation. d: 100L fermentation. e: 200L fermentation. f,g: Two other antibodies purified from 1,001 fermentations of each. f: Antibody B. g: Antibody C. a,h: The pI markers.

of antibody-binding activity and consequently potency. Assay formats can be adapted to determined affinity constants, binding activity relative to a standard, or total antigen-binding units. For routinely monitoring production it is recommended that the relative binding activity of a particular batch be compared with a reference standard that has been characterized by determining its affinity constant as well as its performance in relevant cell-based assays. An ELISA competitive format is the most suitable for this comparison.

There are alternatives to the typical immunoassay approaches described above. They generally involve separation of the antibody–antigen complex from the individual unbound species and quantitative measurement of one of these species. This may be achieved by chromatographic (GPC, HIC, IEC) or electrophoresis (IEF, native gels) techniques. To employ this approach it is essential that the association constant be much greater ($>10^3$) than the dissociation constant for the antibody–antigen complex. GPC appears to be the most useful technique [62], as it involves minimal interaction of the complex with the matrix, can be performed in a relative short time ($<$10 min), and therefore is likely to cause the least disruption. The precision of this technique is much better (% CV $<$2%) than for immunoassays (% CV 2–10%). One of the main disadvantages is that it requires increased amounts of reagents, that is micrograms rather than nanograms.

Concentration. The concentration of purified antibodies can be determined from their absorbance at 280 nm against an extinction coefficient; generally an $E^{1\%}$ of 14.2 has been applied to IgGs [63]. However, this value does vary according to amino acid composition and therefore for accurate determinations it should be calculated for each individual product by amino acid analysis to quantify antibody concentrations. A less accurate method is to calculate the theoretical extinction from the amino acid sequence by the method of Gill and Von Hippel [64].

Isotyping. The subclass of an antibody is an important factor influencing biological activity. Anti-species/subclass antibody reagents are used to confirm identify of product. The Ouchterlony double-diffusion technique is generally used as the assay format. Precipitate bands are recorded and used as a means of identification by reference to appropriate controls in the same assay. The test can be used to support the overall identification of the product as part of a quality control specification.

5.7.2. Process Contaminant Assays

Protein Contaminants. An overall estimate of product purity is usually determined by SDS-PAGE. However, it is necessary to determine the levels of known potential process impurities to lower amounts than those detectable by this general assay. Specific assays are needed that are able to selectively quantify these contaminants at trace levels (nanogram contaminant per milligram MAb). Immunoassays are ideally suited to this purpose.

The protein contaminants of interest are those present in the culture medium (serum proteins or serum-derived proteins such as albumin and transferrin), purification ligands (e.g., protein A), or proteins derived from the host cell. Due to their complexity host cell proteins are most effectively analyzed following an initial fractionation stage such as SDS-PAGE; they are addressed in the section below, which covers western blotting.

ELISAs and RIAs have been employed as assay formats. A limit of detection of around 1 ppm (1 ng/mg antibody) should bc achieved. The assay should be able to detect proteolytic fragments as well as the intact protein contaminant, and therefore the use of polyclonal as opposed to monoclonal antibody reagents may be more suitable. If the contaminants includc serum protein, care should be taken to minimize cross-reactivity with similar proteins present in the polyclonal antisera reagents. Each monoclonal antibody product represents a unique structure with the potential to react nonspecifically with reagents in the assay. Therefore each product should be independently evaluated for interference and the assay should be optimized as necessary.

A problem particular to antibodies is the assay of affinity ligands such as protein A, where they are used as part of a purification scheme. Protein A binds very strongly to the Fc region of IgG molecules and thus has the potential to bind nonspecifically to the assay reagents as well as the product. The challenge for the analyst is to minimize these interactions while maintaining the binding activity of the assay reagents for protein A. This can be achieved by careful optimization of assay solutions (e.g., pH, carrier protein, detergents, and ionic strength) but also by selection of antisera (e.g., chicken anti-protein A does not have a protein A-binding Fc region [65]).

Nonprotein Contaminants. DNA derived from host cells is an important nonprotein contaminant from a regulatory point of view. In products destined for therapeutic use the levels of DNA per milligram MAb are usually demonstrated to be in the low picogram range. This has represented a great challenge to the analyst to develop methods to quantify DNA in the presence of a large excess of antibody protein. The employment of highly sensitive and specific DNA hybridization techniques has formed the main approach to analysis [66].

Protein is generally removed from test samples through digestion with proteinase K, followed by a series of phenol and solvent extractions. DNA is then denatured, immobilized to a membrane, and hybridized with a radiolabeled probe consisting of complementary sequences of the host cell DNA. The amount of radioactivity determined by autoradiography is proportional to the concentration of DNA. A fully optimized assay is capable of detecting around 0.1 pg DNA per 1 mg antibody.

More recently a semiautomated system for measuring DNA contamination has been developed. The "Threshold" system is based on the use of anti-DNA proteins and silicon microchip technology [67]. The limit of detection for the assay is similar to that obtained by the hybridization assay and precision appears to be improved.

The introduction of the polymerase chain reaction (PCR) has provided the ability to develop assays for determining levels of specific sequences of DNA in subpicogram amounts. The technique is most usefully applied to the detection of DNA sequences that represent a particular safety concern, for example, specific viral sequences.

Western Blotting. Western blotting offers a relatively simple method for characterizing proteins separated by gel electrophoresis. It is particularly useful for monoclonal antibodies as the first step to determining the origin of contaminant bands observed by SDS-PAGE. Identification is achieved by probing the blotted proteins with specific antibody reagents.

Anti-species IgG antibodies are readily available from commercial sources. They may be against the whole antibody or particular sections, for example H or L chains. Thoughtful use of the appropriate probes can lead to assignment of the origin of a band to a certain section or sections of the product antibody structure. The specificity and degree of cross-reactivity of each probe should be confirmed. The most important group of non-product-related bands are host cell proteins (HCPs). HCPs represent a mixture of proteins with a wide range of physicochemical properties and therefore a risk of copurification with the product. Polyclonal antisera are raised to antigens prepared from a cell line containing the same genetic material as the product line and grown under the same conditions. In the case of recombinant lines the cell is transfected with the gene construct that lacks the structural gene for the product, while for hybridomas it is often possible to use a similar nonproducing hybrid or the parent myeloma line. The protein composition of the "gene-minus" cell line is compared with the product line by using 2-dimensional gel electrophoresis. Banding patterns should demonstrate $> 90\%$ homology, the product-associated bands having been excluded. It is recommended that probes be raised in two separate species to maximize the range of response against the multicomponent antigens.

Following western blotting and incubation with the primary antibodies (anti-IgG, anti-HCPs) bands are visualized with a suitable anti-species reagent conjugate (e.g., anti-sheep IgG–horseradish peroxidase) and substrate system (e.g., tetramethylbenzidine). Nonspecific cross-reactions with the secondary antibodies must be minimized during assay development. Stained blots are compared with silver-stained gels to assign a label to each band observed by SDS-PAGE. Membranes can be clarified and the visualized bands quantified by densitometry against an appropriate standard.

5.7.3. Further Characterization

Primary Structure. Primary structure of monoclonal antibodies may be investigated by standard protein chemistry techniques, for example, N-terminal sequencing. The H and L chains consist of around 450 and 220 amino acid residues, respectively. Conventional N-terminal sequencing by the Edman process allows the elucidation of the first 20–30 residues. The method cannot be used to evaluate the complete primary structure of monoclonal antibodies but can be used for product identification purposes in support of other routine assays (e.g., potency, IEF). A problem with IgG molecules is that the N-terminal residue of the H chain is often blocked due to the cyclization of a glutamate residue to pyroglutamate, which is not hydrolyzed by the standard Edman reaction. However, this residue can be removed with a specific enzyme, pyroglutamate amino peptidase [68], to enable sequencing to begin on the adjacent residue.

The full primary structure can be confirmed following proteolytic digestion of the antibody and separation of the resulting peptides prior to N-terminal sequencing. This is by no means a simple task for a protein the size of a monoclonal antibody, but it has been reported for an IgG [69]. Once a peptide map has been established, the integrity of future batches of product can be verified against the fingerprint pattern for a given antibody.

Harris et al. [70] demonstrated the value of peptide mapping as a sensitive tool for assessing genetic heterogeneity in Chinese hamster ovary cell lines producing genetically engineered antibody. They were able to detect a variant IgG in which tyrosine was replaced by glutamine at residue 376 in the heavy chain. The variant was present at a level of 1%–27% of total IgG, depending on cell generation number. The authors point out that it is unlikely that nucleic acid analyses would have detected this variant.

Mass spectrophotometry has been used in support of conventional N-terminal sequencing to determine the structure of proteolytic fragments either following isolation or directly interfaced to chromatographic equipment (e.g., LC-MS). This may represent a more practical way of elucidating the primary structure of these types of proteins. The mass of a whole IgG molecule can be determined by laser desorption (LD) mass spectrophotometry (or time of flight MS) and electrospray mass spectrometry (ESMS). The precision of these measurements is of the order of $\pm$ 0.01%, which represents an uncertainty of $\pm$ 15 atomic mass units. The system cannot therefore be used at present to confirm the predicted amino acid composition of IgG molecules.

Carbohydrate Content. Carbohydrate constitutes a variable amount of total antibody mass ranging from around 1% to 10% (w/w). Glycosylation may affect the biological properties of a product and may be altered by changes in the production process [32,71]. It is useful therefore to have a measure of consistency of glycosylation, although it is not usually necessary to define absolute carbohydrate structures.

During the advanced stages of development of a therapeutic antibody it is usual to measure the relative proportion of each monosaccharide unit, the sialic acid content, number of glycosylation sites, and a fingerprint pattern of the oligosaccharide units. This information can be obtained through the application of various conventional types of chromatographic and/or electrophoretic techniques on carbohydrate released from the antibody.

Charge Isoforms. Ion exchange chromatography (IEC) and chromatofocusing (CF) have been used for separating different charge isoforms of monoclonal antibodies [72]. Resolution is inferior to that achieved by IEF but quantitation is more straightforward as the eluent can be directly monitored by a spectrophotometer and the signal can be processed through a conventional integration system.

Capillary electrophoresis (CE) is a rapidly advancing technique that offers resolution similar to that of IEF but with the advantage of shorter times for analysis and direct quantitation of isoforms. A limited amount of data are available on the separation of monoclonal antibodies. However, reports to date [73] indicate the potential this technique has for product characterization.

Hydrophobic Interactions. Evaluation of the hydrophobic nature of the product can provide information on the distribution of such areas on the surface of the protein and detect differences that may occur as a result of changes in conformation. Reverse-phase (RPC) and hydrophobic interaction chromatography (HIC) are the most useful modes of separation. Analysis of the intact product by RPC is often unsatisfactory. Binding to available matrices is fairly strong and requires high organic solvent levels for elution of the protein, which can lead to denaturation and precipitation. The isolated H or L chain subunits are of a more suitable size for direct analysis by RPC. Analysis of these subunits should allow more specific assignment of observed changes.

HIC is more easily applied to the whole antibody molecule. Elution of the protein is under more favorable solution conditions. Generally the level of resolution obtained from these techniques is inferior to that of other methods (IEF, SDS-PAGE, peptide mapping).

Measurement of Secondary/Tertiary Structure. Optical spectrometry techniques have been most commonly used to determine the relative amounts of the different elements of secondary and tertiary structure present in monoclonal antibodies. Measurements of circular dichroism (CD), optical rotary dispersion (ORD), and native fluorescence have been applied to estimate, for example, helix, B-sheet, and random coil content. The precision and accuracy of such techniques is around $\pm$ 2%–4% per structure and therefore is useful only for reference purposes.

Nuclear magnetic resonance (NMR) can also give useful information. Spectra produced for individual antibodies are often very reproducible, but the interpretation of protein structure is less straightforward for molecules of the size of monoclonal antibodies. Nevertheless this approach represents a potentially highly sensitive assay for product characterization. When NMR is employed in a three-dimensional mode, the relative orientation of each amino acid residue within the whole protein can be determine.

5.7.4. Validation of Analytical Methods

Particular attention should be given to demonstrating that the analytical methods applied to monitoring product quality, to product characterization, and to stability-indicating assays are sufficiently reliable. A series of predefined experiments should be performed to validate a technique for its intended application. The degree of validation will depend on the stage of development of the product and scope of the method. For example, it may be enough to define the appropriate assay performance parameters, such as specificity, accuracy, and precision, of a method used to support a product in early clinical trials (phase I/II), while more extensive validation, especially regarding robustness, is required for an assay used to monitor the quality of a product for later clinical trials (phase III) or for the market.

5.8. FORMULATION OF MONOCLONAL ANTIBODY PRODUCTS

Despite their complex nature monoclonal antibodies are relatively stable structures if stored under appropriate conditions. Formulations are designed to minimize the unstable features of a given antibody. The types of factors involved are summarized in Table 5.4 and should be optimized in each specific case.

Solubility and aggregation can normally be controlled by the same factors.

TABLE 5.4. Design of a Solution-Based Formulation

Characteristic	Control Factor
Solubility	Buffer pH at ≥1pH unit from pI Salt concentration/ionic strength Type of buffering agent (e.g., amino acids) Detergents/surfactants (e.g., Tweens)
Aggregation	As above
Oxidation	Antioxidants (e.g., ascorbic acid, butyrate hydroxy toluene)
Deamidation	pH of solution (e.g., pH ≥ 7)
Proteolysis	pH of solution Protease inhibitors (e.g., EDTA)

However, they may not be directly related and, for example, precipitation of an antibody product can be observed without a concomitant increase in soluble aggregate levels. Therefore each characteristic of the product should be monitored independently.

Oxidation of susceptible amino acid residues is not a common problem with monoclonal antibodies but it can occur. Rates of reaction at pH values $\leq$ pH 7 are relatively slow for the modifications associated with proteins (e.g., oxidation of methionine, cysteine, and trytophan). However, if a particular antibody is sensitive to oxidation, then degradation can be controlled by the addition of antioxidants to the formulation.

Deamidation is a commonly observed degradation process for many proteins, including monoclonal antibodies. The rate of deamidation is directly related to the pH of the solution and storage temperature of the product. The simplest way to control the reaction is to maintain pH below neutrality in the formulation.

Proteolysis is another degradative process that affects monoclonal antibodies. It can occur through either enzymic catalyzed reactions or chemical mechanisms. Proteolysis is minimized by choice of pH and buffering agent.

Antibodies are usually formulated in solution or less frequently as freeze-dried preparations. The development of a freeze-dried formulation can produce a product with an extended shelf life compared with a solution and most importantly can facilitate storage at ambient temperature. However, freeze-drying of proteins is a complex science and can require extensive development time. For further information in the field of freeze-drying proteins, the reader is directed to the review of Franks [74]. Product integrity during storage under different conditions can be followed by analytical methods directed at measuring important structural aspects of the proteins without needing to define absolute composition.

5.8.1. Stability-Indicating Assays

Stability-indicating assays are selected on the basis of their ability to detect small changes in antibody structure. The most useful techniques applied to measuring the stability of antibodies are SDS-PAGE, IEF, GP-HPLC, peptide mapping, and a relevant activity assay (ELISA).

Additional stability tests may be directed at aspects of protein structure not monitored as part of routine product quality testing. An example is the use of selected techniques to follow changes in secondary and tertiary structure (fluorescence, CD, HIC, IEC), which may reflect unfolding and exposure of internal regions of the molecule to the external aqueous environment, preceding aggregation of the antibody. An essential aspect of stability studies is the characterization of observed degradation products. This facilitates the definition of degradation processes and provides a basis for further improvement of formulations.

Applying this relatively simple approach to selecting a formulation should provide a shelf life in excess of 12 months for a given antibody stored under standard refrigeration conditions (2°–8°C).

5.8.2. Design and Interpretation of Stability Studies

A large number of diverse formulations are screened by assessing the antibody following storage at elevated temperatures over a relatively short period of time (2–4 weeks), in a selected range of the stability-indicating assays. It is important to monitor the control temperature samples (2°–8°C) as certain effects (e.g., aggregation) may be less extensive at the elevated temperature. Lead formulation(s) are chosen from these studies and the stability is monitored over the intended shelf life of the product (typically more than 12 months). Some of the changes observed in the stability-indicating assays are often a function of more than a single reaction mechanism. The rates of each of these mechanisms may not be directly related to storage temperature; therefore accelerated stability data obtained at elevated temperatures (15°, 20°, 30°, 40°C) cannot routinely be manipulated by the Arrhenius relationship to estimate a shelf life at the lower recommended storage temperature (typically 2°–8°C). However, in some cases the rate of change observed in a particular test method does correlate with zero- or first-order kinetics, and subsequent linear plots of change in product with time can be produced and used to provide an estimate of shelf life at lower temperatures. IEF principally detects changes due to deamidation of exposed asparagine and glutamine residues during the early stages of temperature-stressing of an antibody. This single process produces a shift in the product IEF banding pattern that is quantified by densitometry. A plot of the percentage change in product profile with time for an antibody at various temperatures is shown in Figure 5.6. This can be compared with a similar plot of intact monomer

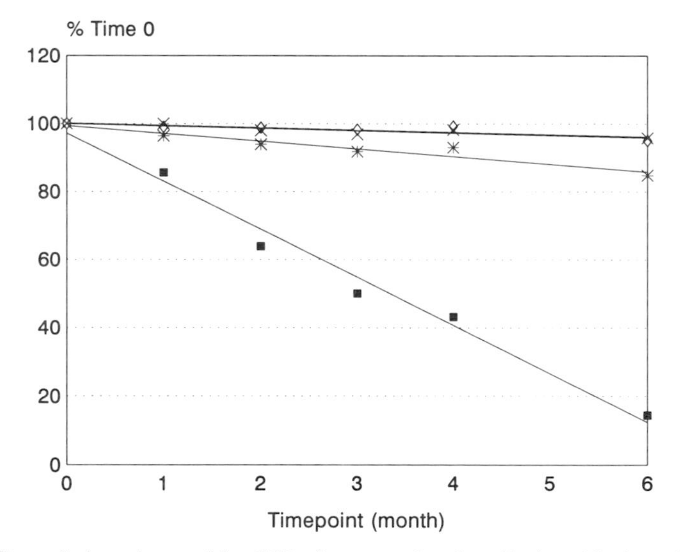

Fig. 5.6. Degradation observed by IEF of a monoclonal antibody with time. The relative amount of intact antibody (percentage of amount at time 0) was measured by IEF for antibody A following storage at four temperatures: −20°C (◇), +4°C (×), +23°C (∗) and +37°C (■), for up to 6 months.

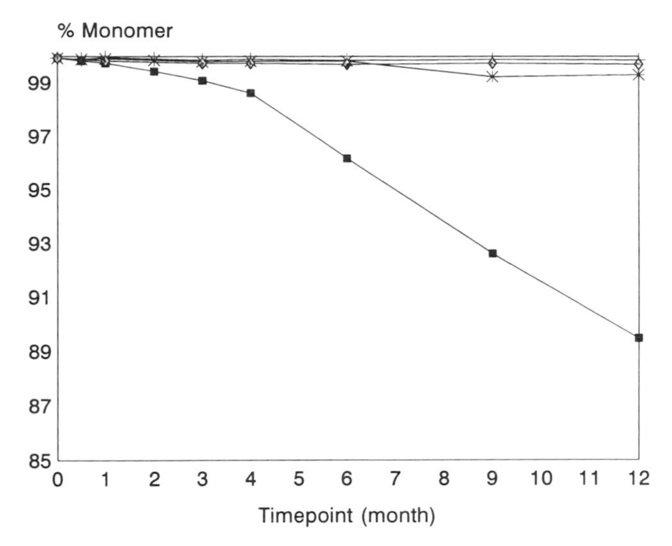

Fig. 5.7. Degradation observed by gel permeation chromatography (GPC) of a monoclonal antibody with time. The relative amount of intact antibody (percentage of monomer) was measured for antibody A following storage at four temperatures: −20°C (◇), +4°C (×), +23°C (*) and +37°C (■), for up to 12 months.

levels for the same antibody in Figure 5.7. In this latter case, there is no obvious order of kinetics for the rate of degradation, and it is therefore most likely a function of a number of effects.

The results from each analytical test can be manipulated in a similar manner to estimate a shelf life for an antibody and the data can be used to support real-time stability data for a product.

5.9. REFERENCES

1. Code of Federal Regulations, Title 21, Part 200–211 (1988).
2. Rules Governing Medicinal Products in the European Community, Vol. 4: Good Manufacturing Practice for Medicinal Products, 1992.
3. Code of Federal Regulations, Title 21, Part 600 (1988).
4. Bader, F.G., et al. Multi-use Manufacturing Facilities for Biologicals Biopharm, pp. 32–40 (Sept. 1992).
5. Lambert, K.J. Problems associated with large-scale manufacture of biotechnology products. BIRA J. 9, 12–14 (1991).
6. Galfré, G., Milstein, C. Preparation of monoclonal antibodies: Strategies and procedures. Methods Enzymol, 73, part B, Immunochemical Techniques, pp. 3–46 (1981).
7. Bazin, H. Production of rat monoclonal antibodies with the LOU rat non-secreting 1R983F myeloma cell line. In Peeters, H.J. (ed.): Protides of the Biological Fluids. Oxford: Pergamon, 1982.

8. Boyd, J.E., James, K. Human monoclonal antibodies: Their potential problems and prospects. In Mizrahi, A. (ed.): Monoclonal Antibodies: Production and Application (pp. 1–43). New York: Alan R. Liss, 1989.
9. Bebbington, C.R. Expression of antibody genes in nonlymphoid mammalian cells. In Methods: A Companion to Methods in Enzymology (Vol. 2, pp. 136–145), 1991.
10. Fouser, L.A., Swanberg, S.L., Lin, -B-Y., Benedict, M., Kelleher, K., Cumming, D.A., Riedel, G.E. High level expression of a chimeric anti-ganglioside GD2 antibody: Genomic kappa sequences improve expression in COS and CHO cells. BioTechnology 10, 1121–1127 (1992).
11. Page, M.J., Sydenham, M.A. High level expression of the monoclonal antibody Campath-1H in chinese hamster ovary cells. BioTechnology 9, 64–68 (1991).
12. Bebbington, C.R., Renner, G., Thomson, S., King, D., Abrams, D., Yarranton, G.T. High level expression of a recombinant antibody from myeloma cells using a glutamine synthetase gene as an amplifiable selectable marker. BioTechnology 10, 169–175 (1992).
13. Gillies, S.D., Dorai, H., Wesolowski, J., Majean, G., Yonag, D., Boyd, J., Gardner, J., James, K. Expression of human anti-tetanus toxoid antibody in transfected murine myeloma cells. BioTechnology 7, 799–804 (1989).
14. Brown, M.E., Renner, G., Field, R.P., Hassell, T. Process development for the production of recombinant antibodies using the glutamine synthetase (GS) system. Cytotechnology 9, 231–236 (1992).
15. Birch, J.R., Lambert, K., Thompson, P.W., Kenney, A.C., Wood, L.A. Antibody production with airlift fermenters. In Lydersen, B.K. (ed.): Large Scale Cell Culture Technology (pp. 1–20). New York: Hansers, 1987.
16. Lewis, A.P., Parry, N., Peakman, T.C., Scott-Crowe, J. Rescue and expression of human immunoglobulin genes to generate functional human monoclonal antibodies. Hum. Antibodies Hybridomas 3, 146–152 (1992).
17. Reis, V., Blum, B., Specht, B.-U. von, Domdey, H., Collins, J. Antibody production in silkworms cells and silkworm larvae infected with a dual recombinant Bombyx mori nuclear polyhedrosis virus. BioTechnology 10, 910–912 (1992).
18. Putlitz, J. zu, Kubasek, W.L., Duchêne, M., Marget, M., Specht, B.-U. von, Domdey, H. Antibody production in baculovirus-infected insect cells. BioTechnology 8, 651–654 (1990).
19. Wiebe, M.E., May, L.H. Cell banking. In Lubiniecki, A.S. (ed.): Large-Scale Mammalian Cell Culture Technology. New York: Marcel Dekker, 1990.
20. Reuveny, S., Lazar, A. Equipment and procedures for production of monoclonal antibodies in culture. In Mizrahi, A. (ed.): Monoclonal Antibodies: Production and Application. New York: Alan R. Liss, 1989.
21. Fazekas de St. Groth, G. Automated production of monoclonal antibodies in a cytostat. J. Immunol. Methods 57, 121–136.
22. Shevitz, J., Reuveny, S., La Porte, T.L., Cho, G.H. Stirred tank perfusion reactors for cell propagation and monoclonal antibody production. In Mizrahi, A. (ed.): Monoclonal Antibodies: Production and Application. New York: Alan R. Liss, 1989.
23. Tyler, J.E. Microencapsulation of mammalian cells. In Lubiniecki, A.S. (ed.): Large-Scale Mammalian Cell Culture Technology. New York, Basel: Marcel Dekker, 1990.
24. Sinacore, M.S., Creswick, B.C., Buehler, R. Entrapment and growth of murine hybridoma cells in calcium alginate gel microbeads. BioTechnology 7, 1275–1279 (1989).

25. Hulscher, M., Scheibler, V., Onken, U. Selective recycle of viable animal cells by coupling of airlift reactor and cell settler. Biotechnol. Bioeng. 39, 442–446 (1992).
26. Runstadler, P.W., Tung, A.S., Hayman, E.G., Ray, N.G., Sample, J.G., DeLucia, D.E. Continuous culture with macroporous matrix, fluidized bed systems. In Lubiniecki, A.S. (ed.): Large-Scale Mammalian Cell Culture Technology. New York, Basel: Marcel Dekker, 1990.
27. Applegate, M.A., Stephanopoulos, G. Development of a single-pass ceramic matrix bioreactor for large-scale mammalian cell culture. Biotechnol. Bioeng. 40, 1056–1068 (1992).
28. Altshuler, G.L., Dziewulski, D.M., Sowek, J.A., Belfort, G. Continuous hybridoma growth and monoclonal antibody production in hollow fiber reactors–separators. Biotechnol. Bioeng. 28, 646–658 (1986).
29. Mizrahi, A. Monoclonal Antibodies: Production and Application. New York: Alan R. Liss, 1989.
30. Lubiniecki, A.S. Large-Scale Mammalian Cell Culture Technology. New York, Basel: Marcel Dekker, 1990.
31. Goochee, C.F., Monica, T. Environmental effects on protein glycosylation. BioTechnology 8, 421–427 (1990).
32. Maiorella, B.L., Winkelhake, J., Young, J., Moyer, B., Bauer, R., Hora, M., Andya, J., Thomson, J., Patel, T., Parekh, R. Effect of culture conditions on IgM antibody structure, pharmacokinetics and activity. Biotechnology 11, 387–392 (1993).
33. Murakami, H. Serum-free media used for cultivation of hybridomas. In Mizrahi, A. (ed.): Monoclonal Antibodies: Production and Application. New York: Alan R. Liss, 1989.
34. Kovar, J., Franek, F. Iron compounds at high concentrations enable hybridoma growth in a protein-free medium. Biotechnol. Lett. 9, 259–264 (1987).
35. Schneider, Y-J. Optimisation of hybridoma cell growth and monoclonal antibody secretion in a chemically defined serum and protein-free culture medium. J. Immunol. Methods 116, 65–77 (1989).
36. Darfler, F.J. A protein free medium for the growth of hybridomas and other cells of the immune system. In Vitro Cell. Dev. Biol. 26, 769–778 (1990).
37. Bell, S.L., Bushell, M.E., Scott, M.F., Wardell, J.N., Spier, R.E., Sanders, P.G. Genetic modification of hybridoma glutamine metabolism: Physicochemical consequences. In Spier, R.E., Griffiths, J.B., Macdonald, C. (eds.): Animal Cell Technology: Developments, Processes and Products. Stoneham, MA: Butterworth-Heinemann, 1992.
38. Carter, P., Kelley, R.F., Rodrigues, M.L., Snedecor, B., Covarrubias, M., Velligan, M.D., Wong, W.L.T., Rowland, A.M., Kotts, C.E., Carver, M.E., Yang, M., Bourell, J.H., Shepard, H.M., Henner, D. High level Escherichia coli expression and production of a bivalent humanized antibody fragment. BioTechnology 10, 163–167 (1992).
39. Buchner, J., Rudolph, R. Renaturation, purification and characterisation of recombinant F_{ab}-fragments produced in Escherichia coli. BioTechnology 9, 157–162 (1991).
40. Davis, G.T., Bedzyk, W.D., Voss, E.W., Jacobs, T.W. Single chain antibody (SCA) encoding genes: One-step construction and expression in eukaryotic cells. BioTechnology 9, 165–169 (1991).
41. Hiatt, A., Caffertey, R., Bowdish, K. Production of antibodies in transgenic plants. Nature 342, 76–78 (1989).

42. Garg, V., Costello, M., Czuba, B. Purification and production of therapeutic grade proteins. In Seetharam, R., Sharma, S. (eds.): Purification and Analysis of Recombinant Proteins. New York: Marcel Dekker, 1991.

43. Ransohoff, T., Levine, H. Purification of monoclonal antibodies. In Seetharam, R., Sharma, S. (eds.): Purification and Analysis of Recombinant Proteins. New York: Marcel Dekker, 1991.

44. Brunner, K.H., Hemfort, H. Centrifugal separation in biotechnological processes. Adv. Biotechnol. Processes 8, 1–50 (1988).

45. Harris, E.L.V., Angal, S. (eds.). Protein Purification Methods, a Practical Approach. New York: Oxford University Press, 1989.

46. Lee, S., Gustafson, M., Pickle, D., Flickinger, M., Muschik, G., Morgan, A. Large-scale purification of a murine antimelanoma monoclonal antibody. J. Biotechnol. 4, 189–204 (1986).

47. Gavit, G., Walker, M., Wheeler, T., Bui, P., Lei, S., Weickmann, J. Purification of a mouse-human chimeric Fab secreted from E. coli. Biopharm, pp. 28–34 (Jan/Feb, 1992).

48. Scott, R., Duffy, S., Moellering, B., Prior, C. Purification of monoclonal antibodies from large scale mammalian cell culture perfusion systems. Biotechnol. Prog. 3, 49–56 (1987).

49. Jehanli, A., Hough, D. A rapid procedure for the isolation of human IgM myeloma proteins. J. Immunol. Methods 44, 199–204 (1981).

50. Goheen, S., Matson, R. Purification of human serum gamma globulins by hydrophobic interaction high performance liquid chromatography. J. Chromatogr. 326, 235–241 (1985).

51. Bukovsky, J., Kennett, R. Simple and rapid purification of monoclonal antibodies from cell culture supernatants and ascites fluids by hydroxylapatite chromatography on analytical and preparative scales. Hybridoma 6, 219–228 (1987).

52. Stanker, L.H., Vanderlaan, M., Juarex-Salinas, H. One step purification of mouse monoclonal antibodies from ascites fluid by hydroxylapatite chromatography. J. Immunol. Methods 76, 157–169 (1985).

53. Nau, D.R. ABx: A novel chromatography matrix for the purification of antibodies. In Seaver, S. (ed.): Commercial Production of Monoclonal Antibodies—A Guide for Scale-Up. New York: Marcel Dekker, 1987.

54. Temponi, M., Kageshita, T., Perosa, F., Ono, R., Okada, H., Ferrone, S. Purification of murine IgG monoclonal antibodies by precipitation with caprylic acid: Comparison with other methods of purification. Hybridoma 8, 85–95 (1989).

55. Kenney, A.C. Large scale purification of monoclonal antibodies. In Mizrahi, A. (ed.): Monoclonal Antibodies: Production and Application (pp. 143–160). New York: Alan R. Liss, 1989.

56. Karlsson, E., Ryden, L., Brewer, J. Ion exchange chromatography. In Janson, J.C., Ryden, L. (eds.): Protein Purification: Principles, High Resolution Methods and Applications. New York: VCH, 1989.

57. Francis, R., Bonnerjea, J., Hill, C. Validation of the re-use of protein A Sepharose for the purification of monoclonal antibodies. In Pyle, D.L. (ed.): Separations for Biotechnology 2. New York: Elsevier Applied Science, 1990.

58. Jiskoot, W., Hoven, A., De Koning, A., Leering, M., Reubsaet, C., Crommelin, D.,

Beuvery, E. Purification and stabilisation of a poorly soluble mouse IgG3 monoclonal antibody J. Immunol. Methods 138, 181–189 (1991).

59. Sasso, E.H., Silverman, G.J., Mannik, M. Human IgA and IgG F(ab′)2 that bind to staphylococcal protein A belong to the V_H III subgroup. J. Immunol. 147, 1877–1883 (1991).
60. Sofer, G.K., Nystrom, L.E. Process Chromatography, A Guide to Validation. San Diego: Academic Press, 1989.
61. Laemmli, L.K. Cleavage of structural proteins during the assembly of the head of bacteriophage T4. Nature 227, 680–685 (1970).
62. Stevens, F.J. Size-exclusion high performance liquid chromatography in analysis of proteins and peptide epitopes. Methods Enzymol. 178, 107–130 (1989).
63. Fasman, G.D. Handbook of Biochemistry and Molecular Biology, Vol. 2: Proteins (3rd ed.). Cleveland: CRC Press, 1976.
64. Gill, S.C., Von Hippel, P.H. Calculation of protein extinction coefficients from amino acid sequence data. Anal. Biochem. 182, 319–326 (1989).
65. Langone, J.J., Das, C., Bennett, D., Terman, D.S. Radioimmunoassay for protein A of *Staphylococcus aureus*. J. Immunol. Methods 63, 145–157 (1983).
66. Per, S.R., Aversa, C.R., Sito, A.F. Quantitation of residual mouse DNA in monoclonal antibody based products. Dev. Biol. Standard. 71, 173–180 (1990).
67. Briggs, J., Panfili, P.R. Quantitation of DNA and protein impurities in biopharmaceuticals. Anal. Chem. 63, 850–859 (1991).
68. Podell, D.N., Abraham, G.N. A technique for the removal of pyroglutamine acid from the amino terminus of proteins using calf liver pyroglutamate amino peptidase. Biochem. Biophys. Res. Commun 81, 176–185 (1978).
69. Bloom, J.W. Reverse-phase high performance liquid chromatographic tryptic digest peptide map comparisons of monoclonal antibodies to human tumour necrosis factor. J. Chromatogr. 574, 219–224 (1992).
70. Harris, R.J., Murnane, A.A., Utter, S.L., Wagner, K.L., Cox, E.T., Polastri, G.D., Helder, J.C., Sliwkowski, M.B. Assessing genetic heterogeneity in production cell lines: Detection by peptide mapping of a low level Tyr to Gln sequence variant in a recombinant antibody. BioTechnology 11, 1293–1297 (1993).
71. Liu, D.T.-Y. Glycoprotein pharmaceuticals: Scientific and regulatory considerations, and the US Orphan Drug Act. Tibtech 10, 114–120 (1992).
72. Silverman, C., Komar, M., Shields, K., Diegnan, G., Adamovics, J. Separation of the isoforms of a monoclonal antibody by gel isoelectric focusing, high performance liquid chromatography and capillary isoelectric focusing. J. Liquid Chromatogr. 15, 207–219 (1992).
73. Costello, M.A., Woittz, C., Deteo, J., Stremlo, D., Wen, L.L., Palling, D., Iqbal, K., Guzman, N.A. Characterisation of humanized anti-Tac monoclonal antibody by traditional separation techniques and capillary electrophoresis. J. Liquid Chromatogr. 15, 108–1097 (1992).
74. Franks, F. Freeze-drying: From empiricism to predictability. Cryoletters 11, 93–110 (1990).
75. McCormick, D.K. Expanded bed adsorption. Biotechnology 11, 1059 (1993).

CHAPTER 6

BIOSAFETY CONSIDERATIONS

GILLIAN LEES AND ALLAN DARLING
Q-One Biotech Ltd., West of Scotland Science Park, Glasgow G20 OXA, UK

6.1. INTRODUCTION

The first therapeutic monoclonal antibody, Orthos's OKT3 anti-CD3, was approved for use in the United States by the Food and Drug Administration (FDA) in 1986. To date, however, few products have been approved, although this year (1993) Cytogen has received approval for its ovarian and colorectal imaging agent Oncoscint.(TM) Centocor and Xoma currently have three product license applications (PLAs) awaiting approval and many more monoclonal antibody products are reported to be in Phase II or Phase II/III clinical trials in the United States [1]. Monoclonal antibodies fall into several broad groups, depending on source and method of generation as described below and include molecules produced in animal or microbial systems. The majority of monoclonal antibodies currently in production are derived from animal cells and we will consider the biosafety issues related to them in detail.

An important consideration in the generation of monoclonal antibodies from cell culture is to ensure that all of the regulatory issues pertaining to the characterization and biosafety testing of potential producer cell lines are addressed at as early a stage in development as possible. It is advisable to conduct a basic schedule of biosafety testing and characterization at the development stage to ensure that the cell substrate is not subsequently found to be unsuitable for production after a lot of time and effort has been expanded on the development program.

Full biosafety characterization of the master cell bank (MCB) and subsequent working and extended cell banks (and/or postproduction cells) should be undertaken as soon as the banks are generated. In addition, it is important to monitor samples throughout the production process to ensure that no adventitious agents are introduced from operators or from culture additives. This monitoring is of particular importance in the generation of biopharmaceuticals from cell culture, as unlike traditional pharmaceuticals these biologicals necessarily involve downstream pro-

Monoclonal Antibodies: Principles and Applications, pages 267–298
© 1995 Wiley-Liss, Inc.

cessing that is relatively "mild," and therefore greater emphasis is placed on the monitoring of the production process.

The biosafety data generated are an essential component of a submission to the regulatory authorities to commence clinical trials (via an IND in the United States or CTX in England), and again there are serious time and cost implications with respect to the overall drug development cycle (5–10 years) for a monoclonal antibody if the biosafety testing requirements are not fully met.

6.1.1. Monoclonal Antibodies From Continuous Cell Lines

Monoclonal antibodies are a relatively recent development. In 1982 the World Health Organization (WHO) produced requirements for the use of continuous cell lines as substrates in the production of inactivated virus vaccines [2]. This was the first such recommendation to industry on the use of continuous cell lines to manufacture products for human use. The acceptance of continuous cell lines for vaccine production was extremely slow, and much of the concern in the use of such "abnormal" cell lines was the potential risk of contamination of the product with tumorigenic agents or pathogenic viruses from the cell substrate [3]. Historically, many of the examples of contamination of a biological product have been associated with, or have arisen from, the use of primary cells or primary biological material in production (Table 6.1). The perceived increased risk associated with the use of continuous cell lines was offset somewhat by the ability to fully test and characterize the cells to an extent not possible with primary material. The production of monoclonal antibodies requires the use of continuous cell lines based on a cell seed system. A cell seed system involves the generation of a master cell bank and a manufacturer's working cell bank (MWCB). The MCB of a continuous cell line must be extremely well tested and characterized to ensure that it is not contaminated with an infectious agent of concern, that the hybridoma is stable, and that the monoclonal antibody produced by the line is of suitable purity and efficacy for therapeutic use.

6.1.2. Recombinant Antibodies

The understanding of the relationship between structure and function of both the genes encoding antibody proteins and the molecules themselves has developed in parallel with an increasing sophistication of *in vitro* techniques of genetic manipulation (see Chapter 3). The functional regions of the molecule are divided into separate domains, which aids the tailoring of antibodies to suit a variety of purposes and

TABLE 6.1. Contamination of Biologicals

Biological	Contaminant
Yellow fever vaccine	Avian leukosis virus, hepatitis B
Polio vaccine	SV40
Factor VIII	HIV
Human growth hormone	Creutzfeld–Jacob disease

the realization of a "second and third generation" of monoclonal antibodies. There are three main areas where genetic manipulation has proved fruitful in the engineering of antibodies: first, in the "humanization" of already existing murine/rat monoclonal antibody antigen-binding specificities [4]; second, in the expression [5] and generation of antibodies in prokaryotic systems, for example, the "coliclonals" [6,7]; and third, in generating novel molecule antigen recognition properties linked to a catalytic moiety [8].

Chimeric Antigen-Binding Molecules. The ability of the antigen-binding activity of antibodies to be retained on a variety of molecules allows the generation of chimeric molecules consisting of antigen-binding linked to enzymatic or other physiological activities [8,9]. Such "targeted" molecules have obvious potential as therapeutic and diagnostic reagents.

Humanized Monoclonal Antibodies. Rodent antibodies are immunogenic in humans and the difficulties in generating stable human monoclonal antibodies have led to intensive efforts to humanize already existing rodent antibodies. When the variable antigen-binding domains are linked to human constant regions, the problems of immunogenicity of the murine antibody can be reduced in the molecule. Alternatively, more extensive humanization includes replacement of the V-region framework with a further reduction in immunogenicity [10]. Recently, the DNA encoding variable regions from antibodies raised in primate hosts have been fused with human antibody constant regions to create "primatized" antibodies, which it is claimed induce little immune response during repeated use [11].

Expression of Recombinant Antibodies. The molecularly cloned antibodies, whether humanized, generated *in vitro,* or chimeric, can be recloned into a number of vectors and expressed in a variety of prokaryotic and eukaryotic systems, the choice of which is dependent on the use and properties of the recombinant molecules. Systems in use include bacteria [12], yeast [13], and mammalian cells such as Chinese hamster ovary (CHO) and mouse myeloma (NSO) cells [14].

6.1.3. Antibody Generation in Prokaryotes

The binding surface of antibodies is in general restricted to six complementarity-determining regions. DNA that encodes such fragments of antibodies can be molecularly cloned into the genomes of filamentous bacteriophage [6,15] or bacteriophage lambda and displayed on the surface of the bacteriophage. The binding properties of the antibody portion can be used to select and purify the bacteriophage carrying the antibody fragment of interest. The use of such systems facilitates the molecular cloning of fragments of antibody genes amplified from immunized animals including humans by the polymerase chain reaction (PCR) using primers encompassing the variable (V) genes. Once cloned, the recombinant proteins can be expressed in a bacterial host and screened for the desired binding activity. Such is the potential of this system or producing large libraries of recombinants that the requirement for the immunization step may be bypassed. Furthermore, once isolated, an antigen-binding fragment can be mutated and the mutants rapidly screened for binding of altered affinity. The sequences can also be molecularly cloned into other desirable molecules including an antibody framework.

6.2. REGULATORY REQUIREMENTS

The regulatory requirements with respect to the biosafety testing of a biological system used to produce a human therapeutic differ to some degree depending on several parameters. For example, the species of the cell substrate, the production system used, the intended use of the product (patient population, dosage and dose regimen), and the regulatory guidelines relevant to the intended market. Since most biotechnology companies wish to market their products internationally, it is important to ensure, as far as possible, that the biosafety testing strategy that is developed satisfies European, U.S., and Japanese recommendations in order that tests do not need to be repeated for each submission [16–19]. Biosafety testing programs should be conducted by specialist laboratories with experience in virus testing and in compliance with Good Laboratory Practice (GLP) [16].

6.2.1. Issues of Concern

The following section is a brief overview of the general regulatory considerations relevant to the production of a monoclonal antibody. The four principal issues to address are (1) the origin of the cells, (2) the seed lot system, (3) the production system, and (4) the purification system.

With respect to recombinant DNA products, data on the genetic stability of the recombinant DNA product will be required in addition to the issues listed above [20–22].

Origin of the Cells. The following should be considered with respect to the identification of the source materials used to produce a hybridoma.

1. The myeloma cell line used to generate the hybridoma must be characterized with respect to demonstrating the purity of the cells and the identity of any immunoglobulin that may be secreted by the cells. It is preferable that the myeloma cell line not secrete any immunoglobulin.
2. The species of animal, the strain, and the immunization procedure used to generate the immune parental cells must be fully described, and details of the source of immunogen must also be provided.
3. The fusion and cloning procedures used to generate the hybridoma must be fully detailed including statistical evidence of cell line clonality, usually by multiple-dilution clonings.
4. The identity of the hybridoma line must be established; and the specificity, class, and subclass, if appropriate, of the immunoglobulin secreted and the

stability of the cell line in terms of antibody secretion must also be demonstrated.

The same careful characterization must be adopted for microbial systems although the test methods employed differ.

Seed Lot System. Whether the biological being produced is a monoclonal antibody derived from a hybridoma line or a recombinant protein from a genetically engineered mammalian cell line, it is essential to demonstrate that a well-defined seed lot system is used in production. A seed lot system involves the generation under Good Manufacturing Practice (GMP) conditions, of a master cell bank (MCB) from which is laid down a manufacturer's working cell bank (MWCB). These banks of vialed cells are maintained frozen in, or in the vapor phase of, liquid nitrogen, and a fresh MWCB vial is used to commence each production cycle. At completion of the production cycle it is necessary to generate postproduction cells (PPC) if an extended cell bank has not been generated. Postproduction cells, as their name implies, are taken directly from the production process at the end of the cycle. An extended cell bank is generated as a parallel culture derived from the MWCB that undergoes at least 10 generations more than the cells used in production. These cell banks are necessary to establish the stability of the system, with respect to production titers and product characteristics and whether the virus status (if applicable) of the cells has altered in any way compared with the MCB.

Production System. The two principal production systems used for the production of monoclonal antibodies are *in vitro* and *in vivo* production systems.

In vitro production systems fall into two categories: finite passage production and continuous or semicontinuous culture production. Of these two, regulatory authorities prefer finite passage production unless continuous culture production is clearly shown to be advantageous. In addition the stability of the continuous culture system must be demonstrated and the yield and quality of the product obtained must be monitored and should not change with respect to specified parameters. With respect to recombinant products the same issues relating to continuous culture pertain, but in addition monitoring of the stability of the expression vector system and integrity of the gene expressed must be undertaken.

In vivo production (e.g., production of ascites) must be based on a defined finite passage system (indefinite passage in animals is not acceptable), and the stability of the hybridoma on *in vivo* passage must be determined. Specific pathogenic-free (SPF) closed-colony animals of a defined genotype and age must be used, and data on the yield and consistency of the monoclonal antibody obtained in the ascites bulk must be generated.

All of the production systems described must be monitored frequently for the presence of microbial contamination. The use of *in vivo* production systems requires rigorous control and monitoring of the animals used for ascites production for the presence of viruses and other adventitious agents and is less favored than *in vitro* production systems in Europe [16].

Purification Systems. Purification systems must be developed that are mild enough to generate a high yield of purified active product but are designed also to remove potential contaminants. As well as contaminating protein from the product that may be derived from the culture medium or from the cells in culture (e.g., from cell lysis), microbial contaminants such as bacteria, mycoplasma, and viruses may be present in the prepurified starting material.

High levels of DNA derived from cell lysis or from microbial contamination may also be present in the prepurified bulk. Contaminating DNA must be removed from a product due to the concern of a risk from potential oncogenic sequences.

Where any bacteria have been present in the production process there is also the problem of contamination from endotoxins (usually lipopolysaccharides) in the prepurified material. Validation studies on the capacity of a down-scaled version of the purification process to remove these potential contaminants should be undertaken and reduction factors for each step of the process should be determined. This will be discussed later in the chapter.

Genetic Stability. The purpose of the testing of the genetic stability of recombinant DNA products is to detect the presence of contaminants that result from changes in protein structure produced by a mutation in the protein coding sequence. A recent supplement [22] to the Center for Biologics Evaluation and Research's 1985 publication *Points to Consider in the Production and Testing of New Drugs and Biologicals Produced by Recombinant DNA Technology* [20] stresses the need for analytical data derived from both nucleic acids testing and protein structural testing to allow a more "complete evaluation of the identity and purity of a recombinant protein product."

The majority of the information required for characterization of the recombinant DNA expression system, including cell expression for production, is the same as is required for all biologicals. However, in addition, details of the method used to prepare the DNA encoding the protein product, including the cell and origin of the source nucleic acid, should be described. The DNA sequence for the whole construct, including promoter sequences, should be included in these data. The host cell system should be compatible with the expression construct and the master cell bank should be derived from a single-cell host containing the expression construct. The stability of the construct in the MCB should be examined, and data on potential rearrangements of the recombined constructs, DNA copy numbers, and the level of expression should be included.

The nucleotide sequence that encodes the protein of interest should be determined at least once for each MCB, these sequence data can be derived from DNA for cells with a single copy of the gene or mRNA from cells with multiple copies of the expression construct. The identity of the MWCB should be assessed at the time of storage by restriction endonuclease mapping of the expression construct for copy number and rearrangements.

A repeat of the characterization of the expression construct, described for the MCB, should be done once on the end-of-production cells for each MCB. There is no requirement for the full characterization to be carried out at the end of each production run.

6.3. BIOSAFETY TESTING

The focus of the remainder of this chapter will be on the biosafety testing that is required on cell banks, in-process samples (including final product), and purification processes to ensure as far as possible the safety of biological products with respect to contamination with microbial agents.

The three principal sources of contamination of a biological are the cells, supplements added during culture, and the operators handling the cell line (Fig. 6.1). A biosafety testing program is designed to detect contaminants by the adoption of three complementary quality assurance testing regimes: (1) the full testing and characterization of the cell substrate to determine which contaminants, if any, are present; (2) the monitoring of in-process samples or bulk harvest and final product to detect contaminants that may have been introduced during the production process; (3) process validation studies, to determine the capacity of the downstream purification process to remove potential (undetected, or as yet unidentified) contaminants (Fig. 6.2). This combined quality assurance triad of checks is necessary, as neither testing nor validation studies alone can assure safety due to the inherent limitations of each regime.

Guidelines and recommendations have been produced by regulatory authorities worldwide to address the issues of concerns in the biosafety testing of biologicals. Specific guidelines exist for monoclonal antibodies [16,18,23] and recombinant products intended for use in humans [20–22] and these are issued by the Committee for Proprietary Medicinal Products (CPMP) in Europe and the Food and Drug Administration (FDA) in the United States. In Japan the Ministry of Health and Welfare have generated a general guideline for production of biologicals from cell culture [21].

Although there are differences between these authorities in the details of the tests required, the underlying principles are the same and are covered by the quality assurance triad of checks discussed above.

One of the major historical differences in requirements between the U.S. FDA and the European CPMP was on which of the cell banks (MCB/MWCB) to undertake the fullest testing. Until recently, the FDA has favored the full testing of the MWCB with a limited set of tests on the MCB, whereas the CPMP has always favored full testing of the MCB with limited testing of the MWCB. At an FDA-cosponsored workshop in January 1992, (Bethesda, Maryland, "Preclinical Safety Testing of Monoclonal Antibodies") it was apparent that the FDA was adopting the preference for full characterization of the MCB, with limited testing of the MWCB, which will greatly reduce the amount of testing that industry will have to undertake on the MWCB in order to satisfy both regulatory authorities.

6.3.1. Biosafety Testing of the Master Cell Bank (MCB)

There are a wide range of tests which have been developed for, and are employed in, a biosafety testing strategy for a rodent master cell bank (Fig. 6.3). The four principal objectives of such a biosafety testing strategy are the following:

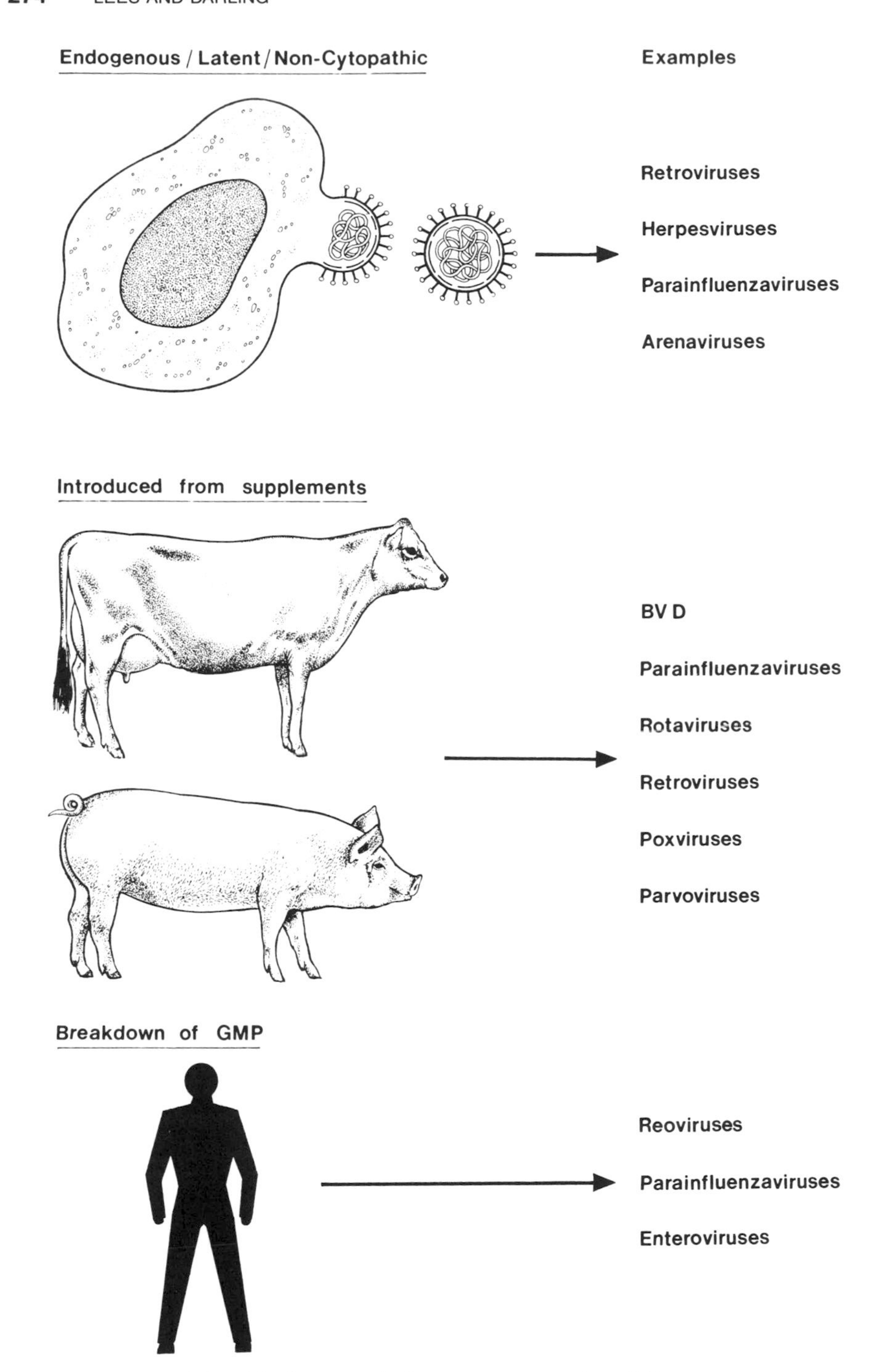

Fig. 6.1. Sources of virus contaminants in cell lines.

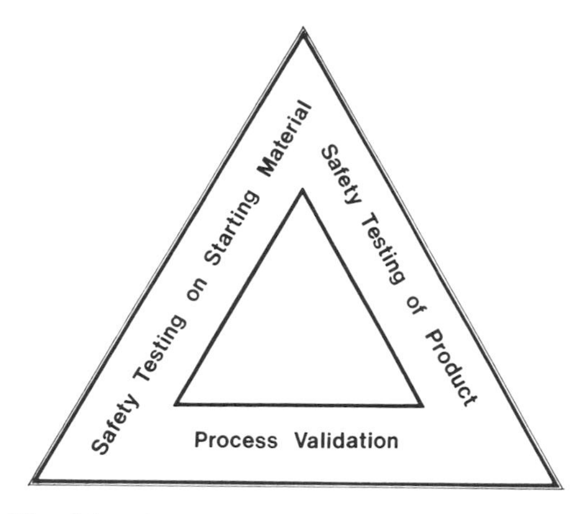

Fig. 6.2. Virus safety depends on a triad of checks.

1. To establish the identity and purity of the cells.
2. To demonstrate the freedom of the cells from contamination with mycoplasmas, bacteria, and fungi.
3. To demonstrate the freedom of the cells from contamination with other adventitious agents.
4. To determine the retrovirus status of the cells.

Identity/Purity of Cells. The identity of the cell line must be demonstrated and the line must be shown to be free from cross-contamination with cells of another species. The biosafety tests involved in the identification of cell lines are isoenzyme analysis and karyology. Karyology is of limited use for polyploid lines such as hybridomas; however, it is recommended for other continuous cell lines like Chinese hamster ovary (CHO) and human diploid cell lines. The U.S. FDA has recently indicated that DNA fingerprinting data can be used as an alternative to karyology studies, in particular with polyploid cell lines [17]. DNA fingerprinting, unlike karyology, is extremely useful for identifying individual hybridoma lines, as each line has a unique DNA fingerprint [24]. This technique enables laboratories developing several monoclonals to be able to identify each hybridoma line separately and to check that no cross-contamination had occurred. Fingerprinting can also be used to establish the identity of bacterial and yeast strains.

For a recombinant cell line there are concerns in addition to those of identity of the host cell, namely: definition of the expression vector system, sequence analysis of the cloned gene, and description of the expression strategy [22].

Detection of Mycoplasmas, Bacteria, and Fungi. To demonstrate the freedom of the cells from contamination with mycoplasmas, bacteria and fungi, it is necessary to conduct both staining and cultivation of the cells in various media. The

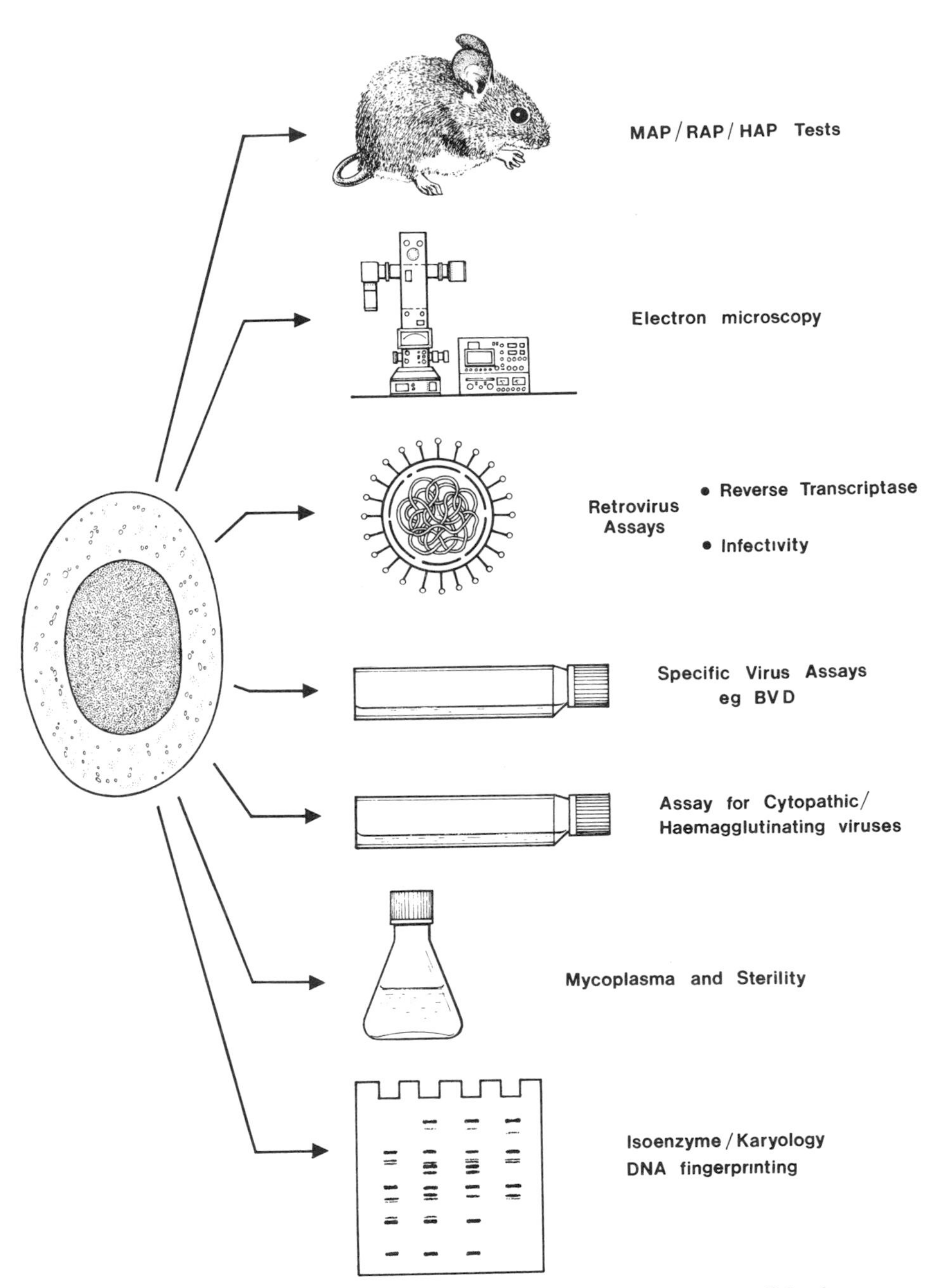

Fig. 6.3. Characterization of a rodent hybridoma MCB master cell bank.

FDA *Points to Consider* documents [17,18] define the test required for regulatory submissions, and this assay also satisfies the European and Japanese requirements. Electron microscopy (EM) studies are conducted and are extremely useful as a general screen for viruses, mycoplasmas, bacteria, fungi, and yeasts. EM studies, however, are only able to visualize a contaminant and are not useful for determining whether it is infectious.

Detection of Other Adventitious Agents. The testing of the cells for contamination with other adventitious agents falls into two broad categories; specific *in vivo* and *in vitro* tests, which detect known viruses of concern either derived from the host species or introduced from additives during culture, and general *in vivo* and *in vitro* tests, which screen for a broad range of animal and human viruses [16–19].

Specific In Vivo *Tests.* For a monoclonal antibody derived from a rodent cell line, rodent viruses derived from the host cell species are obviously of concern. Rodent viruses are detected by inoculating the test cell line into mice, rats, or hamsters depending on the species of origin of the cell line. Serum is collected from the inoculated animals and is screened for the presence of antibodies to a range of viral antigens. For example, 16 murine viruses are detected by the mouse antibody production (MAP) test, including lymphocytic choriomeningitis virus (LCMV), Hantaan virus, Sendai virus, and reovirus-3, which are zoonotic viruses, potentially pathogenic in man. If any of these viruses are detected, the hybridoma is considered unsuitable for production.

Specific In Vitro *Tests.* Many manufacturers use additives of animal origin during the establishment of the cell line and/or during the production process. Two common examples of animal substances used in cell culture are trypsin (porcine origin) and bovine serum or purified bovine protein. It is therefore necessary, when supplements of animal origin are used, to screen for adventitious viral contaminants that may be introduced in these materials.

The specific tests used involve the inoculation of one or two detector cell lines with the test article. The detector cell lines chosen should be sensitive to the viruses of concern in the species of origin of the supplement. These inoculated cultures are examined frequently for evidence of cytopathic effect or gross morphological change. In addition they are subjected to hemadsorption/hemagglutination assays with two or three species of relevant red blood cells and immunofluorescence tests against specific viruses of concern.

Bovine spongiform encephalopathy (BSE) was first identified in the United Kingdom in 1986 and has subsequently been identified in certain other European countries. The potential for contamination of hybridoma cell cultures with BSE introduced from calf serum is considered to be a relatively low risk [25]. The only tests currently available for BSE are *in vivo* assays in murine or hamster systems. These assays are very long and expensive and will not necessarily detect field isolates of BSE. The best approach to take with respect to concerns of cell culture contamination with BSE is to source all calf serum from countries where BSE has not been detected and an official certificate to that effect has been issued by the competent veterinary authority.

General In Vivo *Tests.* Since it is not possible to test for every known animal and human virus, it is also necessary to include general tests that are capable of detecting a very broad range of viruses. The general *in vivo* test that is used involves the inoculation of suckling mice, adult mice, guinea pigs, and embryonated hen's eggs with a test article. The animals and eggs are examined frequently for any signs of the presence of an infectious agent, using indicators such as morbidity or mortality in the animals or deformation of the embryo in the yolk sac of the embryonated eggs. Hemagglutination assays are also conducted on amniotic and allantoic egg samples against chicken and guinea pig red blood cells. The use of such a range of animals and embryonated eggs facilitates the detection of a broad range of viruses from the major virus groups.

General In Vitro *Tests.* The general *in vitro* tests used involve the inoculation of between two and six detector cell lines and the test cell line, with the test article. The FDA recommends that a minimum of two detector cell lines, a human diploid cell line (e.g., MRC-5) and a monkey kidney cell line (e.g., vero) and the test cell line be inoculated. The CPMP recommends, in addition, that a murine cell line (for a rodent hybridoma) and a bovinc cell line be included, and both regulatory authorities suggest that it is wise to include one or two additional detector cell lines to broaden the range of human and animal viruses detected. If an extensive specific bovine virus detection assay is conducted then it may not be necessary to include a bovine line in the general *in vitro* test.

The two principal methods of virus detection employed in the general *in vitro* test are the examination of inoculated cell monolayers over 28 days for gross morphological change or cytopathic effect and hemadsorption assays against human, chicken, and guinea pig red blood cells.

Retrovirus Status. Retroviruses are of concern in the production of biologicals intended for use in humans as they can remain as latent infections (i.e., are not always readily detected); also, they are associated with oncogenic and immunosuppressive conditions. In addition retroviruses have the capacity to recombine either with other retroviruses or with cellular sequences (including proto-oncogenes) to generate novel retroviruses that may have altered host range and/or altered pathogenicity.

Retroviruses are both endogenous (i.e., genetically inherited) and exogenous (i.e., acquired by infection). All mammalian cells harbor endogenous retrovirus sequences and some species, for example, murine and feline cell lines, can express infectious endogenous retrovirus. Consequently the vast majority of rodent cell lines and hybidomas examined show evidence of retrovirus particle expression. Not all retrovirus released from a cell line is necessarily infectious; for example, C-type retroviruses isolated to date from CHO cells are all defective and are consequently noninfectious (Figs. 6.4 and 6.5); however, infectious retrovirus assays should also

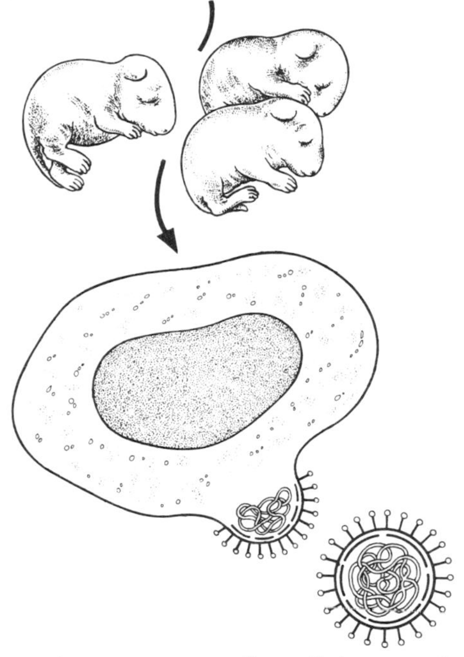

Fig. 6.4. Murine retroviruses.

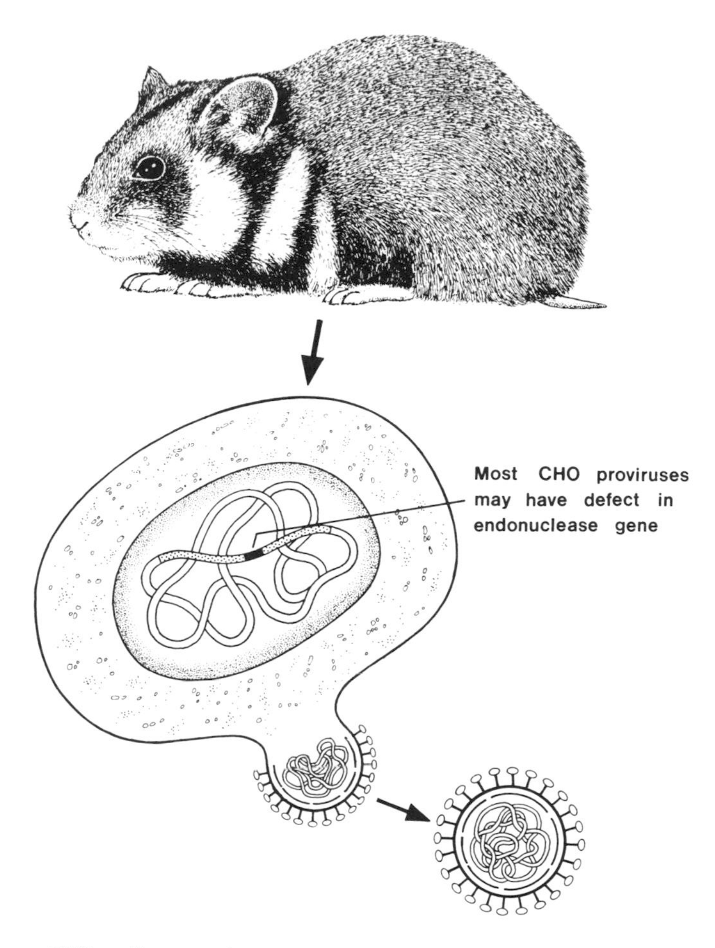

- CHO cells may be reverse transcriptase – positive
- Most virions may be defective for infectivity
- Required to screen by S+L− assay for potential infective recombinant virus

Fig. 6.5. Chinese hamster ovary (CHO) and BHK retroviruses.

be conducted on CHO cells to detect any potential novel recombinant infectious retrovirus. All mice hybridomas and myelomas examined express C-type retrovirus, detected by electron microscopy and reverse transcriptase studies (see below for detail). The retrovirus from these cell lines is also frequently found to be capable of infection (see infectivity assays section below) and may require further characterization at a later stage of development prior to PLA.

The three principal methods of screening for retroviruses mentioned above are electron microscopy, reverse transcriptase assay, and infectivity assays.

Electron Microscopy. As mentioned previously, retroviruses can be visualized by electron microscopy. Although EM can distinguish between intracisternal "A"-type particles and B-, C-, and D-type retroviral particles, it cannot differentiate between an infectious and a defective retrovirus. The original definition of A-, B-, C-, and D-type retrovirus was based on the morphology of the viruses determined by electron microscopy, specifically on the structure of the membrane and the core of the particles. Type A particles have a double shell including an electron-lucent core and they occur intracellularly in two forms. One form, the intracytoplasmic particles, form the cores of type B and type D retrovirus.

Mature extracellular type B viruses contain a type A core surrounded by an envelope derived from the cell membrane. The core is electron-dense and eccentrically placed within an envelope with spikes.

Type C viruses do not contain a preformed core but are assembled entirely at the plasma membrane. Mature extracellular type C viruses have a central condensed, electron-dense core within an envelope with short spikes.

Type D viruses contain a type A core that in the mature extracellular virus has a distinct bar shape within a smooth-surfaced envelope. The retrovirus pertinent to rodent hybridomas are principally A-type and C-type particles (D-type may also be relevant for human hybridomas).

A-type particles are intracisternal particles that do not bud out of the host cell membrane; that is, they do not have an envelope and are noninfectious. They may, however, have a functional reverse transcriptase.

C-type retroviruses bud out of the host cell membrane as enveloped retroviruses and can be infectious (Fig. 6.6).

Reverse Transcriptase Assay. The second method of detection of retrovirus is by assay for the retroviral enzyme reverse transcriptase. This enzyme is an RNA-dependent DNA polymerase that is unique to retroviruses, but once again this assay cannot differentiate between infectious and noninfectious retroviruses. Reverse transcriptase assays are complex to conduct and the correct controls must be included to ensure disruption of the virus and to control for the presence of contaminating DNA polymerase from the cells.

Infectivity Assays. The third method of detecting retrovirus is by inoculating the test article onto susceptible cell lines in an infectivity assay. With respect to rodent hybridomas, murine leukemia viruses are generally noncytopathic on permissive host cells. It is therefore necessary to use a secondary method of detection for infectious retrovirus, principally indicator cell lines and/or reverse transcriptase assay.

Murine leukemia viruses fall into two broad host ranges: ecotropic (i.e., they multiply in cells of murine (or rat) origin) and xenotropic (i.e., they only multiply in cells other than mouse (Fig. 6.4). The direct quantitative assay for ecotropic murine

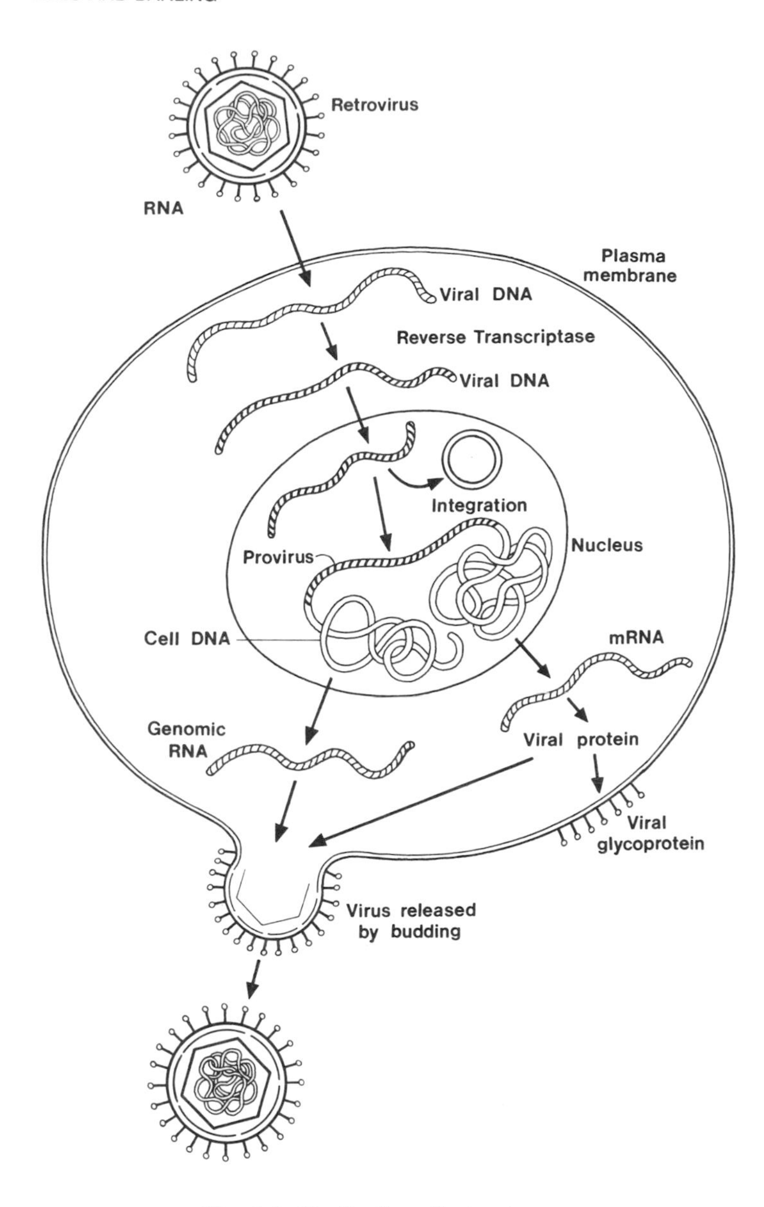

Fig. 6.6. Replication of retroviruses.

leukemia virus is the XC plaque assay and for xenotropic murine leukemia virus is the S^+L^- focus-forming assay. These assays have the advantage of being quantitative but may not detect very low levels of infectious virus. Both direct assays can be extended by subculturing inoculated cells over five passages to amplify any low-level infectious virus present in the test article. Although nonquantitative, this assay is approximately one order of magnitude more sensitive than the direct assay.

The most sensitive assay for the detection of infectious retrovirus is the cocultivation assay. This assay involves the cocultivation of the test article cells with a permissive detector cell line. The detector cell line is passaged five times and reverse transcriptase assays are conducted after passage 1 and passage 5 of culture to detect the presence of retrovirus. If a murine hybridoma is releasing infectious xenotropic retrovirus, cocultivation assays should be conducted to determine the susceptibility of human cells (such as MRC-5 or rhabdomyosarcoma RD) to the contaminating retrovirus. Cocultivation assays are also conducted when no direct quantitative assay is available for a particular retrovirus contaminant or when infectious human retroviruses such as human T-cell leukemia viruses (HTLV) or human immunodeficiency viruses (HIV) are of concern.

As murine retroviruses are ubiquitous contaminants of all murine hybridoma culture systems, the level of expression and the host range of the virus must be established in a hybridoma cell line, and these parameters must be monitored throughout the production process for any change. Assuming that a good level of clearance of retrovirus is obtained by the downstream purification system in virus validation studies, then the presence of an infectious murine retrovirus should not necessarily preclude the use of a hybridoma for the production of a biologic for human use.

6.3.2. Human Cell Lines and Human/Murine Hybridomas

In addition to the tests and considerations previously described for rodent hybridomas, the use of transformed human cell lines or human/murine hybridomas creates extra issues in respect to biosafety testing (Fig. 6.7). The particular additional concerns are principally those of contamination with human retroviruses and herpesviruses. Both of these groups of viruses are of concern as they are associated with the etiology of neoplasia, immunosuppression, and other degenerative conditions and are capable of remaining as latent infection within cells, particularly lymphoid cells.

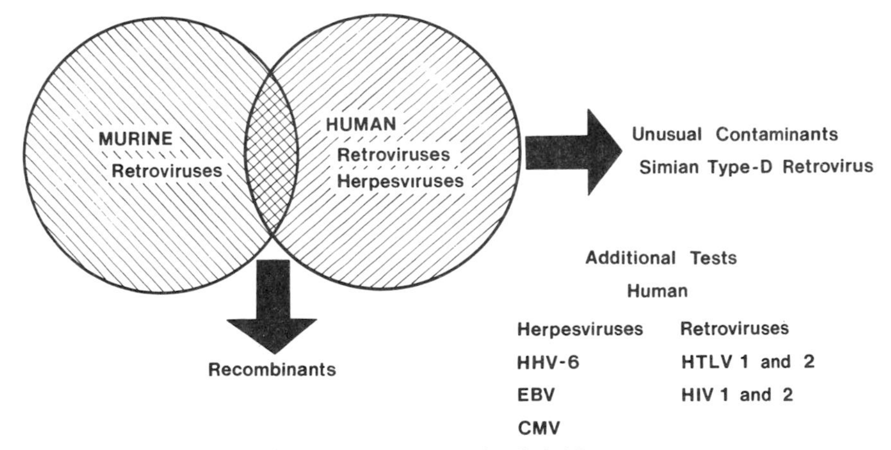

Fig. 6.7. Human–murine hybridomas.

A testing regime should include assays for the known human retroviruses (HTLVs and HIVs) and herpesviruses—Epstein-Barr virus (EBV), cytomegalovirus (CMV), and human herpesviruses-6 and -7 (HHV-6, -7)—and also for unknown or variant viruses (including possible recombinant or pseudotype retroviruses).

Human Retroviruses. The master cell bank should be screened by DNA hybridization or PCR for the known human retroviruses HTLV-1 and -2 and HIV-1 and -2. Virus isolation assays should also be conducted, principally cocultivation assays with appropriate permissive detector cells, to detect variant, recombinant, or unknown viruses that may not be detected by molecular techniques.

Human Herpesviruses. The human herpesviruses cytomegalovirus, HHV-6, and HHV-7 should be screened for, by means of PCR and/or DNA hybridization assays. Virus isolation assays in susceptible cell cultures should also be conducted to detect variant viruses. The human gamma-herpesvirus EBV is often used for the immortalization of human B cells in the generation of human and human/murine hybridomas. Evaluation of these cell types and other human cells requires a hierarchical testing regime in which the cells are screened for the presence of the virus genome; if present, the pattern of genomic expression is subsequently determined.

Other Viruses of Concern. Monoclonal antibodies obtained from human cells raise issues of concern with respect to viruses other than human viruses. For instance, antigen-specific lymphocytes are often prepared by procedures that involve sheep erythrocytes; under these circumstances one should consider conducting general and specific assays for ovine viruses. Screening for human hepatitis viruses should also be considered, as should screening for the human parvovirus B19.

Finally, as previously mentioned, human cells are often immortalized with EBV, which in itself is a potential human pathogen. EBV is commonly produced in the marmoset cell line B95-8, several isolates of which have been shown to be contaminated with squirrel monkey retrovirus (SMRV). SMRV is an endogenous virus in its natural host and is a significant contaminant of several widely used human lymphoblastoid cell lines [26,27]. SMRV can be detected in human cell lines by PCR and/or DNA hybridization techniques. These tests should be conducted in conjunction with isolation assays in Raji cells to detect potential variant forms of the virus.

It is apparent that no defined set of tests and requirements can be laid down that are applicable to every system, owing to the variation in cell lines and processes that are used in the production of monoclonal antibodies. It is essential that at an early stage of development a suitable and efficient biosafety testing regimen be adopted for a biological product. This is best prepared jointly between the manufacturer and the biosafety testing facility to ensure that all issues of concern with respect to regulatory requirements are met. It is also important to approach regulatory authorities with the planned regulatory studies as early as possible for consultation. The EC *Guidelines* and FDA *Points to Consider* documents are flexible, and the regulatory authorities are keen on encouraging industry to approach them at a early stage to discuss any problems that they may have, or envisage, with their particular cell line or process.

TABLE 6.2. Example of a Testing Strategy for a Rodent Cell Bank [1,2]

	MCB	WCB	ECB/PPC
Sterility	×	×	×
Mycoplasma	×	×	×
Virus assays			
In vitro	×	×	×
In vivo	×		×
MAP/RAP/HAP	×		
Bovine viruses	×		×
Retrovirus assays			
TEM	×		×
RT	×	O	×
XC	×	O	×
S^+L^-	×	O	×
Co-cult	O		O
Isoenzyme analysis	×		
Karyology/DNA fingerprinting	×	×	×

[1]MCB, master cell bank, WCB, working cell bank; ECB/PCC, extended cell bank/postproduction cells.
[2]×, recommended; O, may be required.

In summary, the principal tests that should be conducted on a rodent MCB are summarized in Table 6.2. The tests that are conducted on the MWCB are dependent, to some extent, on the results obtained for the MCB. In particular the retrovirus status and any change in that status must be determined. The extended cell bank (or postproduction cells) require basically the same set of tests that have been conducted on the MCB. This is to ensure that no detectable change has occurred in the cell line during production with respect to the stability of the cell line or gene insert (for a recombinant cell line). The retrovirus status must again be assessed to detect any change in the level or characteristics of a retrovirus contaminant of the cell line (Table 6.2).

6.3.3. Testing of In-Process Samples (Table 6.3)

Testing of In-Process Fermenter/Bulk Harvest Samples. The tests that are required on "in process" samples depends to some extent on the results obtained on the master and working cell banks, particularly with respect to the characteristics of the retrovirus detected. It is recommended, however, that a mycoplasma test, certain retrovirus tests, and a limited *in vitro* assay be conducted. Under certain circumstances the FDA may require an *in vivo* assay to be conducted in addition to these other tests. These assays are used to detect the introduction of an adventitious agent during production and to attempt to determine the level of retrovirus present. The FDA currently recommends that the total retrovirus load be estimated at this stage by an appropriate technique. A current technique that is recommended by the FDA for quantitation of retroviral load in bulk harvest is negative stain electron

TABLE 6.3. Example of a Testing Strategy for In-Process Samples/Final Product Derived From a Rodent Cell Bank[1]

	Fermenter/Bulk Harvest	Purified Bulk	Final Product
Sterility	×	×	×
Mycoplasma	×		
Virus assays			
In vitro (limited)	×		
In vivo	O		
Bovine viruses	O		
Retrovirus assays			
RT	×		
XT and/or SL	×		
DNA contamination test		×	
Rabbit pyrogen test			×
Abnormal toxicity (general safety)			×

[1] ×, recommended; O, may be required.

microscopy. There are many problems, however, with the use of quantitative negative stain for bulk harvest samples. Although negative stain electron microscopy is appropriate for the quantitation of specific purified virus preparations, extreme caution must be applied in the interpretation and assessment of negative stain electron microscopy data generated on bulk harvest samples. Negatively stained preparations frequently contain in addition to genuine virus, objects that can be described as "virus-like" but are often attributable to cell debris (X). It is very difficult to differentiate in bulk harvest material between true C-type viral particles and reassociated plasma membranes, vesicles, and any other material of size and diameter similar to those of C-type virus particles. No ultrastructural elements like envelope or virus core are visualized in negative stain electron microscopy studies. Viruses can be identified unambiguously only if these ultrastructural features are clearly demonstrated. It is therefore extremely difficult to discriminate between retrovirus particles and nonviral particles in bulk harvest. This estimated load is then compared with the infectious retrovirus clearance data obtained from the validation of the purification process. It is difficult to relate total retrovirus load in terms of virus particles to the capacity of a purification system to remove infectious retrovirus. It is therefore sensible to conduct an infectivity assay for retrovirus on the bulk harvest material in addition to whatever other, less directly related, type of estimate of load is required. The CPMP currently does not make this comparison but does recommend that tests to detect retrovirus be conducted on the bulk harvest.

Testing of Purified Bulk/Final Bulk Lot. The minimum testing recommended at this stage is for mycoplasma, sterility, and a residual DNA. The latter is conducted by hybridization with species-specific DNA probes. The maximum level of

DNA recommended per dose is 100 pg. However, this number is not rigid due to different dosing regimens, so discussions with the regulatory authorities are recommended if levels exceed the recommendations.

If the cell line is EBV-infected but is not producing active virus, then each lot of final bulk should also be tested for EBV antigens, and if positive the lot may be used only if the purification process has been shown to remove EBV.

Testing of Final Dosage Form/Final Filled Product. In addition to tests to establish the potency, purity, and stability of the product, it is necessary to conduct biosafety tests for sterility, abnormal toxicity (general safety), and pyrogenicity. Currently in Europe the rabbit pyrogen test may still be required, whereas in the United States a validated limulus amoebocyte lysate (LAL) assay is acceptable.

Limitations of Testing. All of the assays described in the biosafety testing regimens that can be adopted are subject to two principal limitations.

First, even the extensive set of assays recommended may not detect an as yet unknown or unidentified contaminant. Although this is covered in part by the inclusion of general *in vitro* and *in vivo* screens as previously mentioned, the sensitivity of these assays with respect to an unknown contaminant clearly cannot be evaluated.

The second limitation of biosafety tests is the limit of detection of the tests with respect to the volume of sample tested. It is clearly not possible or desirable to assay all of the ampoules in a cell bank or all of a bulk or final lot. It is therefore necessary to take small samples from the total volume, which exacerbates the possibility of missing low levels of virus that may be present. For example, if 1-ml samples are tested from a 1,000-ml bulk containing 1,000 virus particles, then statistically 37% of the 1-ml samples assayed would not contain a virus particle (Fig. 6.8).

These limitations, combined with the necessity of demonstrating good clearance of a known contaminant (e.g., murine retrovirus or EBV), demonstrate the crucial role that virus clearance studies play in the overall quality assurance testing of hybridomas or engineering lines producing a biopharmaceutical.

6.4. PROCESS VALIDATION STUDIES

The aim of this section of the chapter is to outline the principles and methods used in performing process validation studies. Process validation is a documented program that provides a high degree of assurance that a specific process will consistently produce a product meeting its predetermined specifications and quality attributes [17]. Performing process validation studies for virus removal or inactivation involves the following basic steps:

1. Scaling down the process.
2. Consideration of virus inactivation.

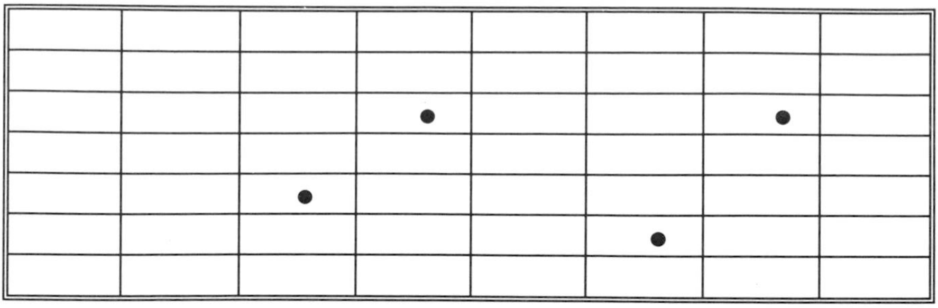

The probability that a sample does not contain infectious virus is:

$$P_{-} = e^{-cv}$$

Where P_{-} is the probability that the sample does not contain infectious virus

c is the concentration of virus particles

v is the volume of sample to be tested

For a 1ml sample, the probability that it does not contain infectious virus at virus concentrations from 10 to 1000 particles per litre are:-

c	10	100	1000
P_{-}	0.99	0.90	0.37

For a concentration of 1000 viruses per litre, in 37% of the sampling 1ml will not contain a virus particle.

Fig. 6.8. Probability of detecting viruses at low concentrations.

3. Choosing the appropriate viruses for the study.
4. Spiking selected steps of the process with high titers of the appropriate viruses.
5. Titrating the output samples and calculating virus reduction factors.
6. Multiplying the virus reduction factors for each step to obtain an overall clearance factor for the whole process.

In Europe, the CPMP has issued a "Note for Guidance" document covering virus validation [25] and the FDA has recently produced a new *Points to Consider* document covering the characterization of cell lines used to produce biologicals that also addresses this area [18].

6.4.1. Scaling Down the Purification Process

One of the most important considerations in the design of these studies is that all the steps in the production process be examined, from initial treatment (i.e., precipitation, clarification) through to the final steps in the production process, such as freeze-drying, as potential virus clearance or inactivation steps. The number of steps to be studied will depend on how comprehensive the validation study needs to be. The steps that are most likely to generate the best virus clearance data should then be downscaled. Downscaling of the purification process is an essential part in performing process validation studies for virus removal/inactivation. For several reasons, including the scale of the production process, it is either impossible or impractical to perform these studies on the full manufacturing scale; therefore the steps to be studied are scaled down to laboratory scale. Although there are no set guidelines governing the size of a scaled-down process, in practice, scale-down factors are usually of the order of 1:100–1:200. It is essential that the downscaled steps of the process mimic as closely as possible the full manufacturing scale process; therefore, the downscaling of the process should be validated to ensure that the yield, purity, buffer compositions, and flow rates are as close as possible to those of the full manufacturing process. This ensures that the virus clearance studies are performed on a process that is as close a representation of the full manufacturing process as possible. In certain cases downscaling is relatively straightforward (i.e., column purification steps) but in other cases, such as ultrafiltration steps, certain problems can arise in sourcing equivalent materials to those used in the full manufacturing scale. These considerations may influence which stages are chosen to be validated. Another consideration in determining the stages to be validated is whether the clearance factors that will be generated by each of the individual stages can be multiplied together to given an overall clearance value for the process. When two stages remove or inactivate virus by independent mechanisms, the clearance factors for each stage can be multiplied together. If the mechanisms of virus removal or inactivation are similar (i.e., where two column steps share a common buffer that is responsible for virus inactivation), then the clearance factors for each step can be considered only additive not multiplicative and this would lead to a lower overall virus clearance factor for the whole process.

The scaled-down study should also be designed to minimize the eluate volumes generated for each stage that will subsequently be assayed for infectious virus. This is due to the considerations mentioned previously concerning the ability to detect infectious virus at low concentrations. Since an aliquot of the eluate from each stage, generally not the total eluate, is titrated, then even if sampling fails to detect infectious virus, the presence of infectious virus in the total eluate cannot be discounted. Therefore a minimal detectable titer based on the Poisson distribution is applied to an aliquot when no infectious virus is detected. Since virus recovery from an individual stage is a function of virus titer and output volume, to reduce the theoretical virus recovery from a stage it is important to keep output volumes to a minimum.

6.4.2. Incorporation of a Virus Inactivation/Removal Step

Whenever possible, inactivation steps should be incorporated into the design of a validation, and both the FDA and the CPMP expect virus inactivation data in addition to clearance data for the process. This can be done by monitoring the inactivation of virus in one of the process buffers (e.g., the low-pH buffers used in the elution of antibody from protein A columns), or it may include the incorporation of a specific virus inactivation step into the process. These studies are usually performed by looking at the kinetics of inactivation, as virus inactivation is rarely linear over time and a persistent fraction is often present. The overall constituents of the buffer can also affect virus inactivation. For example, temperature and pH inactivation steps are influenced by the divalent cation and protein concentration, and these should be adjusted to reflect the levels in the full-scale process. Different viruses also have different responses to different inactivation steps and this should also be borne in mind.

Several specific methods for virus inactivation have been developed, particularly for the blood and plasma fractionation industry, and these are now finding a use in monoclonal antibody production processes. Freeze-drying and heat treatment have been used for several years to ensure inactivation of potential viral contaminants, and several comprehensive studies have been carried out to study the efficacy of these processes (reviewed in Markus-Sekura [26] and Piszkiewicz [27]). The use of organic solvents such as ethanol and β-propiolactone have also been used for the inactivation of viruses [28,29]. Recent developments in the use of solvent inactivation has focused on the use of solvent/detergent combinations [30,31]. The solvent tri-(n-butyl) phosphate (TBNP) in association with detergents such as Triton X-100 or Tween-80 can readily inactivate enveloped viruses. Because the solvent and detergent act synergistically, relatively low concentrations of both reagents can be used and thus the treatment is relatively mild and is unlikely to be detrimental to the product. This treatment is now licensed for use by the New York Blood Centre and is commonly used by manufacturers of products derived from blood.

The addition of a solvent/detergent step into a monoclonal antibody production process can thus generate excellent clearance data for retroviruses and for herpesviruses. However, the main shortcoming of this technique is that infectivity of viruses that do not possess a lipid envelope or do not require their envelope for infectivity is not reduced.

Several studies have shown the effectiveness of UV irradiation, particularly in the presence of chromophore dyes, on virus inactivation [32–34]. This technique is not widely used in production processes, however, but remains a technique that has potential for the future. Similarly, several companies in Europe, the United States, and Japan are now manufacturing specific virus removal filters. These filters have the advantage that virus is physically removed from the solution, so both enveloped and unenveloped viruses can be removed, the efficiency of virus removal being a function of size and shape of the virus. However, there are several inherent problems in relying on filtration as the sole means of providing good clearance data. First, membrane integrity is critical, as any breakage or disruption of the membrane would allow virus to pass through, and this has to be tested before and after use. Thus efficient testing of membrane integrity before and after use will be essential. Second, for certain filters the product has to be recirculated through the filter over a long period of time, which may not be desirable as regards product stability. Filters are also very expensive and prone to fouling; thus they may be of limited use. Consequently, although membrane filtration has a future in production processes, its use is likely to be limited to specific products and is unlikely to be acceptable as the sole virus clearance step in a production process. However, when the overall clearance capacity of a process is low, the incorporation of a specific virus removal filter may generate the necessary extra virus reduction factor needed with minimal disruption to the production process.

By far the most common method of virus inactivation involves the use of pH extremes. Retroviruses, as mentioned above, are the viruses of most concern in hybridomas and are extremely susceptible to inactivation at pH values below 5.0. Therefore the use of low-pH buffers in the purification of monoclonal antibodies from protein A columns is an extremely effective step in ensuring that the product will not contain infectious retrovirus. Herpesviruses, another virus group of concern in hybridomas, are also readily inactivated at low pH values. This treatment unfortunately is not effective for all viruses and small unenveloped viruses such as parvoviruses and polioviruses are not readily inactivated at low pH values.

The incorporation of a specific virus inactivation step or virus removal filter into a process, like filtration, can also help generate the necessary virus clearance values, especially where a purification process may be relatively "mild" and may have been designed to maximize yields and throughput without consideration of the virus problem.

6.4.3. Selection of Appropriate Viruses

Selection of the number and type of viruses to be used in a process validation study depends on a number of parameters. One of the most important considerations is that the viruses to be used must be available at high titer, as it is evident that the theoretical virus clearance factors that can be achieved are proportional to the titer of the virus spikes achieved. The viruses must also be easily assayed to enable accurate and reproducible titration results and should not pose an unnecessary hazard to the operators conducting the validation study. The number of viruses to be used in the study will depend on the results and level of testing, on the master cell bank, the working cell bank, and in-process samples, and on the stage of development of the product. The virus validation data required for a product going into phase 1 clinical trials with terminal cancer patients, for example, are less comprehensive than if the product is to be tested on healthy volunteers.

The viruses selected for the study should also provide a range of physicochemical properties including DNA and RNA viruses, enveloped and unenveloped. In certain cases, the viruses used should be the actual potential contaminants found in the starting material; thus it is obligatory to use HIV when performing validations where blood or blood derivatives are involved in the production process. Similarly, for murine hybridomas and CHO cells, which contain endogenous C-type retroviruses, murine leukaemia virus is used as one of the viruses for the validation studies. For various reasons, however, it is often not possible to use the actual potential contaminant viruses in the process validation study. The viruses may be too dangerous for work without special containment facilities or they may not grow in cell culture, or not to a sufficient titer to use in the studies. The actual number of potential contaminant viruses may also be too numerous to use each potential contaminant in a validation study. To cover such eventualities, model viruses are used in process validation studies. These viruses are chosen either to substitute for specific viruses or, more commonly, to cover a virus group such as herpesviruses, which contains several different pathogenic viruses. New herpesviruses continue to be discovered; therefore validation of a process with a herpesvirus model gives a high degree of assurance that the process being studied will remove or inactivate not only the model herpesvirus being used, but also other identified and as yet unidentified potential herpesvirus contaminants.

In summary, the viruses selected for the validation, whether known contaminants or models, should cover both DNA and RNA viruses, both enveloped and unenveloped, small and large. At least one of the viruses should be highly resistant to physicochemical inactivation, and this virus will provide a severe test of the clearance capacity of the purification system. A retrovirus should always be included in the validation study, as the genomes of these viruses can be inherited through the germ line or, in the case of infection, can be integrated into the genome. An example of viruses that can be chosen to validate a murine hybridoma are shown in Table 6.4.

As mentioned previously, in validation studies for a murine hybridoma, murine leukemia virus (MuLV) is used as a representative of the C-type subgroup of retroviruses. This virus is a medium-sized, enveloped RNA virus that has a low resistance to physicochemical inactivation. A herpesvirus is also included, as these viruses also have the capacity to establish latent infections within cells. Human herpes simplex virus type 1 is an appropriate model to use for murine hybridomas and is a large DNA enveloped virus with a medium to low resistance to physicochemical inactivation. Poliovirus may be included not because polioviruses are potential contaminants of murine hybridomas but because it is a small, tough, unenveloped RNA virus that is very difficult to inactivate. Thus poliovirus in this case is included as a model for other unidentified contaminants that may be highly

TABLE 6.4. Viruses Used in a Virus Validation for a Murine Monoclonal Antibody

Virus	Size	Envelope	Nucleic acid
MuLV	80–120 nM	Yes	RNA
Poliovirus	20–30 nM	No	RNA
Herpes simplex virus type 1	120–150 nM	Yes	DNA
Adenovirus	70–90 nM	No	DNA

resistant to inactivation. To complete the range of viruses in the study a medium-sized, unenveloped DNA virus such as adenovirus should be considered.

For a human hybridoma the choice of viruses is slightly different. It is necessary to show removal or inactivation of the lentivirus subgroup of retroviruses. This group of viruses, which includes HIV, are more resistant to inactivation than the oncovirus subgroup. In this case the ovine lentivirus Maedi-visna virus can be used as a model lentivirus, as it displays many of the physical properties of HIV but does not pose a hazard to operators. Herpes simplex virus type 1 is also used as a model when the hybridoma has been produced with the herpesvirus EBV and to cover for possible contamination with other herpesviruses such as cytomegalovirus, HHV-6, and HHV-7.

6.4.4. Performing the Process Validation Study for Virus Clearance and Inactivation

Once the steps of the purification process to be studied have been decided on, the validation of the downscale demonstrated, and the appropriate viruses selected, the virus validation can commence. One of the first manipulations to be performed is to test whether the samples that will be generated by the steps in the process will be cytotoxic to the indicator cells that will be used to titrate the viruses. Therefore samples from either an unspiked downscaled run or from the actual manufacturing process should first be tested to ensure that they will not interfere with the virus titrations. Cytotoxicity can be a function of the composition or pH of the buffer or can be caused by the product itself. This can often be alleviated by diluting the sample in PBS or tissue culture medium, but care must be taken not to generate large volumes, which would exacerbate the problem of detecting virus present at low concentrations. In these cases ultracentrifugation or diafiltration can be used, but care must be taken to control for these steps, which are not part of the normal purification procedure. When calculating the clearance factors for each stage, cognizance must be taken of necessary correction factors for these manipulations in the results. The process validation experiments are then performed by spiking the starting material for each step of the process with the chosen virus. Only one virus can be added to the starting material, so for a four-virus validation each step to be validated has to be performed four times. It may also be necessary to perform the spiking experiments in duplicate, depending on the stage of development of the product. To ensure that the nature of the starting material is not affected by adding virus in tissue culture medium, the virus spike is added in a volume that is $<10\%$ v/v of the total volume of the material to be spiked. This material is then taken through the purification process and the appropriate fractions are collected for assay of infectious virus. Figure 6.9 gives an example of a two-column validation. In the first column the virus is spiked into the starting material and only the eluate fraction has been collected for virus assay. For the second column virus is again spiked into the starting material, but in this case the flow-through fraction and the wash fraction have been collected as well as the eluate fraction for virus assay. Under certain conditions it is appropriate to collect fractions other than the eluate to examine

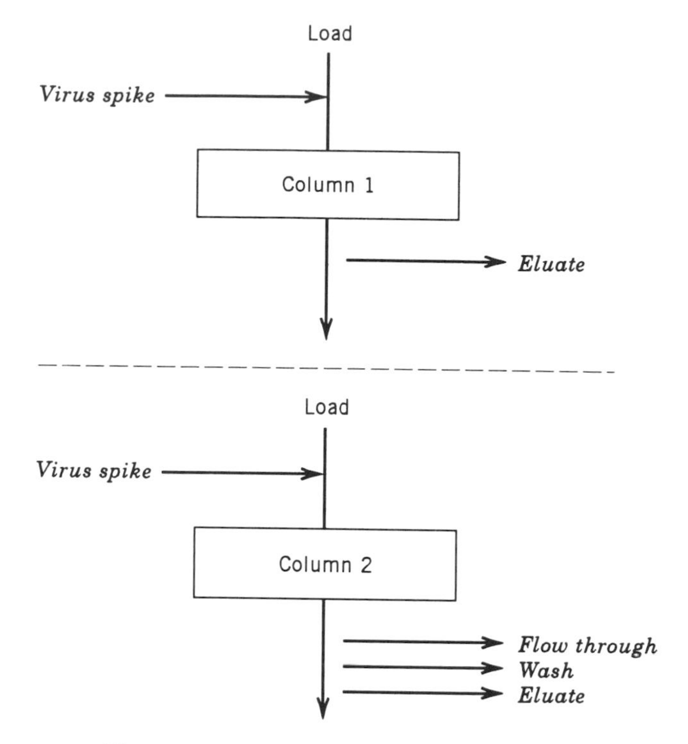

Fig. 6.9. Two-column virus validation study.

where the virus has been distributed after application to the column. The regulatory authorities advise that these fractions be titrated, particularly when virus clearance is dependent on partitioning of the virus rather than inactivation. If chromatography columns are reused, the efficiency of the sanitization regimen used must also be demonstrated, as virus could build up on the chromatography matrix and desorb at any time. These experiments must be performed as kinetic inactivation studies in the sanitization solutions. Another recommendation is that for resins that are continuously regenerated and reused the virus validation should be performed on fresh resin and on resin at the end of the life in the production process. For certain resins this may be upwards of 100–200 cycles of purification and sanitization. The fractions are then assayed for the presence of viruses. All virus assays must be based on infectivity and not on other techniques such as reverse transcriptase activity or PCR detection, as the latter methods do not quantify infectious virus titer.

6.4.5. Calculation of Virus Clearance Factors

Once the virus titrations have been performed, the next stage is to calculate the virus reduction factors for each step tested in the validation. The virus reduction factor for an individual purification or inactivation step is defined as the $\log_{10}$ of the ratio of

TABLE 6.5. Calculation of Reduction Factors[1]

Starting material:	vol v'; titer $10^{a'}$
	virus load $v' \times 10^{a'}$
Final material:	vol v''; titer $10^{a''}$
	virus load $v'' \times 10^{a''}$
	$10^{RI} = \frac{v' \times 10^{a'}}{v'' \times 10^{a''}}$

[1] The virus reduction factor of an individual purification or inactivation step is defined as the $\log_{10}$ of the ratio of the virus load in the prepurification material divided by the virus load in the postpurification material.

the virus load in the prepurification material to the virus load in the postpurification material (Table 6.5). The formula takes into account both the titers and volumes of the materials before and after the purification step. For example, if a protein A–Sepharose step were spiked with 10 ml of murine leukemia virus at a titer of 5.0×10^7 pfu ml^{-1} and the eluate volume were 50 ml that contained virus at a titer of 1.5×10^1, the reduction factor for this step would be

$$10^{RI} = \frac{10 \times 5.0 \times 10^7}{50 \times 1.5 \times 10^1} = 6.7 \times 10^5 = 10^{5.8}.$$

Therefore this step of the process can remove 5.8 logs of retrovirus under the conditions used in the purification step. Because of the inherent imprecision of some virus titrations, an individual reduction factor should be $\geqq 1$ log to be significant. A clearance factor of greater than 5 logs represents an effective virus clearance step.

6.4.6. Calculation of the Clearance Capacity of the Overall Process

Once the individual reduction factors have been calculated, an overall clearance factor for the purification process can be determined. As mentioned previously, reduction factors for individual steps can be multiplied if the steps remove or inactivate virus by different mechanisms. When the steps are similar or identical, the clearance factors can only be additive. Table 6.6 gives an example of a theoretical process for the purification of a monoclonal antibody. The purification regimen illustrated incorporates three column purification steps, a specific detergent virus inactivation step, and an ultrafiltration step. When no virus has been detected after a step and a theoretical minimum titer has been applied, for instance the ion-exchange step, then the clearance factor for that step has been expressed as >3.9 logs. This implies that this step could clear at least 3.9 logs of a virus, which may be an underestimate and higher clearance factors might have been obtained, for instance if a higher virus spike had been used. Summing the individual clearance factors for the individual steps of this process gives an overall clearance capacity of >18.5 logs.

TABLE 6.6. Example of a Typical Retrovirus Removal/Inactivation Study

Purification Step	Log Reduction
1. Affinity chromatography	2.8
2. Detergent inactivation	>5.1
3. Ion-exchange chromatography	>3.9
4. Size-exclusion chromatography	2.5
5. Ultrafiltration	4.2
Cumulative reduction	>18.5

The clearance factor for the overall process gives an indication of the capacity of the purification process to inactivate viruses.

It is important to relate this figure to the assessment of virus risk and the virus load either quantitated or estimated in the starting material. The risk assessment includes how the final product is administered, the dose and frequency of application, and the patient population to be treated (i.e., healthy or terminal patients).

6.4.7. Limitations of Virus Validation Studies

A virus validation study is essential to the assurance that an acceptable level of safety in the final product is established, but does not in itself categorically establish safety. A number of limitations in the design and execution of virus validation experiments may lead to an incorrect estimate of the ability of the process to remove virus infectivity. One of these limitations is the use of model viruses or viruses that have been adapted for growth in tissue culture, which may not behave the same as the viruses present as actual contaminants—for example, in their degree of purity and aggregation. This is important, as aggregation of virus can protect "core" virus from inactivation, and resistant subpopulations of virus can exist that are extremely difficult to remove or inactivate. Therefore a virus that resists a first inactivation step may be more resistant to subsequent steps and as a consequence the overall reduction factor is not necessarily the sum of the reduction factors calculated from individual steps each time with a fresh virus spike-suspension. Virus clearance may have also been overestimated by summing the logarithmic clearance factors for steps in which the virus removal was accomplished by a similar mechanism (e.g., adsorption onto the column matrix). It is also almost impossible to completely miniaturize the full-scale manufacturing process and maintain exactly the equipment and conditions found in the full manufacturing scale. For these reasons it is generally recommended that at least one individual stage of the validation process should remove or inactivate at least 5 logs of retrovirus. This again gives an extra degree of assurance that the virus validation results are meaningful and contribute to the safety of the final product.

6.5. SUMMARY

In summary, the essential features of a biosafety testing program to ensure the generation of a safe biological for human use involve the careful characterization and maintenance of cell substrates and materials used in production. It is important to check for the presence of contaminants and to monitor the levels of contaminants that are inevitable, such as murine retrovirus in hybridoma lines. Virus validation studies are equally important to determine the ability of the downstream purification system to remove and inactivate actual contaminants (such as murine retrovirus) and a range of other potential contaminant viruses, mycoplasmas, and nucleic acid.

The regulatory considerations with respect to the production of monoclonal antibodies are manyfold. The generation of novel biotechnology-derived human therapeutics is an area of rapidly developing technology and consequently both regulatory and biosafety testing bodies must be able to respond to industry with timely, sensible, and scientifically sound recommendations to address new issues of concern.

6.6. REFERENCES

1. Pharmaceutical Manufacturers Association. Biotechnology 9, 947 (1991).
2. WHO Expert Committee on Biological Standardization. WHO Technical Report Series 673, 70 (1982).
3. Cooper, P.D. N.C.I. Monogr. 29, 63 (1968).
4. Riechmann, L., et al. Nature 332, 323 (1988).
5. Boss, M.A., et al. Nucleic Acids Res. 12, 3791 (1984).
6. Huse, W.D., et al. Nature 348, 552 (1990).
7. McCafferty, J., et al. Nature 348, 552 (1990).
8. Chiswell, D.J., McCafferty, J. Trends Biotechnol. 10, 80 (1992).
9. Garrard, L.J., et al. Biotechnology 9, 1373 (1991).
10. Neuberger, M.S., et al. Nature 312, 604 (1984).
11. Winter, G., Milstein, C. Nature 349, 293 (1991).
12. Ueda, H., et al. Biotechnology 10, 430 (1992).
13. Carter, P., et al. Biotechnology 10, 163 (1992).
14. Kotula, L., Curtin, P. Biotechnology 9, 1386 (1991).
15. Bebbington, C.R., et al. Biotechnology 10, 169 (1992).
16. Commission of the European Communities. Guidelines on the Production and Quality Control of Monoclonal Antibodies of Murine Origin Intended for Use in Man (1987).
17. Commission of the European Communities. Guidelines on the Production and Quality Control of Human Monoclonal Antibodies (1989).
18. Office of Biologics Research and Review. Points to Consider in the Characterization of Cell Lines Used to Produce Biologicals (1993).
19. Commission of the European Communities. Guidelines on the Production and Quality Control of Medicinal Products Derived by Recombinant DNA Technology (1987).

20. Office of Biologics Research and Review. Points to Consider in the Production and Testing of New Drugs and Biologicals Produced by Recombinant DNA Technology (1985).
21. Ministry of Health and Welfare Japan. Documents Necessary for Application for Approval on Drugs Manufactured Using Cell Culture Technology (1988).
22. Commission of the European Communities. Guidelines for Minimizing the Risk of Transmission of Agents Causing Spongiform Encephalopathies Via Medicinal Products (1991).
23. Popovic, M., et al. Int. J. Cancer 30, 93 (1982).
24. Oda, T., et al. Virology 167, 468 (1988).
25. Commission of the European Communities. Guidelines on the Validation of Virus Removal and Inactivation Procedures (1991).
26. Markus-Sekura, C.J. Dev. Biol. Stand. 75, 133 (1991).
27. Piszkiewicz, D., et al. Curr. Stud. Hematol. Blood Trans. 56, 44 (1989).
28. Morgenthaler, J.-J. Curr. Stud. Hematol. Blood Trans. 56, 109 (1989).
29. Stephan, W. Curr. Stud. Hematol. Blood Trans. 56, 122 (1989).
30. Horowitz, B., et al. Transfusion 25, 516 (1985).
31. Horowitz, B. Dev. Biol. Stand. 75, 43 (1991).
32. Wallis, C., Melnick, J.L. Photochem. Photobiol. 4, 159 (1966).
33. Chanh, T.C., et al. J. Virol. Methods 26, 125 (1989).
34. Neundorff, H.C., et al. Transfusion 30, 485 (1990).

CHAPTER 7

ANTIBODY PATENTS

PATRICK CRAWLEY
Patents and Trademarks Division, Sandoz Technology Ltd., CH-4002 Basel, Switzerland

7.1. INTRODUCTION

There are numerous commercial applications for antibodies and the technology is undergoing rapid development and change. In this situation many new discoveries and inventions are being made and for commercial reasons these need to be protected. Patents in many cases provide the most effective way to protect these discoveries and inventions, while allowing continued access by the academic research community.

The history of the patenting of antibodies and their uses is relatively recent and opportunities to patent have not always been taken. For example, the discovery of monoclonal antibodies was not patented. Since then experience in the patenting of antibodies has grown. Hopefully, greater awareness of the patent system will improve recognition of future opportunities for patenting. Also, better information on how and when to patent, in order both to protect the invention and to allow publication of the discovery with minimum delay would be helpful. The antibody patent situations that have developed recently are complex, often with a number of overlapping layers of patents relating to any one area. Some knowledge and understanding of patents can help in unraveling these complex situations. An early appreciation of competitors' patent property can help direct research towards less crowded or even patent-free areas, where there is greater opportunity to make and patent inventions and the cost and need to license patented technology from others is likely to be less.

First, a review of the basic principles that underlie the patent system will assist in understanding the sections that follow.

Monoclonal Antibodies: Principles and Applications, pages 299–335
© 1995 Wiley-Liss, Inc.

7.2. ABOUT PATENTS

7.2.1. Why Patent?

Patents often provide the most effective way to protect commercially important inventions, that is, significant technical developments that have broad applicability.

If you do not protect your inventions, you cannot stop others copying them and you cannot secure a reward for your ingenuity and a return on your investment in research. As a result, funding for further research or development of the invention may be jeopardized. Making and developing inventions requires investment in research. This can be lost, and possible commercial development and use may be inhibited, if you cannot prevent unauthorized copying of the fruits of that research. Academic researchers who do not patent may thus lose the possibility of research support from industry and the opportunity to have their discoveries developed as products.

For the inventor in industry the picture is a simple one. An industrial competitor who copies and has no R & D costs can sell products at lower prices than the innovative company, which must recoup R & D investment. Any commercial enterprise that invests in R & D and does not protect its inventions will soon go out of business.

In contrast a well-protected invention can establish a competitive advantage that allows R & D investment to be recouped and an improved market share to be maintained. By their nature inventions provide improved or cheaper products. A competitor who does not have access to the invention is restricted to previous, less-effective products or is not able to benefit from the cost-saving advantages of the invention. Appropriate protection of inventions can make the difference between business success or failure in science-based industries.

The need for effective protection is particularly acute in the pharmaceutical industry, where the costs of developing new products are very high, in particular the costs of obtaining regulatory approval from product-licensing authorities, including the cost of conducting clinical trials. Current estimates for the cost of developing a single new pharmaceutical product, including the costs of obtaining regulatory approval, are in the region of $100 million to $200 million. Many candidate products do not survive the development process, sometimes being discarded at a very late stage of development, which adds to the overall cost. A number of antibody-based pharmaceutical products are currently under development.

Also, having protection for your inventions can provide a valuable source of income or a lever for obtaining access to inventions that belong to others. Patented inventions can be licensed to others in return for royalties or other payments or in exchange for access to their patented inventions.

Similarly, patenting is important for academic inventors, particularly in the current financial climate in which public sector funding for academic research is under considerable constraint. Licensing patented inventions can provide valuable additional income to supplement public sector funding. Also the fact that one has patent protection can encourage industrial collaborators or other sources to fund follow-on work to develop an invention. Without the insurance policy of a patent to protect the investment, it may be difficult to secure further funding from industrial collaborators. The industrial sponsor who has to choose between two project proposals, one of which is protected by patents, will choose the patented one, all other things being equal. Also, having patents from past activities can increase status and credibility with industrial sponsors.

Contrary to popular belief, the existence of patent protection often ensures that inventions are developed, rather than preventing the invention from becoming widely available. Protection safeguards the additional, often very large, investment required to take an invention from the initial ideas stage to a marketable product. Without protection it may not be possible to get anyone to risk the investment required for development, and the rewards will be won by others who protect subsequent follow-on or improvement inventions, as was the case with penicillins and monoclonal antibodies. The existence of patents also acts as an incentive to innovators to make and protect yet further inventions that get around, or add value to, the original patented invention.

Patents cannot prevent inventions from being used for pure, noncommercial research purposes.

7.2.2. What Is a Patent?

A patent is a bargain struck between the owner of an invention and the state by which the state grants to the owner the right to exclude others from enjoying the benefits of the invention for a limited period of time (usually 20 years) in return for a written description of the invention that is published so that the invention is freely available for use by the public at the end of the period of time.

Thus the essence of a patent is *protection in return for disclosure*. This bargain constitutes a legal property right that is granted by the state and is embodied in the patent. As such, a patent is not a positive right: It does not of itself entitle the owner to make or do anything. Rather it is a negative right whereby the owner may stop others from doing things within the area covered by the patent, during the life of the patent. The owner of a patent covering a new pharmaceutical product is not entitled to make and sell the product before having obtained the necessary regulatory approval; even so the way may be blocked because someone else has obtained an earlier patent that covers the whole class of compounds that includes the new product—that is, a generic patent. The ability to make use of a patented invention may be dependent upon other laws, rules, or regulations or may fall under patent or other intellectual property rights belonging to others.

7.2.3. What Is Patentable?

Not every invention or discovery is patentable. There are two types of criteria that must be satisfied before a patent can be obtained. The nature of the invention must be such that it does not fall within a category that is excluded from patentability. Also the invention must have the necessary qualities of *novelty, nonobviousness,* and *industrial applicability* sufficient to justify the grant of a patent.

The requirements for patentability, both in terms of what is excluded and the qualities an invention must have to be patentable, are defined by statute, different national laws paralleling one another except for points of detail. Ultimately questions of what is or is not patentable, or whether a particular activity is covered by a patent, are decided by the courts. In the European patent system, questions of patentability are decided by the European Patent Office Boards of Appeal. In all fields patent law develops in step with changes in technology. However, the application of patent law to biotechnology has presented, and is continuing to present, particular problems, and the situation remains especially fluid in this technical area.

7.2.3.1. Exclusions From Patentability. Some types of invention are judged to be inherently unpatentable. These include inventions that are considered not to be industrially applicable, generally falling into the areas of pure intellectual or artistic endeavor, for example, the discovery or establishment of a new natural law, scientific theory, or method of performing a mental act, or a literary, dramatic, or artistic work of other aesthetic creation. However, some of these latter aesthetic creations may be protected by copyright. Notwithstanding, it is only the non-industrially-applicable inventions *as such* that are not patentable. Industrially applicable products or processes based on these inventions, provided they satisfy the other criteria for patentability, may be patented. For example, the discovery of a new law of nature or scientific theory is not patentable in itself; but a new, nonobvious, and useful product that is based upon the new law or theory may be patented.

A topical example of patenting in which the question of industrial applicability was an issue was the U.S. National Institutes of Health (NIH) patent application relating to partial human genome sequences. In this case, it appears that the functions of the proteins coded by the DNA sequences covered in the patent application were not known when it was filed. The patent applications covering these sequences identified their utility as markers for identifying the position or origin of associated DNA sequences. This is probably not sufficient industrial applicability for patent purposes. The U.S. Patent Office objected to the patent applications on this among other grounds, though the NIH withdrew the application before the point had been decided conclusively.

Further types of inventions that are not patentable are inventions that tend to promote offensive, immoral, or antisocial behavior, as well as certain types of biological and medical invention. Previously patents for contraceptives were refused in the UK on the grounds that they were contrary to morality, though this is no longer the practice. More recently, in the Harvard oncomouse case in the European Patent Office, it was eventually decided after an appeal that a patent covering a transgenic animal was not contrary to morality. The patent relates to a transgenic mouse with increased susceptibility to cancer for use in testing anticancer drugs. However, the patent is apparently being challenged on morality grounds in Germany (Scrip No. 1724, 5 June 1992, page 7) and over a dozen organizations have filed oppositions (see later) to the European patent on these grounds.

Methods of treatment or diagnosis performed *on* a human or animal body are not patentable, although products for use in such methods are. Also essentially biological processes (excluding microbiological processes), such as classical plant or animal breeding procedures, and in Europe "plant and animal varieties," are excluded from patentability. However, patents can be obtained for plants and animals (transgenic animals—see above), provided that the invention involves some measure of human intervention and is not contrary to morality and, in Europe, that the patent covers higher taxonomic groupings than just varieties.

7.2.3.2. Novelty and Nonobviousness. For an invention to be patentable it must be *novel,* that is, new. If the invention is already known, the public is already in possession of the invention and there is no bargain to be struck between the owner of the invention and the state. A patent cannot be granted that prevents the public from doing something that it was previously entitled to do.

Any prior disclosure that makes the invention available to the public destroys the novelty of the invention and renders it unpatentable. Such disclosures may be by written or oral description, by use, or in any other way. *It is of fundamental importance, therefore, to consider patenting and to take any steps necessary before publication or any other disclosure of the invention takes place.* This includes disclosures in the form of posters, or oral presentations at public meetings, or in a PhD thesis.

However, disclosures that are confidential, either by express agreement or in view of the circumstances of the disclosure, are not novelty-destroying, and in some countries such as the United States account is taken only of printed publications, knowledge, and use that are available or take place in the country in question.

In addition, to be patentable, an invention must be *nonobvious.* A patent is a right that can be exercised against the general public, and as such it will be validly granted only for nontrivial inventions that make a real technical contribution to the state of the art. Inventions that are no more than minor developments of existing technology, the next obvious step that workers in the field would readily have thought of doing, are not patentable.

The concept of "obviousness" is one of the more difficult patent concepts to grasp. Many inventions appear obvious once they have been made, but a hindsight analysis is not the correct approach. Also, the inventors are often not the best people to judge the obviousness of their inventions; they are often too close to the invention and may take into account information that is not publicly available. Obviousness is essentially a practical question: The question of whether the invention *would have been obvious* to a skilled person working in the relevant technology (a skilled worker in the art) at the relevant date, given the publicly available information and common general knowledge at that time, that is, the state of the art. The first stage at which the obviousness of an invention is seriously scrutinized is during examination by the Patent Office Examiner, though he is unlikely to be a skilled worker in the art and may not apply the correct criteria. Subsequently, during patent disputes, the question may come to be decided by a court, that is, by a judge or even in some cases a jury, on the basis of conflicting evidence put forward by expert witnesses for both parties.

Clearly the state of public knowledge is constantly changing as the results of research are published in the academic journals, and thus it is necessary to define a fixed point in time against which to judge the novelty or obviousness of an invention. This point in time is known as the *priority date* of the invention and is usually the date of filing of the first patent application that discloses the invention.

As regards priority date, there are two fundamental differences between U.S. patent law and that of most of the rest of the world. The actual date of invention (in the United States) as opposed to the date of the patent application is critical for determining priority between different U.S. patent applications for the same invention. Also there is a grace period of up to 1 year prior to the filing of a U.S. patent application. A publication that occurs during this period does not affect the novelty or obviousness of the invention, provided that the actual date of invention precedes the date of the publication (Fig. 7.2).

7.2.4. How Does One Get a Patent?

Patents are not obtained automatically. It is necessary to file a patent application in each of the territories where protection is required and this application must pass a stringent examination by the relevant patent office before the patent is granted.

7.2.4.1. The Documents. The patent application takes the form of a formal request for the grant of a patent accompanied by a document, referred to as a patent specification, that includes a description of the invention and claims defining the scope of the protection that is being sought. The description of the invention must be sufficiently detailed to enable someone who works in the relevant field to put the invention into practice. The claims (i.e., the numbered paragraphs at the end of the patent specification) define the scope of the invention covered by the patent, so that the public can determine whether their activities fall within the protected area of the patent. The precise wording of these claims is a matter of great importance: If too narrow, a competitor may take the benefit, without falling within the scope of the patent; if too broad, the claims may cover or be too close to something that was known or done previously and may be invalid for lack of novelty or obviousness.

7.2.4.2. The Procedure. The normal procedures by which patent applications are processed to grant are set out in Figures 7.1 and 7.2.

It is necessary to file a separate patent application in each territory where patent protection is required. It is not necessary to file all of these applications immediately. An initial priority application can be filed, usually in the applicant's home country, and corresponding foreign applications can be filed during the next 12 months. Under the terms of an international treaty (the Paris Convention) covering most countries, corresponding foreign applications filed within 1 year of the priority application are entitled to the date of filing of the priority application (the priority date) for novelty and obviousness purposes—that is, publications made after this date do not affect the novelty or obviousness of the invention.

This 12-month period is often referred to as the *priority year* and provides a useful period of time during which the invention can be further developed and perfected. The filing of the priority patent application safeguards the patent position against the effects of any subsequent publication of the invention or later patent

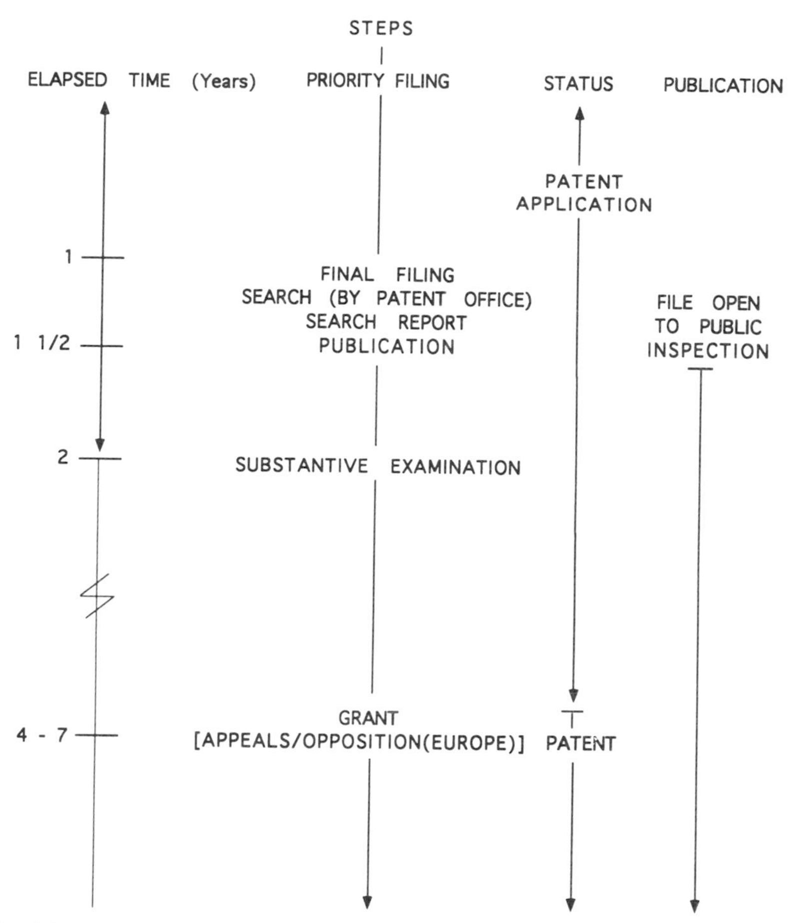

Fig. 7.1. Schema showing the time course and important steps involved in the filing and processing of a patent application in the British and European Patent Offices, in which a final application has been filed 1 year after the filing of a priority application. The patent application is published about 18 months after the filing of the priority application and from this point onwards the patent office file of the application is open to public inspection. The application is usually granted 4–7 years after filing of the priority application. The term of a British or European patent is 20 years starting from the date of filing of the final patent application (not shown).

filings by others. In addition to providing further time to perfect the invention, it also provides an opportunity to gauge the commercial value of the invention before having to invest further in patent filings in other countries. Once the priority application is filed the position is protected and it is possible to make nonconfidential approaches to potential licensees.

In some countries, including the UK and Europe, prior to examination of the

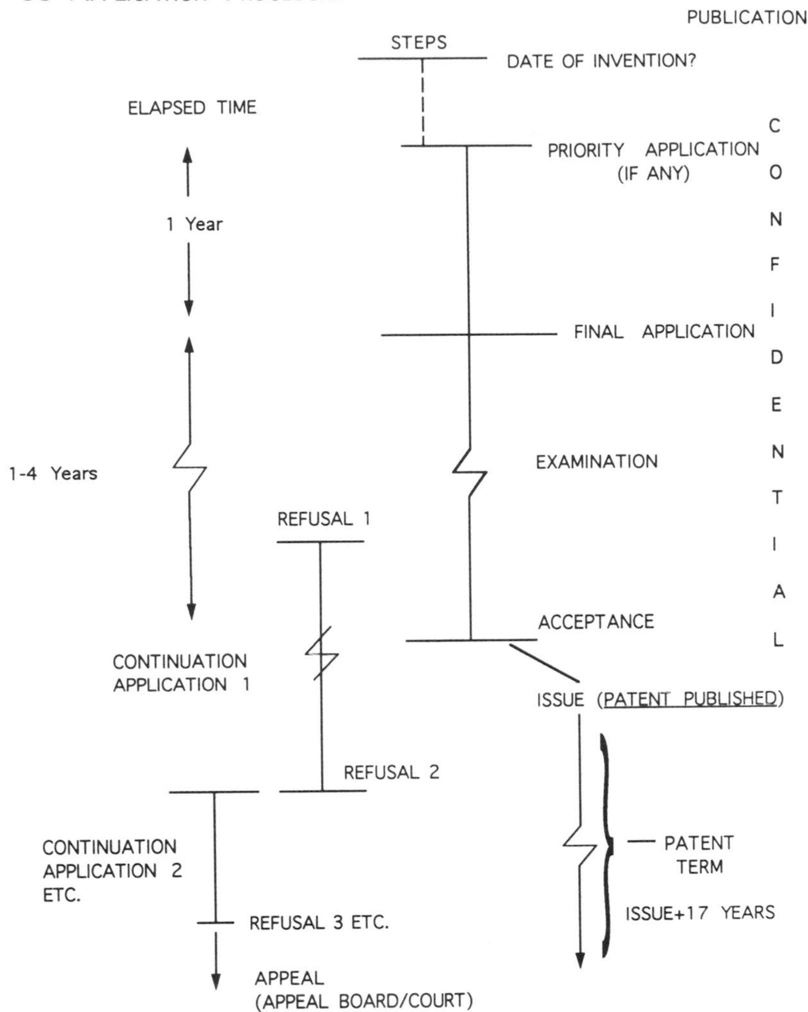

Fig. 7.2. Schema showing the time course and important steps involved in the filing and prosecution of a U.S. patent application, indicating the significant differences between U.S. procedure and British and European patent applications as shown in Figure 7.1, including (1) the fact that proceedings remain confidential until a patent issues; (2) the length of the term of the patent, which is 17 years *starting from the date of grant;* (3) the possibility of keeping a case alive by filing a continuation application if the application is refused, and (4) the importance of the actual date of invention as opposed to the date of filing for determining priority.

patent application, a literature search is carried out by the patent office to identify relevant prior publications, and the results of the search are reported to the applicant to provide the opportunity to abandon the application or amend it before proceeding further. Also in many countries the application is published or laid open to public inspection before examination (i.e., early publication). This early publication usually takes place 18 months after the earliest priority date and includes publication of the search report if any. In this way the public is kept informed about the patent applications that have been filed.

Again, the main country in which the procedures differ from those outlined above is the United States, where there is no publication of pending patent applications, all proceedings between the applicant and the U.S. Patent and Trademark Office being confidential before grant (see Fig. 7.2). In view of this, and the importance of dates of invention as opposed to dates of filing, as discussed above, the United States has a specific procedure, an Interference, for determining entitlement to a patent between copending applications for the same invention (see section 7.6.1).

In most countries an examination of the patent application is carried out through a dialogue between the applicant and the patent office. A patent office examiner examines the application on its merits, identifying objections to the grant principally on the basis of the patent disclosure and the novelty/obviousness of the invention claimed. The applicant then responds to these objections by argument or appropriate amendments to the patent specification (description and claims). This procedure of objections and responses may be repeated a number of times and through this process the applicant and the examiner move towards a mutually acceptable patent specification, which, if reached, results in the grant of the patent. In some countries, where examination is relatively rudimentary (e.g., South Africa), a patent may be granted after only 6 months or so. However, in most countries, where there is detailed examination, for example, in the United States and Europe, it usually takes 2–4 years for a patent to be granted, though it may take longer, particularly for more complex subject matter such as some biotechnology inventions.

In addition to filing separate patent applications in each of the countries where protection is required, it is possible to file a single European patent application and to obtain grant of a European patent covering all or some of the 17 countries of the European Patent Convention (EPC). On grant, this European patent must be converted into a bundle of national patents in the countries designated in the application. It is possible also to file a single "International Patent Application" designating many of the world patent territories. However, such an international patent application must be converted to separate national and regional (e.g., EPC) applications after the initial stages of the procedure if it is to proceed further (Fig. 7.3).

The costs of filing an initial priority patent application are relatively low, about £1,000–3,000 for a simple case, including the cost of employing the services of a patent attorney to prepare the patent specifications and other documents. Clearly patent professionals' charges are not cheap but there is no safe alternative to engaging a patent attorney to help avoid the difficulties that can arise. The costs of proceeding with corresponding foreign patent applications is high, about £1,000–

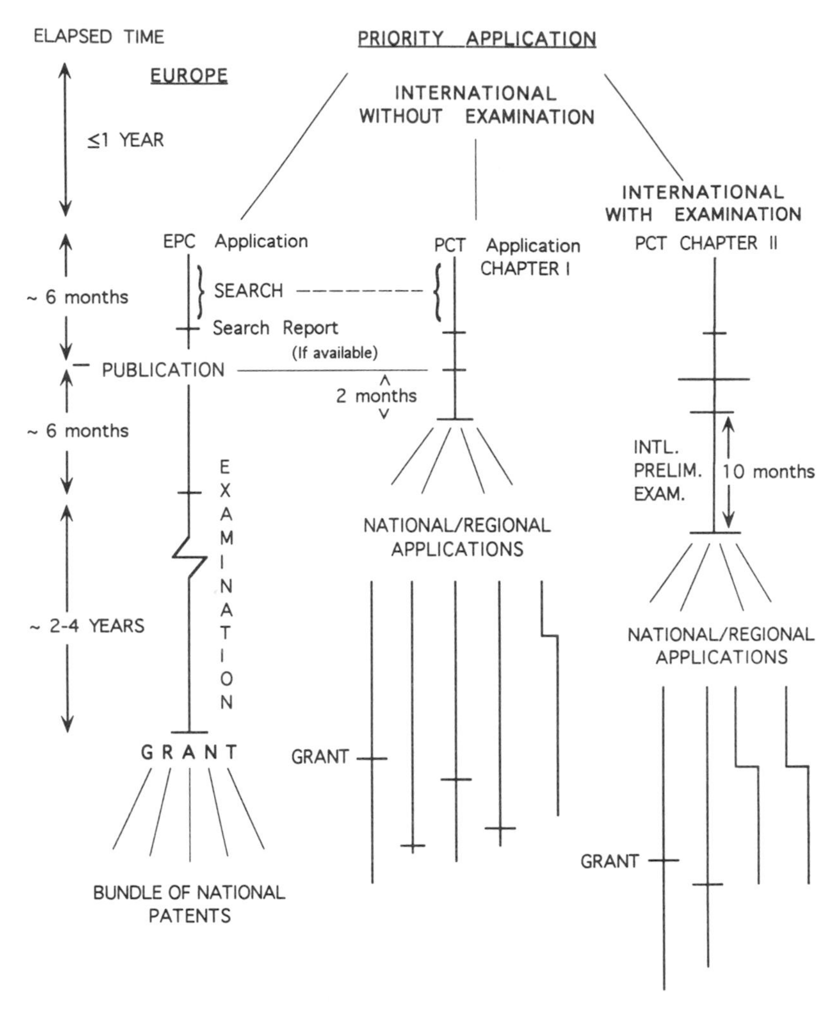

Fig. 7.3. Schema showing the time course and important steps involved in the filing of patent applications under the provisions of the European Patent Convention (EPC) (a European patent application) and the Patent Cooperation Treaty (PCT) (an international patent application). An EPC application follows the procedure set out in Figure 7.1. A PCT application is conducted as a single consolidated procedure only during the earlier stages, which consist of search, search report, and publication for an application proceeding under the provisions of Chapter I of the PCT; or, in addition, an international preliminary examination, for an application proceeding under the provisions of Chapter II of the PCT. After these early stages the PCT application must be converted into corresponding national (e.g., UK and U.S.) and regional (e.g., EPC) applications, which then follow the procedures set out in Figures 7.1. and 7.2.

3,000 per country to file, followed by prosecution costs of about £2,000–3,000 per country to process the application through to a granted patent. When the patent is granted there are still more costs; the renewals fees, often payable annually, start at a £100 or so per year, and range up to £1,000 per year during the final years of patent life, for each country.

7.2.5. Ownership and Exploitation of Patents

In the simple case a patent is owned by the person or persons who made the invention. Often, however, it is the inventor's employer who owns the patent, provided the invention arose during the course of the normal duties of the employee inventor. For most scientist employees in industrial research this is explicitly covered by a clause in the contract of employment by which inventions made by the employee belong to the employer. The position as regards academic inventors is usually the same and the patent is usually owned by the academic institution. The disposition of the rights in an invention may be varied by agreement; for instance, in the case of academic inventions when there is industrial sponsorship the rights may be licensed or even assigned to the industrial sponsor.

As noted previously, the patent right is not a positive right to do something oneself, but a negative right to exclude others from the patented area. This right becomes effective on grant of the patent, although the patent owner may be entitled to compensation for acts done by others back to the date of early publication of the patent application. The patent right lasts for a finite period of time, which varies from country to country. In many countries the life of a patent is 20 years reckoned from the date of filing of the patent application. As it may take 4 years or more before a patent is granted, the useful patent life may be much shorter. Again the United States is anomalous: The patent life is 17 years, though reckoned from the date of issue of the patent, not the filing date.

A patent owner can benefit from the patent by using it to prevent competitors making use of the patented invention. In this way the patent owner can maintain a competitive position and safeguard R & D investment. Alternatively the owner may obtain financial rewards by licensing others to use the invention in exchange for royalty or other payments.

Ultimately, however, it is the responsibility of the patent owner, or possibly the licensee if so required, to enforce the patent rights against anyone who makes use of the patented invention without permission. *The patent office is not responsible for policing the patent.* Enforcement involves the civil courts, where it is necessary to take action against the offending party for infringement of the patent rights. Putative infringers may counterclaim for revocation of a patent on the grounds that the invention is not patentable, as well as argue that their activities are not covered by the patent. All of this is very costly, and thus the threat of litigation is often enough to get the parties to negotiate a solution. The cost of enforcing patents through the courts is very high, usually reckoned in tens of thousands if not hundreds of thousands of pounds, and may even cost over £1 million in a complicated contested case.

7.2.6. Patents Versus Secrecy

The main alternative to patenting, if an invention is to be protected, is to keep it secret. Indeed in some cases secrecy may be the only way of protecting the invention, for example, when it falls into one of the categories that are excluded from patentability. However, there are some inventions that cannot be kept secret—for instance because the nature of the invention is instantly apparent, or can be analyzed, on examination of the product.

Also in the academic world, where the main intention of research is to discover new things and make the information available to the academic community at large, there is generally no real alternative to patenting if inventions are to be protected.

In practice a combination of the two modes of protection are used. For instance the structure of a new pharmaceutical compound may be covered by a patent, but the fine detail of the production process and the formulation procedure may be maintained as trade secrets.

In industry the main alternatives are patenting and secrecy, whereas in the academic world the main alternatives are patenting before publishing or publishing without patenting.

7.3. A BRIEF HISTORY OF ANTIBODY PATENTS

Although antibodies have been known for over a 100 years, it is only in relatively recent years that they have become common subject matter in patents.

7.3.1. Immunoassay Patents

Early antibody patent applications, in the late 1960s, were those relating to immunoassay procedures. Notable were those relating to enzyme-labeled immunosorbentassays (ELISAs). Akzo/Organon obtained a series of patents relating to ELISAs. In the area of homogeneous, that is, single-phase enzyme immunoassays, key patents were obtained by Miles Laboratories covering the use of enzyme-labeled reagents and enzyme inhibitor-labeled reagents.

7.3.2. Monoclonal Antibodies

The development of monoclonal antibody technology, following the seminal work of Kohler and Milstein in the mid 1970s, gave rise to a very significant increase in the number of patent filings on antibodies. Although the original monoclonal antibody invention was not patented, attempts were made by others to obtain broad patents covering generic classes of antigens. For example, Koprowski et al. have obtained U.S. patents covering monoclonal antibodies to viral (USP 4196265) and antitumor monoclonal antibodies (USP 4172124), though these attempts have not been successful in most countries. However, in the immunoassay area, Hybritech obtained broad patents, the Tandem Assay patents, relating to the use of monoclonal antibodies in sandwich immunoassays (see Chapter 2). Subsequently there have been a great number of patents relating to more narrowly defined classes of monoclonal antibody products, monoclonal immunoassay procedures, methodologies for obtaining monoclonals, and the use of monoclonal antibodies as immunopurification reagents.

7.3.3. Recombinant Antibodies

The next technological development to be patented was the development of recombinant antibodies, by procedures in which genes coding for antibody heavy and light chains have been obtained, originally from monoclonal antibody-producing cell lines, and reexpressed in host cells. This has opened up the possibility of making altered antibodies, including chimeric and humanized antibodies (e.g., humanized CDR-grafted antibodies), as well as antibody fragments, such as Fv's, and novel antibody molecules such as single chain antibodies (see Chapters 1 and 3). More recently patent applications have been filed on methods for cloning human antibody genes and transferring human antibody repertoires to transgenic animals—that is, to produce animals capable of generating human antibodies in response to antigen challenge. Suffice it to say, however, that the patent situation in this area is highly complex, with many layers of overlapping patents. A detailed review and analysis of recombinant antibody patents is given in section 7.6 below.

7.3.4. Therapeutic Antibodies

Gamma-globulin, obtained from fractionated donor human serum, has been used for a number of years as a therapeutic antibody reagent. Initially in the 1930s and 1940s relatively crude material was used in an intramuscularly injectable form as a panacea for many ills, though no patenting was done at this stage. Subsequently, in the 1960s, technology was developed to further purify the gamma-globulin and provide an intravenously injectable product. The lack of early patent filings may reflect the attitude at that time against patenting medical inventions as well as the fact that at that time product patent protection was not available.

The development of monoclonal and recombinant antibody technologies provided a ready source of antibodies for therapeutic use, and since the late 1970s many patent applications have been filed in this area, including those relating to the use of antibodies to target drugs and other agents to specific sites *in vivo,* such as cancer cells—the so-called magic bullet principle. Also there are numerous patent applications and some granted patents covering the use of antibodies to block or otherwise modify the activities of various natural biological molecules *in vivo,* including in particular the modification of the activity of lymphokines and cell surface receptors. These are dealt with in greater detail in section 7.5.

7.4. ANTIBODY PRODUCTION (METHODOLOGY)

There are relatively few patents in the areas of antibody production technology and the methodology for obtaining antibodies generally. The established methods used to raise antibodies by immunizing a suitable animal such as a sheep with the desired antigen, waiting for an immune response to develop in the animal, and bleeding the animal and separating the desired antiserum are generally not patentable because they are essentially biological processes.

7.4.1. Antibody Production Processes

Once monoclonal antibodies had been invented, the next step was to devise methods to produce sufficient quantities of these homogeneous reagents for their use, initially as immunoassay reagents and subsequently as therapeutic products. The first methods used for production of monoclonal antibodies involved growth of the antibody-producing cells as tumors in mice and harvesting the antibodies from ascites fluid recovered from the animals. Again there was no significant patenting in this area in view of the essentially biological nature of the processes being used.

Although the ascites method is relatively efficient for production of small quantities of antibody, its use for production of larger quantities, or products for use in humans, has obvious drawbacks. Consequently, cell culture techniques were developed; however, again relatively few patent applications were filed initially, though more recently a steady stream of patent applications are being filed in this area, particularly from U.S. applicants. The reasons for this slow start to patenting in this area are probably twofold. Often the processes used were adaptations of existing cell culture techniques and thus the question arises whether the processes were sufficiently nonobvious to support the grant of a patent. Also, generally the inventions were processes and it is difficult to enforce patents that protect processes, since it is often not possible to detect that the patented process has been used from an examination of the product. Not only is it difficult to know when the patent is being infringed but also special steps may have to be taken to obtain evidence of the infringement, since the process is usually carried out behind closed doors. It is often necessary to obtain a court order to inspect the suspected premises and this is available only if a good prima facie case can be made.

Given these uncertainties, it is not surprising that many antibody production process inventions are protected by secrecy rather than by patenting. The dual risks that patents might not be granted or enforceable because the inventions might not be sufficiently nonobviousness, and that even if they are, they might suffer from policing difficulties, make it questionable whether patenting is worthwhile, since patenting requires publication of the invention. If the patent is not granted, or is liable to be invalidated because of prior publications, patenting may give away more than it achieves.

Notable exceptions to the nonpatenting of antibody production processes are patents relating to processes that have particular cell culture equipment requirements, such as the microencapsulation patent filings of Damon Biotech (British patent application GB 2153830 A) (see also Chapter 5). If it is necessary to use special equipment to apply the patented invention, the problems of policing a process invention may be less severe. General information about the type of process or equipment being used is often given in press releases and company information handouts. Also if it became evident that competitors had started ordering costly specialized equipment suitable for use in the patented process, they could be required to prove that they are not making use of the patented process.

7.4.2. Recovery and Purification

As is the case for production processes, antibody recovery and purification processes had until recently generally not been the subject of patent filings. Again the inventions tend to be process inventions based upon established protein recovery and purification procedures. One antibody purification patent that illustrates some of the points discussed above is a patent belonging to Bio-Rad (GB 2160530 and U.S. 4,704,366) that covers the use of protein A to affinity-purify immunoglobulins (in particular IgG) under high-salt buffer conditions during the adsorption step. The use of high pH combined with high-salt conditions gives much improved binding of certain antibody classes compared with use of lower salt conditions. The fact that an antibody producer is a heavy user of protein A could indicate that the Bio-Rad process is being used. However, this patent has been the subject of litigation in the United States between Bio-Rad and Pharmacia and the U.S. patent has been invalidated on the ground of obviousness at a jury trial in the San Francisco federal court. This litigation is discussed later in section 7.7.

7.5. APPLICATIONS AND THERAPEUTIC USES

7.5.1. Diagnostic and Assay Uses

As discussed above, many of the earlier patents related to the use of antibodies as specific binding reagents in immunoassays, often for diagnostic purposes (e.g., for diagnosis of thyroid dysfunction). The first assays used were radioimmunoassays and there does not appear to have been any general patenting of such assays.

Subsequent developments of this technology were aimed at improvement of the sensitivity of the assays and avoiding use of radioactive labels and the possible attendant health risks.

7.5.1.1. Enzyme-Labeled Immunoassays. One area in which there was significant patent activity was that of enzyme-labeled immunoassays. An expected additional benefit of using enzyme labels was that the enzyme could provide a built-in signal amplifier. Instead of there being only a fixed quantity of radioactive material present to provide the signal as in radioimmunoassay, the enzyme label could, with sufficient substrate, provide a large signal by reacting with substrate to produce, for example, a colored product, without itself being used up.

There were two approaches, the development of heterogeneous and of homogeneous immunoassays. The former involved use of solid phase reagents and separation of solid and liquid components, such that the extent of the binding reaction between the antibody and analyte could be determined by separating the solid and liquid phases and measuring the colored product, for example, in the liquid phase. The most commonly used form of such assays are the ELISA assays (see Chapter 2). Such assays and variants of them were covered in a series of patent filings from Akzo/Organon (e.g., granted U.S. patents 3,654,090; 3,791,932; and 4,016,043).

A further technical development from these heterogeneous assays (two-phase) were homogeneous (single-phase) enzyme immunoassays. These assays depend upon use of soluble reagents and modification of the activity of an enzyme or enzyme inhibitor used as a label attached to the antibody reagent (e.g., enzyme inhibitor-labeled antibody) when it binds to its corresponding antigen (analyte). The advantage of such assays is that a solid/liquid separation step is not required. For example, on binding of the inhibitor-labeled antibody with the corresponding analyte, the activity of the inhibitor is abolished and an enzyme, present in the reaction mixture, remains active. Thus the presence of a colored signal, produced by reaction of the enzyme with a substrate, indicates the presence of the analyte in the sample. There are two relevant patent families in this area, one from Syva relating to the use of enzyme labels (USP 3817837) and the other from Miles relating to use of enzyme inhibitor labels (USP 4134792).

Most of these broad enzyme-labeled immunoassay patents have now reached the end of their patent lives or are soon to do so though they have withstood some challenges.

7.5.1.2. Monoclonal Antibody Immunoassays. More recently with the development of monoclonal antibodies, broad patents have been obtained by Hybritech relating to the use of monoclonal antibodies in sandwich immunoassays (U.S. patent 4,376,011 et al.). These patents cover the use of monoclonal antibodies having antigen-binding affinities greater than 10^{-8} M. For practical purposes this limitation on the scope of the claims appears of little significance, as affinities of this level are usually required to achieve satisfactory assay performance. These patents have been the subject of well-publicized litigation in the United States and will be discussed in greater detail in section 7.7 below.

In addition to the Hybritech patent property there are also patent filings from a number of other companies, including Akzo, Hoffman La Roche, and Unilever, directed to various forms of sandwich or two-site immunoassay involving use of two monoclonals. In the United States the situation is fairly straightforward, with Hybritech holding the dominant position with broad generic patent claims that have withstood challenge in two infringement actions. However, in Europe matters are still being sorted out through oppositions and appeals in the European Patent Office (see section 7.7.2.3 below).

7.5.2. Therapeutic Uses

Recently there have been two main areas of patent activity: (1) the use of monoclonal antibodies to bring about site-specific delivery of therapeutic or cytotoxic entities to particular tissue or cell sites *in vivo* (the so-called magic bullet principle); and

(2) the use of antibodies to block or otherwise modify the activity of endogenous biologically active molecules *in vivo*.

7.5.2.1. *Site-Specific Delivery.* There are a number of patent filings with broad claims in this area, including ones from Hybritech, Gansow, and Immunomedics (Goldenberg). Follow-on patent filings in the area have focused on particular types of cytotoxic agent and the technology used to link the cytotoxic agents to the antibody. Thus, a number of groups have developed technology based on the use of radioactive metals as cytotoxic agents. Gansow and Immunomedics make broad claims in this area.

The Immunomedics patent filings claim the use of radioisotopes or magnetic resonance contrast agents attached to monoclonal antibodies or antibody fragments that bind selectively to normal tissue. When these agents are used, areas of disease are not imaged and thus show up as unlabeled spots on the scan. These patent applications were originally filed in the early 1980s and a patent has been allowed recently on appeal by the U.S. Patent and Trademark Office. However, certain claims of the corresponding European patent have recently been found to be invalid in opposition proceedings (see section 7.6 below).

It is interesting that the Immunomedics application as originally filed contained claims directed to the labeled antibody reagents irrespective of their mode of use. These do not appear to have been granted yet, though in the United States continuation applications containing such claims may still be in the pipeline. (In the United States the claims in a patent application can be packaged out into one or more further patent applications, known as continuations or divisionals, and these can then be processed to grant as separate patents.)

Patent applications have also been filed on particular ligands for use in binding radioactive metal atoms to antibodies or fragments. It is of fundamental importance that the radioactive metal remain firmly bound to the antibody. If not, not only is the desired targeting effect not achieved but undesirable generalized cytotoxic effects may result.

7.5.2.2. *Antibodies as Biological Modifiers.* There are two main types of molecular targets against which antibodies are being developed to modify biological activities *in vivo:* soluble biologically active molecules, such as cytokines, and biologically active molecules, which are associated with cell surfaces, such as cell surface receptor molecules (see Chapter 2). The underlying principle of such uses is that the antibody by binding to the molecule will interfere with the usual interactions of the molecules (e.g., cytokine–receptor interactions) and thereby modify the physiology or disease process associated with the molecule. For example, antibodies to lipopolysaccharide of bacterial endotoxin have been proposed for use in the treatment or prevention of endotoxic shock associated with bacterial infection (see section 7.7.1.4), as has anti-TNF to neutralize the toxic effects of the TNF induced by lipopolysaccharide.

Another well-known example is that of patents relating to antibodies to T-cell markers (see Chapter 1) for use as immune response modifiers; Ortho has a series of patent filings covering a wide range of T-cell surface antigens and monoclonal antibodies to them, the OKT series. This Ortho patent property has been involved in litigation with Becton Dickinson in the United States, which was settled at an early stage. The corresponding European patents have been opposed by a number of companies; this is discussed in greater detail in section 7.7.

The patent situations in the therapeutic antibody area are highly complex. Usually there is at least one, often many, patent filings relating to antibodies to each molecule of interest. In addition to broad patent filings relating to the antibodies themselves and general therapeutic use claims, there may also be patent filings on specific narrow therapeutic uses.

7.6. ANTIBODY ENGINEERING PATENTS

The area that has generated the most antibody patent activity in recent years is that of recombinant antibodies, or antibody engineering, where parallel development of monoclonal antibody and recombinant DNA technologies come together (Fig. 7.4). Monoclonal antibody-producing hybridoma cell lines provide a ready source of genes that code for antibody proteins with specific binding properties. Recombinant DNA technology had already been used to express DNA sequences coding for single chain mammalian proteins. The first proteins to be prepared by recombinant DNA techniques were relatively small single chain molecules, such as growth hormones and insulin chains. Thus the application of recombinant DNA techniques to the production of antibodies, which are larger, multichain products, was not obvious.

A diagrammatic representation of the structure of a typical antibody molecule is given in Figure 7.5, being made up of two heavy (H) chains and two light (L) chains, each of which has variable (V) and constant (C) domain parts as illustrated. The various humanized antibody fragments referred to later are also indicated (refer also to Chapters 1 and 3).

The use of recombinant DNA techniques has opened up the possibility of making altered products in which the amino acid sequences of the natural antibody molecules are altered at the genetic level by making changes in the DNA sequences coding for these proteins. A very important impetus in this regard has been the desire to provide products that are "human" in character for use for *in vivo* treatment of humans (see Chapter 1).

In addition recombinant (rec) DNA technology may be used to produce fragments of whole antibody molecules, such as Fv, FAb′, and $F(Ab)_2$ products, as well as totally new antibody products such as single chain antibodies.

Fig. 7.4. List of prominent recombinant antibody patent filings in the order of their priority filing dates. For each filing a representative patent or patent application number is given, together with the priority date, the name of the patent owner, and, in parentheses, the first-named inventor and a brief indication of the subject matter covered.

RECOMBINANT ANTIBODY PATENTS

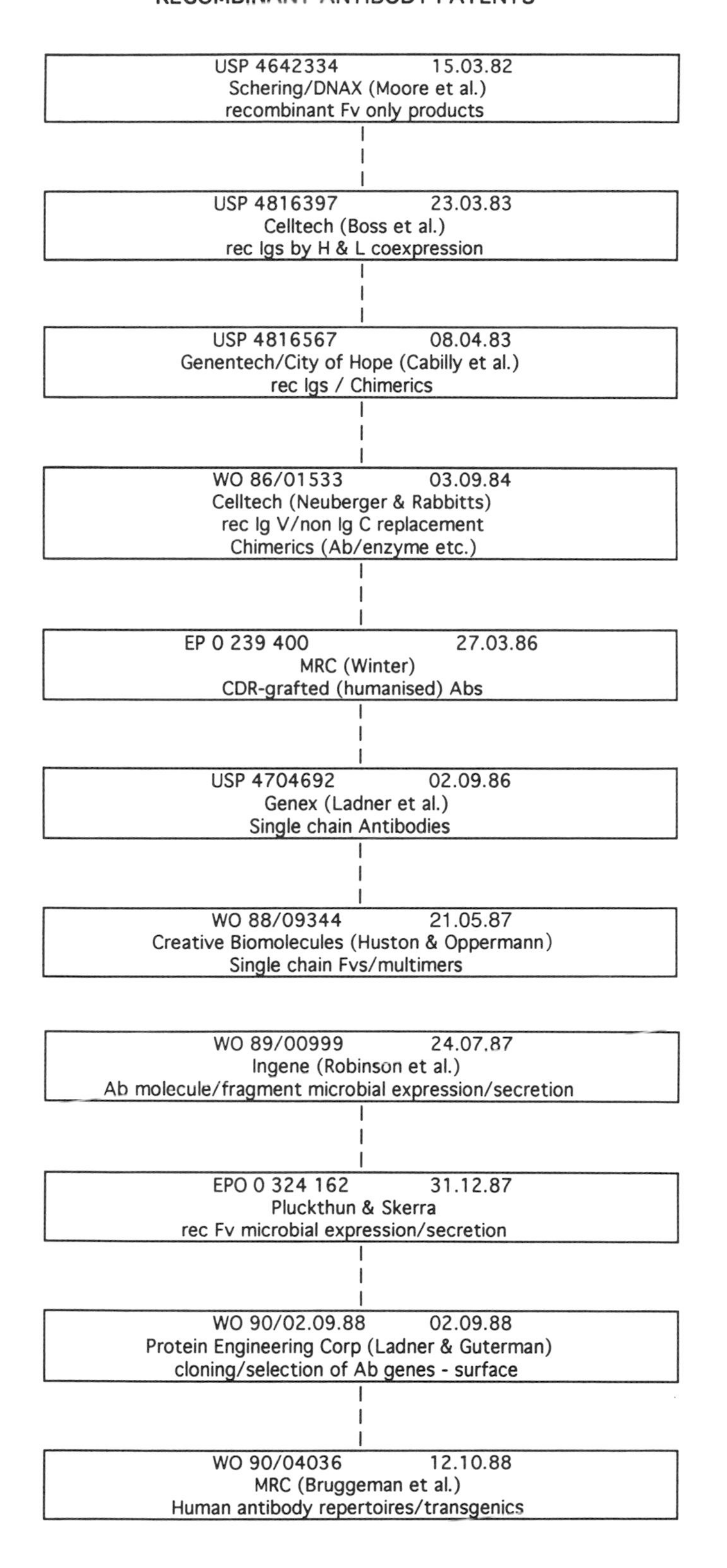

Each type of altered antibody product or fragment has its patent filings. In some cases a number of patent filings from different applicants overlap with one another. The patent filings that seem to dominate the key areas are summarized below and in the accompanying diagram (Fig. 7.4).

One of the first recombinant antibody patent applications was filed by Moore et al. (Schering/DNAX; granted U.S. patent 4,642,334) in March 1982; it deals specifically with variable domain only antibody fragments, and is dealt with under the Antibody Fragments (section 7.6.3) below.

7.6.1. Basic Recombinant Antibody Patents

The next antibody patent filings, about 1 year after the Schering/DNAX filing, were by Celltech (Boss et al.; U.S. patent 4,816,397 and European patent 120694) followed shortly afterwards by filings from Genentech and City of Hope (Cabilly et al.; U.S. patent 4,816,567 and European patent 125023).

The Celltech patent filings broadly cover processes for rec DNA production of antibody products involving coexpression of the antibody heavy and light chain polypeptides in the same host cells. The molecules covered include altered antibodies, provided they contain at least heavy (H) and light (L) chain variable (V) domains. Celltech has been granted patents in the UK, the United States, and Europe. This patent property covers all recombinant antibody production processes involving heavy and light chain coexpression in the same cell. Coexpression of heavy (H) and light (L) chains has turned out to be the production method of choice. In particular, coexpression is required when the product is expressed in cells, such as mammalian cells, that process the heavy and light chain polypeptides and secrete them as active antibody molecules.

The Genentech patent filings cover similar subject matter and overlap with Celltech's, although during prosecution the U.S. and European applications have been focused on the production of altered antibody molecules, including in particular chimeric antibody products—that is, antibodies in which the parts that determine antigen-binding specificity (the variable domains) are derived from a different source than the remainder of the antibody (the constant domains). Patents have been granted in the United States, Europe, and some other countries with claims directed to chimeric antibodies. The patent applications as filed also included claims covering coexpression of H and L chains.

Clearly there is significant overlap between the Celltech and Genentech patent filings. This overlap will be resolved in favor of either Celltech or Genentech on the basis of the dates that each can establish for their inventions. In most countries this is a simple case of determining the filing dates of the relevant patent applications; the party having the earlier filing date will prevail. In the United States, however, the situation is more complex, as it is the actual date on which the invention was made or introduced into the United States that counts. Non-U.S. inventors have to rely upon the date of filing of their priority patent application or specific acts of introduction into the United States to establish an invention date. U.S. inventors are

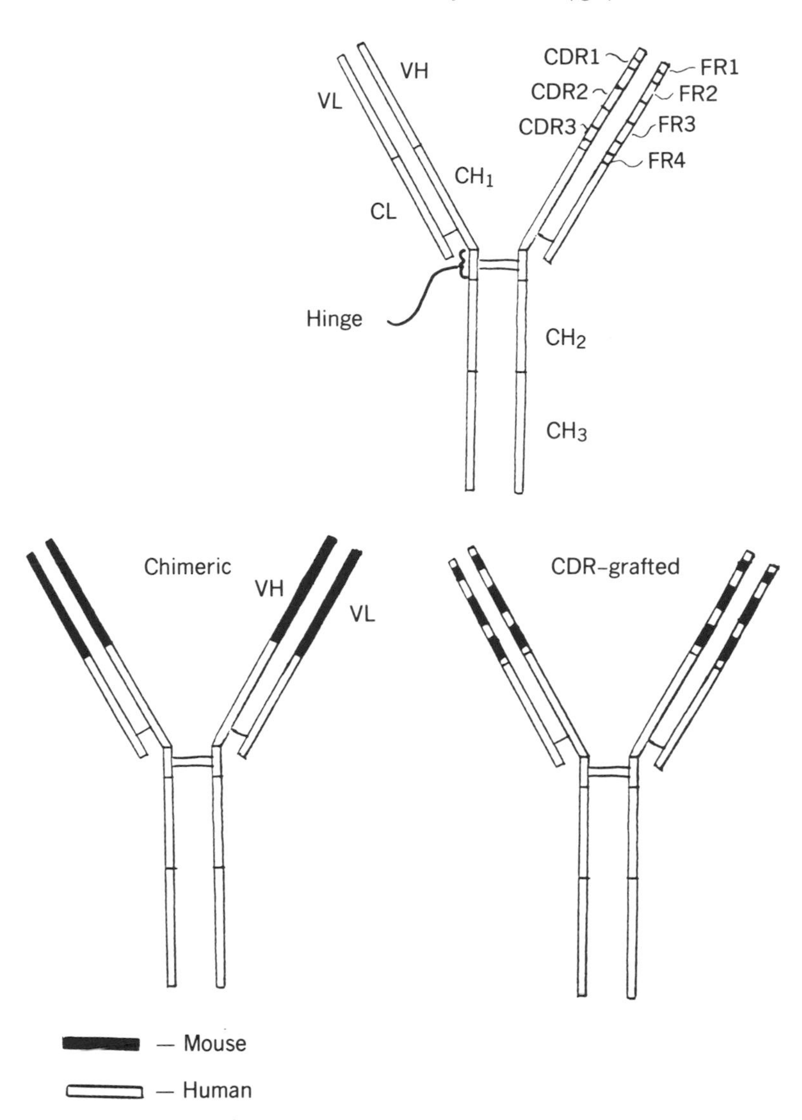

Fig. 7.5. Schematic representations of human, and mouse–human chimeric and CDR-grafted IgG antibody molecules including the variable domains of the light (VL) and heavy (VH) chains, their complementary-determining regions (CDR1, CDR2, and CDR3) and framework regions (FR1, FR2, FR3, and FR4), and the constant domains of both the light (CL) and heavy (CH1, CH2, and CH3) chains.

able to rely upon their experimental activities in the United States to establish a date of invention, which explains the importance of laboratory notebook records for U.S. researchers. The determination of who made the invention first involves a special procedure in the U.S. Patent and Trademark Office known as an *interference*. In this procedure the parties can bring forward evidence of actual experimental activity in, or introduction into, the United States to establish their date of invention, in addition to relying on the patent application documents. Currently such an interference is ongoing between Celltech and Genentech to determine who has priority in the United States for the invention of H and L chain coexpression. The granted Celltech and Genentech European patents are currently under opposition in the European Patent Office (see section 7.7.2).

7.6.2. Humanized Antibodies

As discussed above, the desire to make humanized versions of murine monoclonal antibodies has acted as an important incentive to the development of recombinant antibody technology. There have been two stages in the development of humanized antibodies: chimeric antibodies and complementarity-determining region-grafted (CDR-grafted) antibodies.

7.6.2.1. Humanized Chimeric Antibodies. Chimeric antibodies are altered antibody molecules in which the variable domains (i.e., those parts of the antibody molecule that determine its antigen-binding specificity) are derived from one antibody and the constant domains are derived from a different antibody (see Fig. 7.5 and also Chapters 1 and 3). Thus as a first stage humanized chimeric antibodies have been provided in which variable domain coding sequences from a murine hybridoma that produces a monoclonal of desired specificity are joined to constant domain coding sequences of a human antibody. Both Celltech (Boss et al.) and Genentech (Cabilly et al.) have broad claims directed to such chimeric antibodies. There are numerous subsequent filings from other groups including Stanford (Oi and Morrison; EP 0 173 494), Research Development Corporation of Japan (Taniguchi; EP 0 171 496), Celltech (Neuberger and Rabbitts; WO 86/01533), Centocor (Schoemaker et al.; EP 0 256 654), Ingene (Robinson et al.; WO 87/02671). However, it is likely that these applications will be granted with narrower claims in view of the general disclosure of chimerics in the earlier Celltech and Genentech filings. Such narrow claims are likely to be of less value since, being more specific, they may not cover useful alternatives.

A possible exception to this are the later Celltech (Neuberger and Rabbitts; WO 86/01533) filings, which in addition to simple variable domain/constant domain chimeric antibodies also cover chimeric antibody molecules in which constant region domains have been replaced by functional polypeptides of nonimmunoglobulin character. For example, the nonimmunoglobulin polypeptide may be a toxin or enzyme and the antibody variable domains may be from a site-specific antibody, such that the resultant molecule is a site-specific toxin (e.g., a tumor-specific toxin) or provides a means of localizing an enzyme at a particular site.

Humanized chimeric antibodies have been prepared and tested in *in vivo* studies and shown to be less antigenic in humans than the corresponding murine antibodies from which they were derived. However, these chimeric antibodies may still elicit some antibody response in humans directed against their nonhuman variable domains (see Chapter 2).

7.6.2.2. Humanized CDR-Grafted Antibodies. A further refinement, beyond the chimeric antibody approach, for preparation of humanized antibodies is CDR-grafting (see Chapters 1 and 3). The heavy and light chains of an antibody molecule both contain variable domains, which carry the antigen-binding site (see Fig. 7.5), the specificity of which is uniquely determined by certain regions of the variable domains, the CDRs. An alternative to making a chimeric molecule is to use only the minimum essential parts of the variable domains required to transfer the antigen-binding specificity, the CDRs, and provide "CDR-grafted" antibodies with considerably less mouse antibody sequence than the corresponding chimeric antibody.

The earliest patent filing in this area belongs to the UK Medical Research Council, naming Greg Winter as the inventor (European patent application EP 0 239 400 and U.S. patent 5225539). These patents, filed in 1986, broadly cover the basic concept of CDR-grafting and describe initial experiments in which the CDRs were derived from murine monoclonal antibodies specific for relatively simple hapten antigens (NIP), though they also include work on CDR-grafting of an antilysozyme antibody-binding specificity.

As CDR-grafting techniques have been applied to an increasing number of antibodies, it has become apparent that simple transfer of the CDRs alone is often not sufficient to regenerate satisfactory antigen-binding properties in the grafted product. In order to obtain a product with binding characteristics similar to or approaching those of the original monoclonal antibody it has been necessary to make changes in the framework regions near the CDRs (see Chapter 3).

A Medical Research Council (MRC) patent application relating specifically to Campath-1 (anti-T-cell) antibodies, filed in early 1988, covers Campath-1-specific CDR-grafted antibodies in which amino acid changes are made outside the CDRs. However, the changed residue 27 of the heavy chain, although falling outside the definition of the CDRs according to Kabat et al. [1], falls within the structural loop adjacent to CDR1.

Subsequently, criteria for identification of the framework changes that are required to regenerate full antigen binding have been established and patent applications have been filed by Protein Design Labs (Queen et al.; WO 90/07861) and Celltech (Adair et al.; WO 91/09967). The Protein Design patent application is based on work with a single antibody only, an antibody to the IL-2 (anti-TAC) receptor. The Celltech patent application is based on work done with a number of different antibodies and sets out a hierarchy of framework positions at which changes can be made to improve the binding characteristics of the grafted antibody.

7.6.3. Antibody Fragments

Antibody fragments have potential advantages for various uses, including *in vivo* therapy: They are smaller than whole antibody molecules and thus are potentially less antigenic and more able to penetrate tissue.

Prior to recombinant DNA technology antibody fragments had been generated by controlled enzymic digestion of whole antibody molecules, and $F(Ab)_2$, Fc, FAb′, and Fv fragments had been prepared (see Fig. 7.6 and Chapters 1 and 3). However, such enzymic digestion processes are unpredictable and are likely to give heterogeneous products unsuitable for use in pharmaceutical applications. Recombinant DNA techniques provide an alternative, more predictable route to antibody fragment production that is capable of providing uniform, homogeneous products in large quantities.

These potential advantages of recombinant antibody production were recognized in one of the first recombinant antibody fragment patent filings made by Moore et al. (Schering/DNAX; USP 4,642,334 and EP 0 088 994) in 1982. These patents are concerned specifically with the production of recombinant Fv fragments. Although the patent specification contains considerable and detailed description of the cloning and manipulation of the antibody genes, there are no data on expression of the Fv product. The broad Celltech (Boss et al.; USP 4,816,397 and EP 120694) and Genentech (Cabilly et al.; USP 4,816,567 and EP 125023) filings, which followed shortly after, also cover antibody fragment production in general terms.

Subsequently there were a number of patent filings covering different aspects of recombinant antibody fragment production. Potentially important ones are those from Ingene, Pluckthun, Genex, and Creative Biomolecules (see below).

7.6.3.1. Antibody Fragment Secretion. Patent filings from Ingene (WO 89/00999; priority date 24/07/87) and Pluckthun (EP 0 324 162; priority date 31/12/87) have particular relevance for the production of antibody fragments in bacterial host cells. In most cases mammalian host cells are preferred for recombinant antibody production, in view of the ability of such cells to assemble process and secrete the antibody molecules. Also in the case of whole antibody molecules, it may be important for the product to be glycosylated and even the form of the glycosylation may be critical. Mammalian cells are likely to produce products having glycosylation which is closest to that of natural antibodies. However, the productivity (amount of product produced per liter of culture per day) of mammalian cell expression systems is generally not as high as bacterial expression systems, though the latter systems are not normally able to process and secrete expressed products (see Chapter 3).

It has been found, however, that it is possible to express antibody fragments in bacterial cells and secrete functional antigen-binding products from the cells if appropriate secretion sequences are attached at the amino termini of the heavy and light chain polypeptides. Such processes are potentially of great benefit for particular therapeutic applications, for instance, for preparation of antibody fragments used to deliver cytotoxic radioactive isotopes to tumor cells. Some of the first patent disclosures relating to such bacterial antibody expression/secretion are in the Ingene

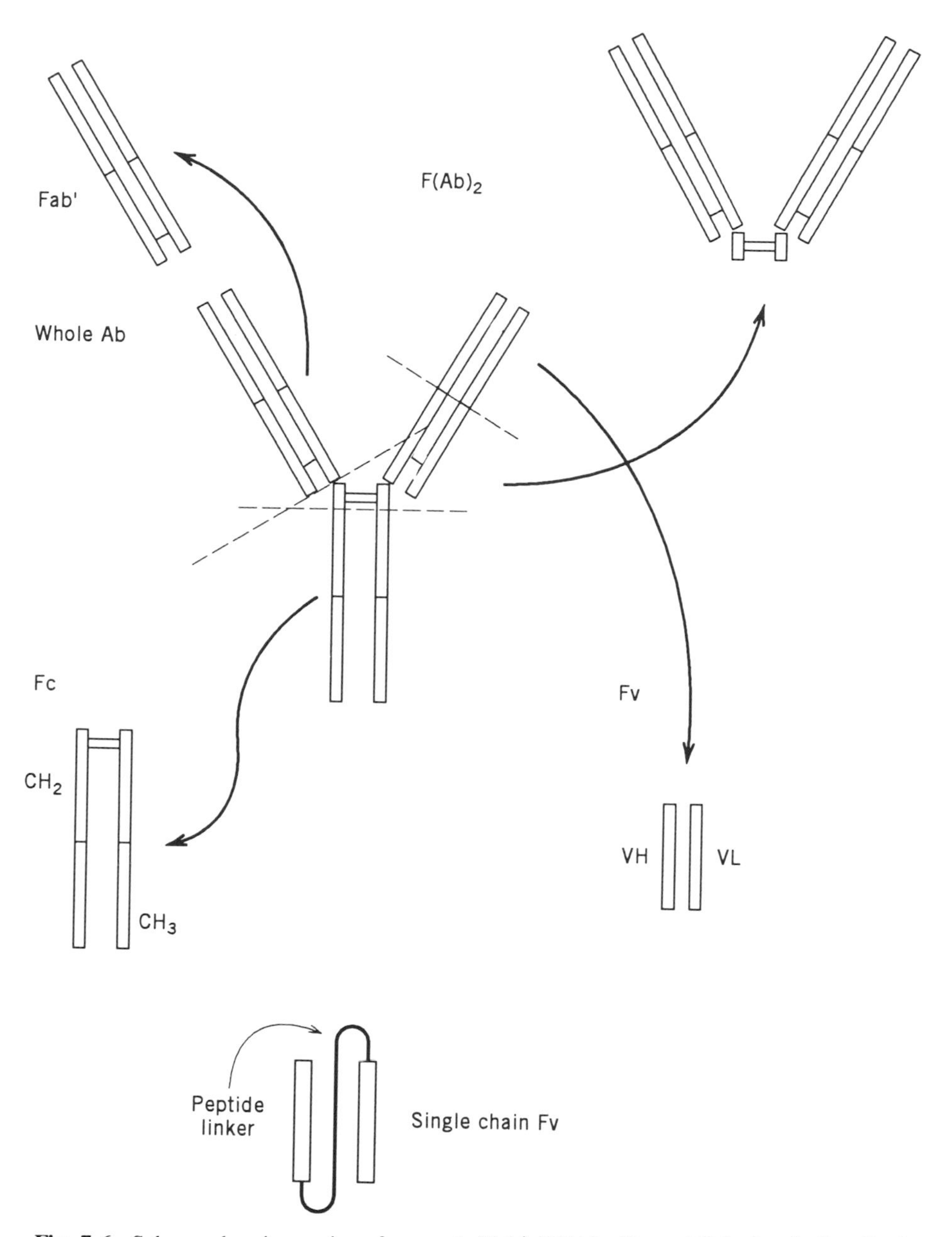

Fig. 7.6. Schema showing various fragments (Fab′, F(Ab)$_2$, Fc, and Fv) of an IgG antibody molecule and how these fragments relate to and may be derived by cleavage from the whole antibody molecule.

patent filings (WO 89/00999) made in July 1987, which disclose the expression in, and secretion from, bacteria of antigen-binding FAb fragments. The Pluckthun patent application (EP 0 324 162), initially filed in December 1987, claims the use of secretion sequences to achieve secretion of antibody products in general from bacteria, though has data only on recombinant Fv expression and secretion.

7.6.3.2. Single Chain Antibodies. Single chain antibodies are a particular form of novel antibody-based molecule that can be made by recombinant DNA techniques. Each antigen-binding site of an antibody is made up of one heavy chain variable domain and one light chain variable domain (see Fig. 7.6), which in the complete antibody molecule are held together by disulfide bonds between constant domains of the heavy and light chains. Thus, in the case of Fv fragments the natural disulfide bonding that holds the heavy and light chain variables domains together is not present and the only forces that are available to hold them together are relatively weak electrostatic forces. As a result Fv fragment products can suffer from instability problems. Single chain antibodies provide a solution to this problem. In single chain antibodies the carboxy terminus of one antibody peptide chain, for instance, the carboxy terminus of the light chain variable domain, is connected via a peptide linker to the amino terminus of the other antibody peptide chain, for instance, the amino terminus of the heavy chain variable domain. This linkage between the heavy and light chains helps to hold the two chains together and does not appear to interfere with the antigen-binding activity, provided a linkage of appropriate length and sequence is used.

There are two potentially important patent filings in this area, the first filed in September 1986 from Genex (Ladner et al.; WO 88/01649 and granted U.S. patents 4,704,692 and 4,946,778) and a subsequent filing from Creative Biomolecules (WO 88/09344 and EP 0 318 554) in May 1987. Both filings cover essentially the same ground, though the later Creative Biomolecules patent applications contain more data on multiple single chain antibody products, that is, products in which two or more single chain antibody units are joined together or joined to other polypeptide moieties. An interference may be declared between the Genex patents and Creative Biomolecules patent applications in the United States (see section 7.6.1 above for an explanation of interferences).

7.6.4. Human Antibody Genes

The chimeric and CDR-grafting approaches to the preparation of humanized antibodies are based upon the use of nonhuman hybridoma cell lines as the source of the DNA that encodes the desired antigen-binding specificity. Hence chimeric and CDR-grafted products always contain some, however little, nonhuman amino acid sequence and thus carry some risk, however slight, of being antigenic when administered to humans (see Chapter 1). A more satisfactory solution might be to use human antibody genes, but there are ethical and practical considerations which make it difficult to obtain human hybridomas for all desired antigens (see also Chapter 1 and 3).

As a response to this and other needs, techniques have been developed recently to clone and screen antibody genes (e.g., human antibody genes) to select for ones coding for antibodies with the desired specificity and binding properties. Patent applications have been filed by Protein Engineering Corp. (Ladner and Guterman; WO 90/02809), the Medical Research Council (Winter et al.; EP 0 368 684), Scripps/Stratagene (Lerner et al.; WO 90/14430), and Molecular Affinities Corp. (Wigler; WO 91/10737). Protein Engineering has the earliest filing date (02/09/88). Their method relies upon intracellular expression and surface presentation of the antibodies on the cells for screening and thus may be the dominating patent property if such surface presentation is used. The MRC and Scripps/Stratagene filings, first filed in November 1988 and May 1989, respectively, both cover processes that involve PCR cloning of V_H/V_L libraries and clearly have considerable overlap.

Patent applications have also been filed by the Medical Research Council (Bruggeman et al.; WO 90/04036 and EP 0 438 474), Genetics Institute (Wood et al.; WO 91/00906), Cell Gensys (Kucherlapati; WO 91/10741), and Yeda (Reisner; EP 0 438 053) covering the generation of transgenic animals that contain a foreign (e.g., human) antibody gene repertoire, such that when the animal is challenged with an antigen it produces foreign antibodies, not its own antibodies. Such animals could be used for preparation of hybridoma cell lines that secrete human antibodies (i.e., antibodies coded by human genes). The patent position is clearly complex, though the MRC appears to have the earliest filing date (October 1988), followed by Genetics Institute (July 1989).

7.7. ANTIBODY PATENT LITIGATION

Given the level of scientific activity in the antibody area over recent years and the potential commercial importance of the products under development, it is not surprising that there have been patent disputes and some litigation. This has involved *litigation* of granted patents, mainly in the United States, and *oppositions* against granted patents in the European Patent Office.

7.7.1. Litigation in the United States

In the United States the litigation has taken the form of infringement actions in which owners of granted patents have attempted to assert their rights against others whom they believe to be using the patented inventions without permission. These include actions between Hybritech and Monoclonal Antibodies and Hybritech and Abbott on the Hybritech tandem assay patents; between Scripps and Genentech and Scripps and Baxter Travenol over Scripps patent property relating to Factor VIII and monoclonal antibodies to Factor VIII; and between Xoma and Centocor over a Xoma patent (exclusively licensed from the University of California) that concerns monoclonal antibodies to bacterial endotoxin. There were also early reports of litigation between Ortho and Becton Dicksinson over Ortho patents relating to anti-T-cell monoclonals, although apparently this litigation was settled at an early stage (however, see section 7.7.2.1. below). In the process area there has been litigation between Bio Rad and Pharmacia on a Bio Rad patent covering particular processes for affinity purification of antibodies using protein A.

7.7.1.1. Hybritech v. Monoclonal Antibodies Inc. and Hybritech v. Abbott. These were patent infringement actions in which Hybritech took action against Monoclonal Antibodies Inc. and Abbott for infringement of their U.S. tandem assay patents (David et al.; USP 4,376,110).

This patent property covers the use of monoclonal antibodies in two-site immunometric assays that have potential advantages over competition-based immunoassays, such as radioimmunoassays (see Chapter 2). However, immunometric assays require use of monospecific antibody reagents. Monoclonal antibodies clearly satisfy this reagent requirement and it is no wonder that the use of immunometric/sandwich assays has increased very significantly since monoclonal antibodies became generally available. A number of companies developed and started to sell monoclonal antibody-based immunometric assay products, including Monoclonal Antibodies and Abbott.

Initially Hybritech took action against Monoclonal Antibodies in the California District Court, claiming that their patent was infringed. Monoclonal Antibodies counterclaimed that the patent was invalid for lack of novelty and obviousness, and the claim was successful in the district court, the judge finding the patent invalid and not infringed. However, Hybritech appealed this decision to the specialist U.S. patent appeal court, the CAFC, where the decision of the district court was reversed and the patent was found to be valid and infringed. As a result Monoclonal Antibodies was required to stop their manufacture and sale of double monoclonal antibody-based immunometric assay products.

Subsequently Hybritech also initiated an action against Abbott for infringement of this U.S. patent property. Abbott attempted to show that the patent was invalid, but again it was found to be valid and infringed.

The main issue in both cases was the validity of the Hybritech patents, both Monoclonal Antibodies and Abbott having conceded at an early stage that their products would infringe if the patent were valid. No publication of sufficiently early date could be found that disclosed the Hybritech invention and thus the critical question was one of obviousness. Sandwich assays had been known for at least 10 years before, and monoclonal antibodies had been invented more than 5 years before, the Hybritech U.S. patent application was filed in August 1980. Also monoclonal antibodies had already been used in radioimmunoassays. Hybritech argued that it was by no means obvious at their date of invention that monoclonal antibodies could be used successfully in immunometric assays, in part because it was not clear at that time that monoclonal antibodies of sufficiently high affinity could be obtained. It appears that these arguments were given some weight and the validity of the Hybritech patent was upheld.

However, a number of potentially damaging disclosures were not available for attacking the novelty of the Hybritech patent because of the special provisions of U.S. patent law. Very shortly before the Hybritech filing date there had been some disclosures at conferences and in doctoral theses that at least suggested, if not disclosed, the use of monoclonal antibodies in immunometric assays. However, the latter disclosures had taken place within 1 year of the Hybritech filing date and it was established during the litigation that Hybritech was entitled to a date of invention in the United States that predated any of these disclosures. As a result the latter disclosures were not available as references against the Hybritech claims. In the United States there is a 1-year grace period prior to the U.S. filing date during which publications do not affect validity, provided the invention was made before the publication took place. Thus in the very special circumstances of U.S. patent law it was possible for the Hybritech patent claims to be found nonobvious.

If the disclosures that took place shortly before the Hybritech filing date could have been taken into account, the matter might have been decided differently. Indeed in Europe and Japan these publications are available to attack validity of the Hybritech claims. No European or Japanese patent has yet been granted to Hybritech on this case.

Lessons: This case highlights the fact that the patent situation for a particular case can sometimes differ between the United States and the rest of the world because of the grace period and date of invention provisions of U.S. patent law (see previous sections 7.2.3.2. and 7.6.1). Also it is a salutary warning against making a too hasty assessment of obviousness: It is always necessary to consider this question in respect to the correct point in time, the priority date of the invention, and to guard against hindsight. Many inventions appear obvious once they have been made, but that is not the correct approach for assessing patentability.

7.7.1.2. Scripps v. Genentech. Scripps has patent property, U.S. Patent 4,361,509, now licensed to Rhone-Poulenc Rorer, that relates to the use of monoclonal antibodies to immunopurify Factor VIII, a natural factor involved in blood clotting and used for treatment of blood clotting disorders such as hemophilia. The Scripps patent includes claims directed to purified Factor VIII. Genentech was developing a pure recombinant Factor VIII as a product to treat hemophilia and subsequently licensed this product to Miles. Obtaining pure, virus-free Factor VIII was becoming a significant issue in view of the contamination of donor blood samples with HIV.

Scripps took action against Genentech in the California District Court for infringement of their patent and as an initial step applied to the U.S. Patent and Trademark Office (USPTO) for reissue of the patent. Reissue is a USPTO procedure by which the validity of a patent is reconfirmed or its claims adjusted in light of prior publications or errors that have been discovered since the patent was granted. In these reissue proceedings Scripps sought successfully to introduce claims into their patent that cover purified Factor VIII irrespective of its method of production, that is, product *per se* claims.

The main issues to be decided were whether the Scripps claims directed to purified Factor VIII covered Genentech's recombinant product and whether the reissued patent was valid in view of prior disclosures and certain aspects of Scripp's conduct in relation to the obtaining of the original patent and the reissued patent.

The district court found the patent to be invalid because of prior disclosure of purified Factor VIII in a college dissertation and because the inventor had not disclosed the best mode for carrying out the invention.

Under U.S. patent law it is required that the inventor disclose the best mode of carrying out the invention; it would be inequitable for the inventor to disclose a less than best way of carrying out the invention to the public while keeping the best mode secret and enjoying the benefits of patent protection. In this case it was argued that the best mode required a particular monoclonal antibody that had not been made available to the public by deposit of the hybridoma cell line which produced it.

Scripps appealed this decision to the federal court and the decision was reversed. This court found that the patent was valid and that the claims directed to a purified product, when defined in terms of the process used to obtain it, were not limited only to products when prepared by the process designated in the claims—that is, the claims were not restricted to natural Factor VIII but included recombinant Factor VIII. In particular the holding that the inventor had failed to disclose the best mode was reversed.

Lessons: This case illustrates the immense value of patents covering products when the claim covers the product however produced, including the case in which the product is defined in the claims in terms of a process used to obtain it. Also this case highlights the extreme caution that must be exercised, in particular in the United States, to ensure that the disclosure of a patent application is clear, contains all the necessary information to enable the reader to carry out the invention, and in the United States discloses the best mode, even to the point of providing a deposit of an essential organism or cell line. In this case the point was decided in Scripps's favor but only after an appeal.

7.7.1.3. Scripps v. Baxter Travenol. After commencement of the action against Genentech, Scripps also filed an action against Baxter Travenol (subsequently Baxter International, Inc.) in the Delaware District Court for infringement of their reissue patent. Baxter are licensees under a Genetics Institute patent relating to recombinant Factor VIII.

Initially the court held the patent to be invalid following the district court decision in the *Scripps v. Genentech* litigation. However, as a result of the federal circuit decision reversing the latter decision, the *Scripps v. Baxter* litigation was restarted. This case was eventually settled by an agreement between the parties in May 1993 under which it is reported that Baxter has agreed to pay Rhone-Poulenc Rorer $105 million as well as royalties and to grant Rhone-Poulenc Rorer the right to sell concentrates of the product licensed to Baxter [2].

7.7.1.4. Xoma v. Centocor. Xoma is the exclusive licensee of a University of California patent (U.S. patent 4,918,163) that relates to monoclonal antibodies to gram-negative bacterial endotoxin. Both Xoma and Centocor are involved in the development of such antibodies for use in the treatment or prevention of septic shock in human patients caused by gram-negative bacterial infection. Lipopolysaccharide endotoxins released from the cell walls of infecting bacteria play a major role in the pathogenesis of bacterial infection. The resulting endotoxemia can develop into a life-threatening disease: It is estimated in the United States that about 50,000 deaths per year are attributable to endotoxic shock. It is hoped that antibodies to these endotoxins will provide effective therapeutic products for treatment or prevention of bacteremia and endotoxic shock.

Xoma filed suit against Centocor in the district court of the Northern District of California for infringement of USP 4,918,163. In a jury trial the validity of the patent was upheld and Centocor was found to infringe the patent. Following the jury decision the judge issued an order entitling Xoma to receive payments as well as use and sales reports from Centocor. There are various interesting aspects to this case, including the fact that neither of the parties had received marketing approval for its product in the United States (indeed both parties were subsequently refused marketing approval by the FDA). Thus, as in the *Scripps v. Genentech* Factor VIII litigation, a matter to be decided was whether premarketing activities were an infringement, which in the circumstances of the case they were deemed to be.

The claims granted in USP 4,918,163 appear at first glance to be very narrow in scope, being defined in terms of a particular murine monoclonal antibody from a deposited cell line (XMMEN-OE5) and anti-lipid A IgM antibodies which competitively inhibit binding of this monoclonal to purified lipid A—that is, in terms of the class of antibody (IgM) and one epitope of the antigen (LPS). The Centocor antibody, like the Xoma antibody, is of class IgM and thus this was not an issue. However, the Centocor antibody was a mostly human monoclonal antibody (i.e., had human light and heavy chains but, being produced by a mouse–human hybridoma, contained mouse J chain), and thus the issues to be decided were whether the Xoma claims covered antibodies from other mammalian sources or only murine monoclonal antibodies, and whether the Centocor antibody inhibited binding of the Xoma antibody to an extent sufficient for it to fall within the Xoma claims. Expert evidence was provided by both sides, the Xoma expert witness showing experiments that demonstrated competitive inhibition and the Centocor expert showing experiments by which he was apparently unable to demonstrate such inhibition. The jury decided to weigh Xoma's results more heavily and decided in their favor.

An appeal to the U.S. Court of Appeals for the Federal Circuit against this decision was filed by Centocor. However, the case was settled by agreement between the parties in July 1992, Centocor reportedly [3] agreeing to pay royalties to Xoma and both agreeing to forego further litigation on the patents. However, it is not clear how much money Xoma will receive by way of royalties as neither product has yet received approval from the FDA.

A further interesting feature of this case is that Centocor also has patent property relating to antiendotoxin monoclonal antibodies and their use in the treatment or prophylaxis of bacterial endotoxin-related disorders (Pollack and Hunter; USP 5,057,598). Although this patent is based on an initial patent application filed in the United States in May 1983, it was not issued until October 1991, just as the trial on the Xoma patent was finishing. The Xoma/University of California patent is based on a U.S. patent application filed only in April 1986, 3 years after Centocor, but it has been issued in April 1990 18 months before the Centocor patent.

Lessons: Claims directed to antibodies defined in terms of the epitope specificity of a particular antibody may be interpreted broadly to cover all antibodies having that specificity regardless of the method of obtaining the antibody.

It can pay to get a patent granted early. In the United States applicants are often suspected of delaying the grant of their patents to prolong the effective life, as in the United States the life of the patent runs from the date of grant, not the date of filing.

7.7.1.5. Bio-Rad v. Pharmacia. Bio-Rad had obtained a patent (USP 4,704,366) relating to an improved method of affinity-purifying antibodies by using protein A as the adsorbent, in which high salt concentrations were used during the adsorption step in order to greatly increase the efficiency of the purification procedure and make it usable with several classes of immunoglobulins. Protein A is a cell wall protein of staphylococcal bacteria that has a specific high-affinity binding interaction with the constant domains of antibodies, particularly immunoglobulin G.

Pharmacia had been supplying protein A-based assay kits and affinity adsorption products for a number of years and is currently the major supplier of protein A for purification of antibodies. Indeed Pharmacia had earlier patent property relating at least to assay uses of protein A.

Bio-Rad filed suit against Pharmacia in the federal district court in San Francisco, accusing them of publishing information that encouraged the purchasers of their protein A affinity purification products to infringe the Bio-Rad patent. The jury found that the invention was obvious and the patent invalid. This decision is currently being appealed by Rio-Rad.

7.7.2. European Patent Office Proceedings

In the European Patent Office (EPO) a few of the key antibody patent applications are now being granted and thus oppositions are beginning to be filed. When a European patent is granted, it is possible for interested parties to file an opposition against grant of the patent, giving reasons why the patent should not have been granted in the first place. This procedure takes place in the EPO. Both the opponent and the patentee present written arguments and then have the right to be heard at an oral hearing, after which a decision is taken by the EPO. This decision may be appealed to the European Patent Office Boards of Appeal. There is no further appeal from the decision of the appeal boards (see Fig. 7.7 for an overview of the procedure).

In effect the opposition proceeding is an extension of the EPO examination of the patent application, though as a proceeding involving third parties rather than the normal procedure of examination, which involves only the applicant and the patent office. A few oppositions on patents relating to antibodies have been decided so far. These include oppositions against an Ortho patent relating to anti-T-cell monoclonal antibodies, against a University of Texas patent relating to bispecific antibodies, against an Akzo patent relating to use of monoclonal antibodies in immunoassays, and against a Unilever patent relating to use of monoclonal antibodies in immu-

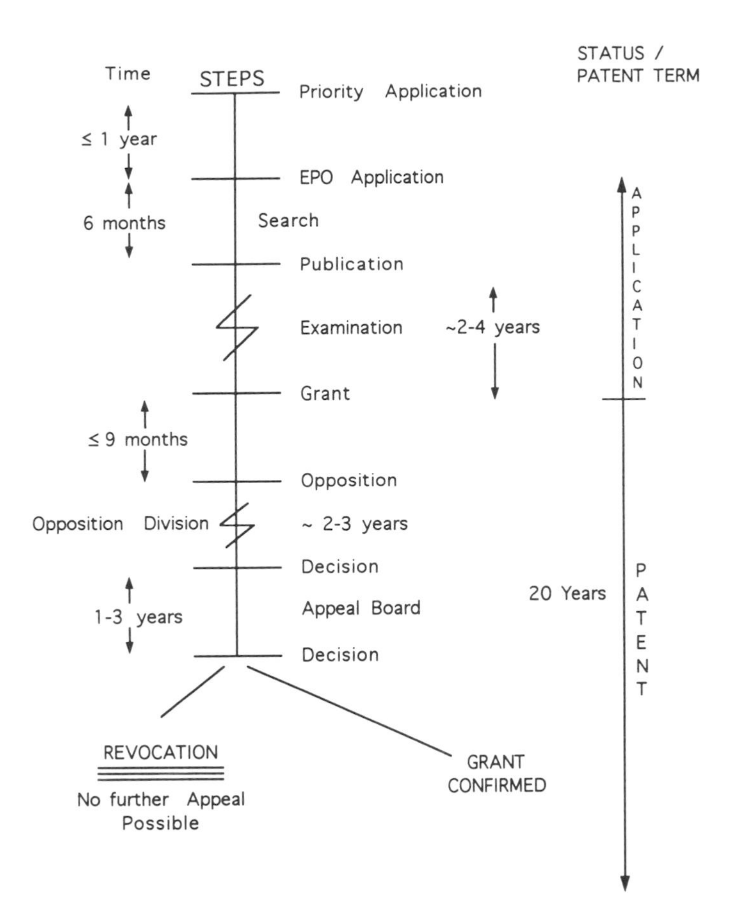

Fig. 7.7. Schema showing the time course of a European patent application as in Figure 7.1 but additionally including the time course for the European Patent Office opposition procedure following grant of the European patent. An opposition to grant may be filed by an interested party within 9 months after the grant of the patent. This procedure takes place in the EPO and involves an initial stage before the Opposition Division with the possibility of an appeal to an appeal board, and it may last from about 3 to 6 years, all of this within the 20-year patent term.

nopurification. A number of other oppositions have been filed against antibody patents, including oppositions against other Ortho patents relating to anti-T-cell antibodies and the Celltech (Boss et al.; EP 0120694) and Genentech (Cabilly et al.; EP 0125023) recombinant antibody patents, though these oppositions are still not finally decided. The details of the decided oppositions are as follows.

7.7.2.1. Ortho Oppositions. Ortho has patent property, granted in the United States and Europe, relating to the OKT series of monoclonal antibodies, which have specificity for T-cell surface antigens. Such anti-T-cell antibodies are therapeutic products useful for modifying the immune responses of patients, for example, to protect transplant patients against graft rejection. The antibodies function by blocking or otherwise modifying the activity of the subset(s) of T cells that carry the antigen in question.

The Ortho patent claims define the antibodies by their binding specificities for particular subsets of cells of the immune system. In this way Ortho has sought to define the antibodies in terms of the antigens for which they have binding specificity. The object of such claims is to cover for each antigen all monoclonal antibodies that bind to that antigen. At the time at which Ortho patent applications were filed (1978–1979) the various T-cell antigens had not been characterized in molecular terms, and reference to the binding characteristics of the antibody to a collection of cells was a convenient way of defining the antigen.

Ortho's European patents have been opposed by a number of parties. Two of these oppositions, to the patents covering OKT 1- and OKT 4-like antibodies, have now reached a final conclusion. These patents were found to be invalid after appeals to the European Patent Office Technical Board of Appeal, even after the validity of the patents had been originally upheld by the Opposition Division. Both cases were decided on substantially the same issues. The main claim of the OKT-4 patent read as follows:

> Mouse monoclonal antibody which (i) reacts with essentially all normal human peripheral helper T cells (being about 55 per cent of all normal human peripheral T cells), but (ii) does not react with any of the normal human peripheral cells in the group comprising non-helper T cells, B cells, null cells and macrophages.

The patent also included a claim directed to a particular monoclonal antibody (OKT-4) produced by a deposited cell line. In essence the main claim defines any murine monoclonal antibody that has specificity for the antigen that is characteristic of normal human peripheral helper T cells. It has subsequently turned out, however, that the antigen bound by the OKT-4 antibody *is* present on certain macrophage cells (monocytes). The European Patent Office Technical Board of Appeal decided that the patent was invalid because it would not be possible for someone to repeat the experiments described and obtain the correct antibody following the instructions set out in the patent specification. Even the claims relating to the deposited cell line were found to be invalid, because only the basic techniques of monoclonal antibody production were described, without special instructions on how to get the particular antibody specificity claimed. The patentee's arguments that the wider occurrence of the OKT-4 antigen could only be detected with the development of improved assay techniques that became available after the publication date of the patent application did not carry weight with the appeal board.

(The date of publication of the patent application is the notional point in time at which the judgment is made of whether the patent specification contains sufficiently clear and comprehensive instructions to repeat the invention—that is, the date on which the sufficiency of the disclosure is judged.)

The appeal on the OKT-1 antibody patent was decided on very similar facts. Claim 1 defined the class of antibodies in terms of its binding to a panel of cells, but the deposited cell line produced a monoclonal that did not have the identical binding pattern.

Lessons: Defining antibodies in terms of the antigens to which they bind can be useful, but extreme care must be exercised when doing so to ensure that the definition does not contain any inaccuracies. If it is possible to define the antigen in molecular terms (e.g., sequence, size, mobility on gels, etc.) this should be done. The best level of definition available should be used.

7.7.2.2. University of Texas Opposition. The University of Texas has patent property, including a granted European patent, relating to "trioma" and "quadroma" technology. A trioma is the product of the fusion of a monoclonal antibody-producing hybridoma cell line with an antibody-producing B cell; a quadroma is the product of a hybridoma–hybridoma fusion. Both types of cell lines produce, among other things, a bispecific antibody having the dual specificities of the antibodies of the two parent cells. Such a bispecific antibody is a hybrid molecule having one heavy chain/light chain heterodimer derived from each of the antibodies of the parent cells. Some of the product claims granted to the University of Texas, however, were not limited to products produced by such cell lines but covered similar products obtained by other means. Bispecific antibody products had been described previously in the literature: Experiments had been carried out by others some years previously in which two antibodies had been subjected to partial enzyme digestion followed by reconstitution to give crude products containing molecules having dual specificity.

The University of Texas European patent was opposed by a number of companies. During the opposition the University of Texas attempted to maintain a claim in which the bispecific antibody was defined as being "derived from monoclonal sources" though without requiring that it be the direct product of a trioma or quadroma cell line. Eventually, after an appeal a more limited claim, in which the antibody is defined as one "produced by cultivation of a quadroma cell . . . and/or a trioma cell" was held to be patentable.

There was a corresponding Hybritech European patent application directed to bispecific antibodies but this had been allowed to lapse. However, it has been reported recently that an interference in the U.S. Patent and Trademark Office between the equivalent Texas U.S. patent and Hybritech U.S. patent application has been settled by an agreement by which both parties have licensed each other to their patents.

Lessons: A convenient way to define an antibody for patent purposes may be in terms of the process used to obtain it; however, claims that extend beyond the direct products of the process (i.e., "obtain*able* by" claims) may not be tenable if there is close prior art, and only claims in the "obtain*ed* by" form may be possible.

7.7.2.3. Akzo Opposition. Akzo had obtained a European patent (EP 0 045 103) covering the use of two different monoclonal antibodies in immunoassays in which

the antigen is directly bound by at least two antibodies. The subject matter of this patent closely parallels the Hybritech tandem assay patent property and was opposed by a number of parties on grounds of lack of novelty. After an appeal by Akzo, procedure specified in a modified claim covering such double monoclonal immunoassays, in which sandwich-type immunoassays (i.e., solid-phase capture antibody/soluble labeled antibody) were specifically excluded from the scope of the claim, was found to be novel. Sandwich-type immunoassays involving use of monoclonal antibodies had been described in another European patent application of earlier priority date, and thus the Akzo patent would have lacked novelty if it had continued to cover this assay format.
Lesson: This case illustrates the way in which priority is sorted out between patents having overlapping subject matter in countries other than the United States. In non-U.S. countries priority is determined simply on the basis of the filing dates of the applications, the earlier applicant getting the patent to the overlapping subject matter.

A number of cases have also been decided in the European Patent Office, in which replacement of prior art polyclonal antibodies by monoclonal antibodies was not deemed to involve an inventive step.

7.7.2.4. Unilever Opposition. Unilever obtained a patent relating to the recovery of immunoglobulins from milk, the invention claimed being the use of low-affinity monoclonal antibodies to immunopurify the antibodies from the milk. The patent was revoked after an appeal, for lack of inventive step. The previously published literature included a description of the immunopurification of antibodies from colostrum by using monospecific polyclonal antibodies as the immunoadsorbent. The appeal board held that the replacement of monospecific polyclonal antibodies in an immunopurification process by monoclonal antibodies is the desired, logical, and obvious step in solving the problem to improve a purification process as described in the closest prior art.

7.7.2.5. Goldenberg Opposition. This case related to an appeal against the maintenance of a patent by the Opposition Division. The patent related to the use of radioactive labeled monoclonal antibody fragments for tumor imaging. The closest prior art described the use of polyclonal antibody fragments for the same use but indicated that the method had limitations and had not yet been proven to be clinically useful. The prior art also included a description of the use of radioactively labeled monoclonal antibody whole molecules to locate tumors in which the advantages of using monoclonals were emphasized. The relevant claims of the patent were found to be invalid for lack of inventive step.

7.7.2.6. Bogoch Application. This case was an appeal from a decision of the Examining Division of the European Patent Office that rejected a patent application that included a claim directed to the use of anticancer monoclonal antibodies to detect cancerous or malignant tumor cells. Polyclonal antisera had been used previously for this purpose but were not suitable for quantitative analysis because of cytotoxic properties. The EPO Appeal Board upheld the rejection of this claim, finding it obvious to replace cytotoxic polyclonal antisera with noncytotoxic monoclonal antibodies.

Lesson: These cases give a clear indication of how at least the European Patent Office is likely to deal with cases claiming analogy products or processes in which polyclonal antibodies are simply replaced by monoclonal antibodies.

7.8. REFERENCES

1. Kabat et al. Sequences of Proteins of Immunological Interest. Washington, DC: USDHHS, NIH, 1987.
2. Biotechnology Law Report 12, 377–378.
3. Biotechnology Law Report 11, 565–567.

INDEX